# DATA ANALYSIS
## The Ins and Outs of
## Solving Real Problems

## Competitive Methods in Operations Research and Data Analysis

Series Editor: J. Janssen, *Free University of Brussels, Brussels, Belgium*

Editorial Board: S. Osaki, *Hiroshima University, Hiroshima, Japan*

R. L. Disney, *Virginia Polytechnic Institute and State University, Blacksburg, Virginia*

---

DATA ANALYSIS: The Ins and Outs of Solving Real Problems
Edited by J. Janssen, F. Marcotorchino, and J. M. Proth

# DATA ANALYSIS

## The Ins and Outs of Solving Real Problems

**Edited by**

### J. Janssen
Free University of Brussels
Brussels, Belgium

### F. Marcotorchino
IBM Scientific Center
Paris, France

**and**

### J. M. Proth
I.N.R.I.A.
Le Chesnay, France

SPRINGER SCIENCE+BUSINESS MEDIA, LLC

Library of Congress Cataloging in Publication Data

International Symposium on Data Analysis, the Ins and Outs of Solving Real Problems
    (4th: 1985: Brussels, Belgium)
    Data analysis.

    (Competitive methods of operations research and data analysis)
    Includes bibliographical references and index.
    1. Multivariate analysis—Data processing—Congresses. 2. Factor analysis—Data pro-
cessing—Congresses. I. Janssen, Jacques, 1939–    . II. Proth, Jean-Marie, 1938–
III. Title. IV. Series.
QA278.I57   1985                        519.5′35                        87-10967
ISBN 978-1-4615-6792-9        ISBN 978-1-4615-6790-5 (eBook)
DOI 10.1007/978-1-4615-6790-5

Proceedings of the Fourth International Symposium on Data Analysis:
The Ins and Outs of Solving Real Problems, held June 10–12, 1985,
in Brussels, Belgium


© 1987 Springer Science+Business Media New York
Originally published by Plenum Press, New York in 1987
Softcover reprint of the hardcover 1st edition  1987
A Division of Plenum Publishing Corporation
233 Spring Street, New York, N.Y. 10013

PREFACE


        This book is the result of the fourth International Symposium
on Data Analysis held on June 1985 at the Université Libre de
Bruxelles with the help of the European Institute for Advanced
Management.

        As the preceding ones, the organization of the Symposium started
with a call for real life problems from which an International Com-
mittee selected six topics and asked for several solutions.   These
topics are :
1) Multivariate and longitudinal data on growing children
2) Prehistoric assemblages and lithic artifacts from a small West-
     european area
3) A comparison of results of European elections
4) Classification of heterogeneous data related to microcomputers
5) Group technology in production management
6) Juvenile delinquency
They are covered by the six chapters of this book in the following
systematic way :
a) firstly, a presentation of the problem is given in the original
     context of the relevant discipline (Medicine, archaelogy, politics,
     marketing, production and education);
b) Secondly, we present the solution found by people who presents
     the problem;
c) thirdly, we find the other retained solutions among the most
     significative ones;

d) finally, a short conclusion compares the different approaches.

The diversity of the six selected problems clearly shows that Data Analysis can be used for solving a wide variety of problems. Moreover, the fact that each problem is approached by several different way — at least two — also shows that, in general, a "universal" statistical method does not exist. This is renforced by the strong complementary of proposed solutions, aside the fact that any "good" or judicious approach must detect the basis of the internal structure of the data.

Finally let us mention that the strong interest from people both from University and Industry for the Symposium stimulates the Editors to reorganize, in an enlarged version in the topics not only in Data Analysis but also in Stochastic Modelling, such a meeting in the future.

Finally, the Editors should like to mention the strong help given by IBM-France and the CADEPS (Centre d'Analyse des Données et Processus Stochastiques, Ecole de Commerce SOLVAY, Brussels) for taking in hands the local organization and in particular the one of Mr P.CULLUS (CADEPS) for the help with final preparation of this volume.

Jacques Janssen<br>
F. Marcotorchino<br>
J.M. Proth

CONTENTS

CHAPTER 1. MULTIVARIATE AND LONGITUDINAL DATA ON GROWING CHILDREN

SECTION 1.1

PRESENTATION OF THE FRENCH AUXOLOGICAL SURVEY

Michel Sempé[1] and Groupe d'Auxologie Médico-Sociale[2]

[1] Laboratoire de Biométrie, Université Claude Bernard
   Villeurbanne, France
[2] Hôpital Debrousse, Lyon, France

An investigation about the growth, development and maturation
of normal newborns, children and teenagers was supervised for 22
consecutive years (from 1953 to 1975). Individuals were observed
every six months.

This investigation was initiated by Professor R.Debré in 1951,
in relation with representatives of several European nations. Prof.
R.Debré was the President and Founder of the International Childhood
Center in Paris.

It began in Parisian region in 1953, with the collaboration of
Dr. N.Massé, who managed the education section of the International
Childhood Center, and Dr. M.-P.Roy-Pernot, who followed the study
until 1975 when he studied individuals became adults.

As the investigation has an international impact, thirteen
coordination meetings were organized between 1954 and 1975 in Paris,
Stockholm, London, Brussels, Zurich, Davos and Rennes. After each
meeting, proceedings were published.

A final report of the French team was published in 1976. Histo-

3

rical features, results, analyses based on those results and complete
literature, are presented in this report.

Besides the publication of the French size curves and weights
curves established and brought up to date since 1963, three articles,
the subject of which was somatic and radiographic data for clinical
and teaching use, were published in 1979.

The set of observed facts has been studied in the framework of
several medical theses.  Presently, these results are also being
used for two, and probably three, scientific theses.

The existing data set is organized following three principal
orientations, linked altogether by their origin (longitudinal observations of the same individuals) :
- a prevalent one : somatic and radiographic;
- psychological investigations;
- periodical sociological investigations.
The results of the two last investigations are virtually unexploited.

In France such a work is uncommon, because the same individuals
were concerned from the age of one month to the adult stage, making
it possible to analyse statistical variability precisely.

Obviously, the size of the sample decreased during the consecutive years of measurement : at the beginning there were 588 selected individuals, 497 were examined at least one time, 194 were
periodically examined until 11 years old, 116 until 16 years old.
For all the subjects, observations were stopped after 20 years.
Some additional information is available from irregular examinations.

However, it is remarkable that the equilibrium in terms of numbers of children of each sex was maintained;  moreover, the experi-
mental design enabled us to complete the set of data after acciden-
tal absences.

So, about fifty boys and fifty girls were followed from their birth until the end of their adolescence, at regular intervals of six months (birthday and half-year). In the cases of sporadic absences, missing data were estimated by interpolation (there was no extrapolation).

Among the available auxometric measurements we have :
- length, measured in lying position,
- stature, measured in standing position,
- weight
- head circumference,
- crown-rump length, measured in lying position,
- chest circumference,
- left upper-arm circumference,
- left calf circumference,
- maximum pelvic width,
- etc.

Available maturative data are
- dentition (number of teeth),
- external genital characters,
- X-rays of left wrist and hand.

It will be noticed that the whole of the 5,011 X-rays thus collected were analysed using several methods. They are at present being coded so that they may be processed by computer, the purpose of this operation being an application of our particular method to the interpretation of the above-mentioned X-rays.

The final report of the survey gives more details about collected data. The major points concerning this French survey, which was carried out from 1953 to 1975 and is now being analysed, are the following :

- This survey was strictly sequential; data were collected at six-month intervals, the problem of sporadic missing data being solved by interpolation.

- This survey makes possible the study of correlation between anthropometric measurements and maturation information (dental, sexual and skeletal); every maturation indicator is described, defined so as to be easily processed by computer.

- This survey allowed the gathering of much other information on the psychological and socio-economic levels of the individuals studied. This information is available, and has never been analysed.

References

M.Sempé et al. (1976). Rapport d'activité terminal de l'Equipe Française. (Centre International de l'Enfance, XIIIème Réunion de coordination des équipes chargées des Etudes sur la Croissance et le développement de l'Enfant Normal; 60 p. + ann.)

M.Sempé, G.Pédron, M.-P.Roy-Pernot (1979). Auxologie. Méthode et séquences. (Paris, Laboratoire Théraplix éd., 165 p.).

M.Sempé, C.Pavia (1979). Atlas de la maturation squelettique - ossification séquentielle du poignet et de la main / Atlas de la maduracion esquelética - osificacion secuencial de la muneca y de la mano / Atlas of skeletal maturation - sequential ossification of the wrist and hand. (Lyon, SIMEP S.A., 241 p.).

M.Sempé (1979). Une étude auxologique française : méthodologie séquentielle, mensurations somatiques, maturation squelettique, in Pédiatrie Sociale Préventive, t.7, XI, 1979 (Lyon, Association Corporative des Etudiants en Médecine de Lyon, 249 p.).

SECTION 1.2


INTRODUCTION TO THE PROPOSED SOLUTIONS

Pieter M. Kroonenberg

Department of Education, University of Leiden

In this introduction an attempt is made to give an overview of
the analyses made of the multivariate growth curves.  The detailed
description of the data is contained in Sempé *, and will not be
repeated here.  In short, the data consist of 12 yearly measurements
on 8 morphological variables for 30 normal French girls.

All methods used to analyse these data are descriptive, and all
authors but one (Mineo) use in one way or another linear combinations
of the variables, years and/or individuals to analyse the data.  None
of the papers deals explicitly with multivariate time series or standard multivariate statistical techniques based on multivariate
normal distribution theory.  No author uses the analysis of covariance structures approach to longitudinal data (see, for instance,
Jöreskog & Sörbom, 1977; Goldstein, 1979; Swaminathan, 1984; and
their references).

In his paper Mineo searches for clusters of girls with specific
characteristics.  To describe these characteristics he primarily
presents the separate means and the rate of growth per cluster for

---

* References without a date refer to papers included in this volume.

7

each of the variables.  Furthermore, logistic curves are used to fit
the means of the variables for each cluster.  It should be noted
that none of the authors either link their contributions to the
basic literature on physical anthropology (see e.g. Borow et al,
1984) and/or the more commonly used methodology in that field (for
an overview of generally used functions to fit growth curves and
other related procedures with respect to growth curves, see
Goldstein, 1979).

The approach taken by Pontier & Pernin is in a sense somewhat different from most other papers using some kind of linear
combinations as their search for indices rests on a motivation ex-
terior to these particular data, and is derived from the subject
matter itself.  In particular, they look for indices which accentuate
either the relative state of growth at a particular age compared to
all other ages and relatively independent of the individuals, or
accentuate the relative growth of individuals compared to all other
individuals and relatively independent of age.

With these indices they intend to assist the relative assessment of new individuals compared to the present (training) sample
by using these two indices, and eventually other ones produced by
the method (see their section 1.5).  In each of the other methods,
however, such assessment is also possible by using the group information, and performing some form of multiple regression, or treating
the new individual as an 'individu supplémentaire'.  For instance,
in STATIS (Lavit & Pernin) this would entail using $X_k X_k' DY \Lambda^{-1}$
(section 1.3), where $X_k$ are the data for the new subject $k$.
Similarly in SPECTRAMAP (Lewi & Calomme) and in TUCKALS (Kroonenberg;
using $T'X_k U$) such an appraisal could be carried out, be it not
as compact as in LONGI (Pontier & Pernin).

Setting the Mineo and Pontier & Pernin study aside for the
moment, the other three studies may be ordered by the simultaneity

of analysis of the entire data set. Lewi & Calomme start with analysing the averages per variable and time point over individuals,
and after having found a satisfactory representation, they continue
by adding more detail of the subjects. In particular, they examine
individuals at specific time points and investigate specific individuals over time. This approach identifies growth patterns in length
and circumference measurements. After eliminating the above mentioned averages, Lavit & Pernin first perform a global analysis concentrating on the developments of the variability over time, and
the relationships between the measurements over the years. Instead
of simple averaging to find what they call a compromise solution for
the individuals, they use a special kind of 'optimal' averaging of
the matrices of scalar-products between individuals at each occasion.
This compromise solution is the basis for interpreting the trajectories or evolutions over time for both variables and individuals.
After eliminating the same averages, the Kroonenberg study attempts
one single analysis of the entire data set to find simultaneously
optimal component (or compromise) spaces for both variables and individuals, and their relationships at each age. From the basic
parameters of the solution various quantities are then derived to
investigate the patterns in more detail. In a sense the increase in
simultaneity has to be bought by an increase in complexity of interpretation, which is not necessarily a good thing. It should also be
mentioned that Lewi & Calomme use the centred average solution as a
reference point and discuss the individual characteristics by por-
traying them together with the (condensed) average solution, while
both Lavit & Pernin and Kroonenberg only display the deviations from
the averages. This leads to far greater visual similarity between
the figures of the latter two studies, compared to the former.

    Two further aspects of the analyses deserve special attention
and consideration. The first is a global one. Only the curve
fitting of Mineo takes explicitly into account the most salient
factor of the design, namely time. In their discussion and conclu-

sion Pontier & Pernin see the non-inclusion of time as an asset
rather than a liability of their technique, and their conclusions
and arguments could be directly extended to the other analyses.
One may, however, question their point of view as an analysis which
does not include a design variable should be inherently less powerful than one that does. Whether techniques exist that deal adequately with large amounts of multivariate longitudinal data and explicitly make use of the sequential information is not clear from
the present collection of papers.

The second aspect which is crucial in the analyses is the way
the data are handled before the analysis proper. Pontier & Pernin
base one index on per age centred and standardized variables, using
$\bar{X}_{.jk}$ and $s_{.jk}$, and the other index on variables centred and scaled
over all individual-age combinations using $\bar{X}_{.j.}$ and $s_{.j.}$. Lewi
& Calomme indicate that their analyses use 'logarithmic expression,
row- and columnwise centring of the data and global standardization'
before the analysis proper. This results in an analysis of ratios
such as between the various measurements of length and circumference.
Lavit & Pernin indicate that the data were standardized to mean
zero and unit variance at each age, which means that they removed
$\bar{X}_{.jk}$ and $s_{.jk}$. Finally, Kroonenberg uses yet another standardization, namely the data are transformed by first subtracting $\bar{X}_{.jk}$,
and then scaled by $s_{.j.}$. Without going into further discussion
of this issue (see some remarks by Pontier & Pernin, section 1.5,
and Kroonenberg), it is clear that the standardization used is of
vital importance, and is far more complex for three-way data than
for two-way data. Standardization critically influences what is
exactly analysed by a technique and how the numerical results may
be interpreted. The proper standardization should, therefore, be
carefully considered.

With respect to the results it is evident that in a general
sense the analyses agree, and come to similar general conclusions,

showing that the data are well-structured.  If a choice needs to be
made, the details provided by  an analysis and the ease of interpretation are crucial, while also the simplicity of the analysis
itself plays a rôle.  From the latter point of view, considering
the methodological rather than the substantive or practical orien-
tation of the contributions, one may wonder with Sempé in his final
comments at the Symposium if an auxologist, or even more your country
physician, might not prefer just the average curves per variable of
the group under discussion, and plot the position of a girl brought
in for consultation for each variable separately, as was done in
Figures 4 and 5 (lefthand panels) of Lavit & Pernin, and leave the
nice technicalities for us methodologists to ponder about.

## References

J.Borow,R.Harespie, A.Sand, C.Sussanne, H.Hebbelink(Eds). (1984).
    Human growth and development. New York, Plenum.
H.Goldstein (1979). The design and analysis of longitudinal studies.
    London, Academic Press.
K.G.Jöreskog & D.Sörbom (1977). Statistical models and methods for
    the analysis of longitudinal data. In D.J.Aigner & A.S.Goldberger
    (Eds.), Latent variables in socioeconomic models. Amsterdam,
    North Holland.
H.Swaminathan (1984). Factor analysis of longitudinal data. In
    H.G.Law, C.W.Snyder Jr, J.A.Hattie & R.P.McDonald (Eds.),
    Research methods for multimode data analysis. New York, Praeger.

SECTION 1.3

SOLUTION USING STATIS

Christine Lavit [1] and Marie-Odile Pernin [2]

[1] Unité de Biométrie, INRA-ENSA-USTL, Montpellier, France

[2] Laboratoire de Biométrie, Université Claude Bernard
Cilleurbanne, France

Introduction

In this article, we present a real life problem which concerns
physical growth of normal children.  In pediatrics, the physical
development of the body is essential to the judgement of the child's
growth quality and health.  This subject, which has often been studied, is well-known when the parameters are considered one by one :
particularly weight, height, and head circumference, especially when
compared to the average growth of a sample population of children.
A very complete univariate analysis was done, starting from this
data, by M.Sempé (1979).

Nevertheless, human growth is a biological phenomenon which "in
itself" is dynamic, which concerns the entire individual and happens
to each human being in a specific way.  Therefore in a real life
problem we must consider the knowledge of growth in a multivariate
way and then refer back to each individual case.

In this particular paper, we consider a multivariate problem :
the morphological development of children, summarized by eight
parameters during 12 years of growth.  We have tried to provide

13

answers concerning the following points :

 * describing morphological evolution of each child,

 * finding out any pattern in variables or individuals and for
example finding out typical morphologies, or characteristic ages in
morphological evolution.

 To achieve the above purposes, we use STATIS, a multidimensional
data analysis method.

## 1. Data

 Data were collected from 1953 to 1975 in the framework of a
prospective, longitudinal, and interdisciplinary study of growth and
development of children.  This study was coordinated with other
European longitudinal studies (in Brussels, London, Stockholm, and
Zürich) by International Children's Center in Paris, and supported
by INSERM with the collaboration of M.Sempé * .  The children included
in this study were selected according to three main criteria :
  - Parents of French nationality and living in the suburbs of
    Paris;
  - Weight at birth between 2.5 and 4.7 kg;
  - Absence of any anomalies at birth.

 In this paper, we work on a sample of 30 girls selected from
this survey because the data is complete.  We consider eight morphological parameters measured each year between their fourth and
fifteenth birthdays :
PD : weight(dcg)
TA : length (mm), i.e., measured in lying position
SS : crown-coccyx length (mm)

---

* Michel Sempé, Dr.Pediatrician,
   * Laboratoire de Biométrie, Université Claude Bernard - Lyon 1,
     69622 Villeurbanne cedex, France
   * Hôpital Debrousse, 29, rue Soeur Bouvier, 69322 Lyon, France
INSERM : Institut National de la Santé et de la Recherche Médicale

PT : chest circumference

PB : left upper arm circumference (mm)

PG : left calf circumference (mm)

AB : maximum pelvic width (mm)

PC : head circumference

## 2. The STATIS method

### 2.1. General description

STATIS is a DATA ANALYSIS method available to simultaneously analyse K individuals x variables data matrices $X_1, \ldots, X_K$, each $X_k$ containing some different information about the same phenomenon. Though in most real cases k is a time parameter, the STATIS technique does not take into consideration any order on the $X_k$. STATIS's graphic output gives a global description on a plot where each point represents a matrix, a plot of compromise individuals in two or three axes which can be explained by the variables and, in addition, we can follow on this plot each individual's evolution around its compromise point.

STATIS has been worked out and developped by the team of Professor Escoufier : mathematical results are gathered in Glaçon (1981).

The method has been programmed in Fortran language for the following kind of data : each $X_k$ is an observed matrix $n \times p_k$ of $p_k$ quantitative variables taken on n individuals, the individuals are the same along the K studies and we are interested in describing each individual's evolution. This program belongs to MODULAD*, a library of DATA ANALYSIS programs in Fortran 77, conceived as short subroutines and available on IBM, MULTICS CII, VAX, UNIVAC... Another version solves the symmetrical problem: the variables are the

* INRIA Domaine de Voluceau-Rocquencourt, BP 105,
    78153 Le Chesnay Cedex

same along the  K  studies and we are now interested in describing
each variable's evolution.   This last program can be asked to Lavit *.

The program output has been interpreted in Bernard and Lavit (1985)
on sociological data and Lavit and Perez-Hugalde (1985) on economics data.

Other French researchers have worked on closely related techniques : Escofier and Pages (1985), Foucart (1983) and Le Foll (1982).
Lately, different approaches of simultaneous analysis of data matrices have been applied on the same data in a special issue of "Statistique et Analyse de Données" (1985).

## 2.2. Usual notation in Principal Component Analysis

Let  X  be a  $n \times p$  data matrix of  p  quantitative variables
observed on  n  individuals weighted by a diagonal matrix  D.  First,
each column of  X  is standardized to have mean  0  for the metric  D.
The sample covariance matrix  $V = X'DX$  (or correlation matrix
if the variables are scaled to have unit variances) and  $WD = XX'D$
have the same eigenvalues, and if we write the eigenvectors of  V
as columns of a matrix  U, the columns of the matrix  $Y = XU$  are
the eigenvectors of  WD  normalized by the corresponding eigenvalues.
If we note  $\Lambda$  the diagonal matrix of the decreasing eigenvalues
of  WD  or  V, we have :

$U'U = I_{p \times p}$   (identity matrix)

$Y'DY = \Lambda_{n \times n}$ .

The columns of  U  are the principal axes and a column of  Y
contains the coordinates of the individuals on the corresponding
principal axis.  The columns of  $Z = X'DY\Lambda^{-1/2}$  contain the coordinates of the variables in the D-orthonormal basis of the columns
of  $Y\Lambda^{-1/2}$ .

---

* Unité de Biométrie, place Viala, 34060 Montpellier Cedex, France.

## 2.3. Statis method

## 2.3.1. Global analysis

Here, we will explain the theory in a particular case : each matrix $X_k$ is observed on the same population of individuals weighted by the metric $D$. The variables need not be exactly the same from one study to another. This possibility authorizes the treatment of missing data : if the measure of a variable fails for the $k^{th}$ step, you can drop this variable in $X_k$.

First, each column of the $n \times p_k$ matrix $X_k$ is standardized, according to $D$, to have mean $0$. The square $n \times n$ matrix $W_k = X_k X_k'$ is symmetric and positive semi-definite. The operator $W_k D$ is a matrix of scalar products between the individuals for the $k^{th}$ study. We can now compare the $K$ operators. We want to plot the $W_k D$ on a graph in order to provide an overall description of the $K$ studies. So we need to define a distance between operators and a scalar product for that. We can choose (Robert and Escoufier, 1976) :

$$COVV(W_k D, W_\ell D) = Tr \ (W_k D W_\ell D)$$

or

$$RV(W_k D, W_\ell D) = \frac{Tr \ (W_k D W_\ell D)}{\sqrt{Tr \ (W_k D)^2 . Tr \ (W_\ell D)^2}}$$

$$= Tr \left( \frac{W_k D}{\|W_k D\|} \cdot \frac{W_\ell D}{\|W_\ell D\|} \right)$$

The $K \times K$ matrix $C$ with elements $C_{k\ell} = COVV(W_k D, W_\ell D)$ used as a covariance matrix between the $W_k D$ provides their graphic representation as in PCA. The $W_k D$ coordinates are the eigenvectors' coefficients of $C$ normalized to the corresponding eigenvalues. When $C_{k\ell} = RV(W_k D, W_\ell D)$, $C$ can be interpreted as a correlation matrix between the $W_k D$, and the plotted points lie inside the unit circle.

If the variables are only standardized to have mean  0 :

$$Tr \; (W_k D W_\ell D) = \sum_{i=1}^{P_k} \sum_{j=1}^{P_\ell} \; Cov^2((X_k)^i, (X_\ell)^j),$$

and if, in addition, the variables are scaled to have unit variances :

$$Tr \; (W_k D W_\ell D) = \sum_{i=1}^{P_k} \sum_{j=1}^{P_\ell} \; Corr^2((X_k)^i, (X_\ell)^j).$$

Thus, the more neighbouring  $W_k D$  and  $W_\ell D$  are on the plot, the more closely related  $X_k$  and  $X_\ell$  are.

2.3.2. Compromise

All the elements of  C  are positive, hence all the coefficients $\alpha_1, \ldots, \alpha_K$  of its first eigenvector can be chosen positive (Frobenius theorem). We take this eigenvector standardized :

$$\sum_{k=1}^{K} \alpha_k^2 = 1.$$

Then  $W = \sum_{k=1}^{K} \alpha_k W_k$, which is a linear combination of positive semi-definite matrices with positive coefficients, is positive semi-definite. WD  is called the "compromise matrix" and represents a dummy matrix of scalar products between individuals which summarizes most of the  $W_k D$. Actually, WD  lies on the first axis of the previous graph and if, for example, a  $W_k D$  were orthogonal to this axis, its  $\alpha_k$  would be weak in the compromise definition.

WD  has the following maximization properties : for every standardized vector  $\beta' = (\beta_1, \ldots, \beta_K)$,

$$\| \sum_{\ell=1}^{K} \beta_\ell W_\ell D \|^2 \leqslant \| WD \|^2 = \lambda_1 \quad \text{(first eigenvalue of  C),}$$

$$\sum_{k=1}^{K} < \sum_{\ell=1}^{K} \beta_\ell W_\ell D, W_k D >^2 \leqslant \sum_{k=1}^{K} < WD, W_k D >^2 = \lambda_1^2 \; .$$

SOLUTION USING STATIS

## 2.3.3. Close-up analysis

We are going to describe the initial data as compared to the compromise.

With this kind of data, statisticians are used to doing the Principal Component Analysis of

$$(X_1,\ldots,X_K) \quad \text{or} \quad \begin{pmatrix} X_1 \\ \vdots \\ X_K \end{pmatrix}$$

But in STATIS we consider the $n \times (\Sigma\, p_k)$ matrix $X = (\sqrt{\alpha_1}X_1,\ldots,\sqrt{\alpha_K}X_K)$ and the compromise we have just seen (2.3.2) equals $XX'D$.

Let $\Lambda$ denote the diagonal matrix of the decreasing eigenvalues of $WD = XX'D$ and $Y$ the matrix whose columns are the eigenvectors of $WD$ normalized by $\Lambda = Y'DY$.

The Principal Component Analysis of $X$ gives a simultaneous representation of the $\Sigma\, p_k$ variables in the D-orthonormal basis of the columns of $Y\Lambda^{-1/2}$ by $X'DY\Lambda^{-1/2}$ (the columns of $\sqrt{\alpha_k}X_k'DY\Lambda^{-1/2}$ are the coordinates of the variables of the $k^{th}$ study). In addition, the columns of $Y$ are the coordinates of the compromise individuals. In the best cases we can explain the axes with the correlated variables and consequently interpret the position of the compromise individuals in these axes.

But this PCA does not solve the problem of the simultaneous representation of the $nK$ individuals as they are seen by each study.

We define the coordinates of the individuals, as they are seen by the $k^{th}$ study, by the columns of $W_k DY\Lambda^{-1}$. Thus each individual's evolution can be read in the compromise axes on a trajectory around its compromise position ($WDY\Lambda^{-1} = Y$ is the barycenter of the $W_k DY\Lambda^{-1}$).

How can we read a trajectory ? The $k^{th}$ point of an individual's trajectory situates this individual as compared to the mean individual of the $k^{th}$ step. And the segment between two points $k$ and $k + 1$ gives an idea of this individual's variation from $k$ to $k + 1$ as regards to the mean individual's variation. If we know how to interprete the compromise axes with the variables, the reading of each individual's trajectory will be quite easy.

## 3. Results and interpretation

We applied STATIS to the above mentioned data, standardized to have mean zero and unit variances. The results are analysed in the two-fold steps of STATIS : global analysis, using the COVV scalar-product between the $W_k D$, and close-up analysis.

## 3.1. Global analysis

The graphic representation of the 12 ages by 12 points on the first-second axes global analysis plane (Figure 1) can be analysed in terms of "0 to age" vector norms, and angles between these vectors.

This plane represents 88 percent of the overall variability. In addition, all squared cosines of angles between "0 to age" vectors and this plane lie between 0.86 and 0.94, except for 3 ages : 4, 14, and 15 (Table 1). Therefore, we consider this graphic representation to be good and to allow the following interpretations.

Although the STATIS method does not include time in itself, points representing ages appear as arranged in an orderly manner on the plane. In addition, the angles between two consecutive "0 to age" vectors are almost regular. Thus, this plot clearly reveales the chronological and dynamic aspect of growth; it must be understood that the child is not a miniature adult, but a being who evolves and changes.

Nevertheless, the ages of 7 and 11 are set apart from the other ages by their position on the first axis : their abscissa are maxi-

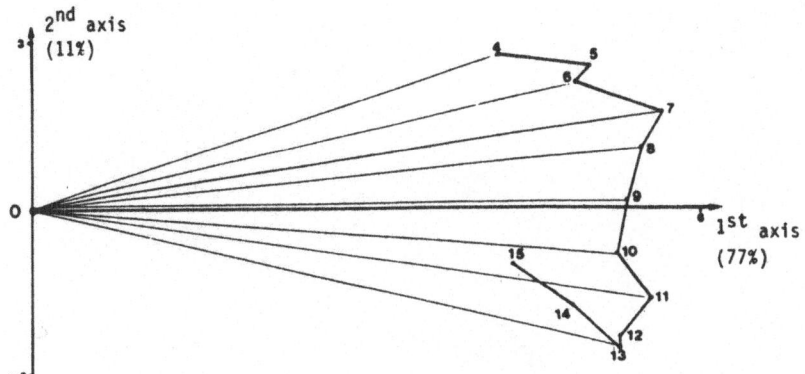

Fig.1. Global analysis first - second axes : the twelve points re-
       present the twelve ages, from 4 to 15.  7 and 11 have the
       maximum abscissa and thus appear as peaks on the age curve.

Table 1. Squared cosines of angles between each "0 to age" vector
         and the compromise plane.  "0 to age" vector norms.

| ages | 4 | 5 | 6 | 7 | 8 | 9 | 10 | 11 | 12 | 13 | 14 | 15 |
|---|---|---|---|---|---|---|---|---|---|---|---|---|
| squared-cosines | .838 | .939 | .939 | .942 | .867 | .864 | .869 | .903 | .926 | .932 | .807 | .686 |
| norms | | | | | 5.63 | 5.52 | 5.29 | 5.25 | 5.62 | | | |

mum, and thus they appear as peaks on the age curve.  Besides, for
each of these two ages, the "0 to age" vector norm is relatively
high compared to those of ages 8, 9 and 10 which means the correlations between variables at 7 (or at 11) are, globally, higher, than
at the other ages.  Therefore, 7, and 11 seem to be particular ages
in children physical development.

    The close-up analysis, a study of evolution of each child and
each variable, attempts to provide more details.

### 3.2. Close-up analysis

    The close-up analysis is concerned with, on the one hand, va-
riables and on the other hand, individuals.

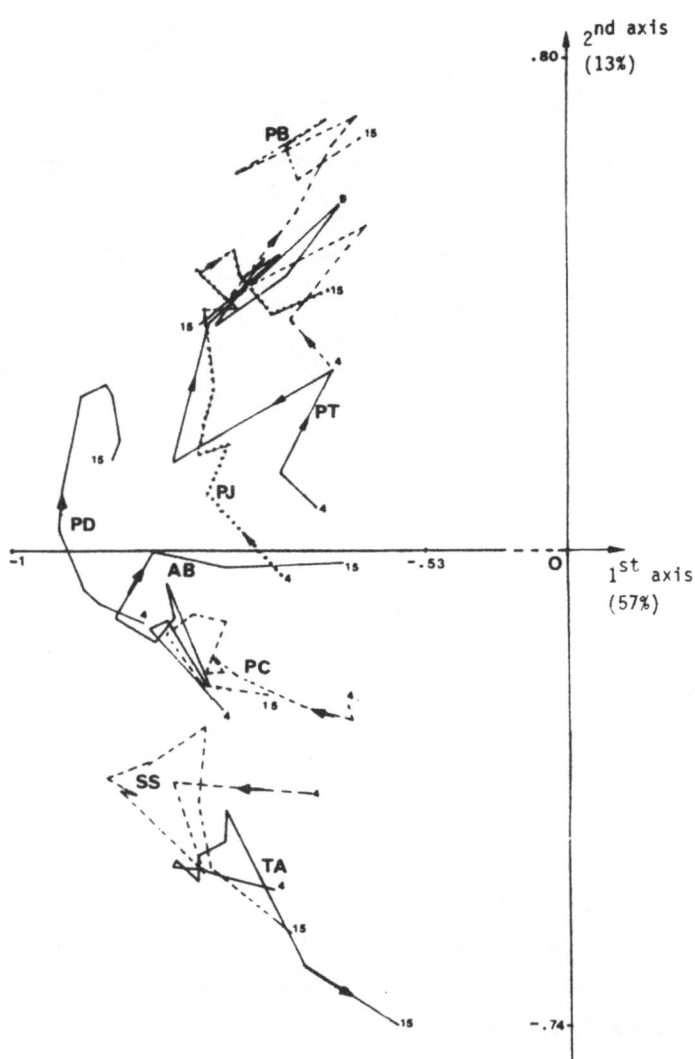

Fig.2. First – second axes of close-up analysis of variables (cf.
       Table 2). The second axis opposes two biological groups of
       variables : stoutness parameters (PD, PT, PJ, and PB) which
       have vertical ascending trajectories, and skeletal parame-
       ters (TA, SS, PC, and AB), which have "turning trajectories".

Table 2. Correlations between each variable at each age and the
compromise first axis; correlations between each variable
at each age and the compromise second axis.

Correlations variables / first axis          variables / second axis

| ages | | PD | TA | SS | PC | PT | PB | PJ | AB | PD | TA | SS | PC | PT | PB | PJ | AB |
|---|---|---|---|---|---|---|---|---|---|---|---|---|---|---|---|---|---|
| | 4 | -.83 | -.68 | -.64 | -.59 | -.63 | -.61 | -.67 | -.75 | -.12 | -.53 | -.38 | -.24 | -.07 | -.28 | -.04 | -.24 |
| | 5 | -.89 | -.74 | -.80 | -.58 | -.67 | -.66 | -.76 | -.83 | -.08 | -.51 | -.36 | -.26 | .12 | .36 | .08 | -.12 |
| | 6 | -.91 | -.77 | -.76 | -.65 | -.61 | -.57 | -.73 | -.81 | -.06 | -.49 | -.50 | -.24 | .28 | .50 | .16 | -.11 |
| | 7 | -.94 | -.80 | -.86 | -.72 | -.80 | -.73 | -.77 | -.76 | .03 | -.49 | -.37 | -.19 | .14 | .40 | .15 | -.20 |
| | 8 | -.93 | -.80 | -.84 | -.75 | -.75 | -.65 | -.75 | -.76 | .09 | -.48 | -.38 | -.17 | .35 | .53 | .25 | -.21 |
| | 9 | -.91 | -.77 | -.88 | -.74 | -.60 | -.61 | -.76 | -.81 | .24 | -.51 | -.35 | -.19 | .54 | .63 | .37 | -.05 |
| | 10 | -.91 | -.77 | -.83 | -.76 | -.66 | -.58 | -.72 | -.80 | .24 | -.47 | -.33 | -.19 | .43 | .67 | .38 | -.10 |
| | 11 | -.91 | -.74 | -.83 | -.74 | -.75 | -.72 | -.77 | -.82 | .24 | -.45 | -.33 | -.11 | .35 | .59 | .43 | -.14 |
| | 12 | -.88 | -.74 | -.76 | -.78 | -.69 | -.61 | -.72 | -.87 | .26 | -.40 | -.27 | -.10 | .45 | .67 | .47 | -.11 |
| | 13 | -.87 | -.70 | -.77 | -.81 | -.71 | -.66 | -.72 | -.82 | .24 | -.51 | -.39 | -.13 | .43 | .63 | .43 | -.00 |
| | 14 | -.86 | -.64 | -.75 | -.76 | -.67 | -.65 | -.68 | -.74 | .17 | -.64 | -.49 | -.21 | .46 | .58 | .37 | -.03 |
| | 15 | -.87 | -.53 | -.66 | -.69 | -.76 | -.57 | -.62 | -.60 | .14 | -.74 | -.59 | -.22 | .36 | .64 | .40 | -.02 |

## 3.2.1. Evolution of the variables

As in Principal Component Analysis we obtain a representation
of the correlations between each variable at each age and the compromise axes (Fig.2). Thus we can describe the evolution of the variables, find out patterns in variables and deduce a biological interpretation of the compromise axes.

The first axis (57% of the overall variability) is a size factor
dominated by weight, although all the variables are well correlated
with this axis (having the same sign). The second axis (13%) seems
to be a form factor. It opposes several variables at two levels
(Table 2 and Fig.2) :

- globally, one can distinguish two large groups of variables :
PT, PB, PJ and PD which are positively correlated with the second
axis; TA, SS, PC and AB which are negatively correlated with this
axis. In fact PC and AB have nearly zero correlations and only TA
and SS are really opposed to PT, PJ and especially to PB.

    - more closely observed, two forms of trajectories occur :
vertical ascending trajectories for variables PD, PT, PJ and PB;
trajectories which can be called "turning trajectory" for variables
TA, SS, PC and AB.

    Therefore, this plot reveals two biological groups of variables :

    - stoutness parameters (PD, and especially PT, PB, and PJ),
which show irregular development and are very sensitive to environment (diet, physical exercise etc).

    - skeletal parameters (PC, AB, TA, SS), which can be separated
into two subgroups : parameters of skeletal width (PC and AB) and
parameters of skeletal length (TA and SS), but both have a regular
evolution and present very few fluctuations.

    The importance of stoutness parameters on the second axis ap-
pears negligible ($\simeq 0.20$ Table 2) at early ages (i.e. from 4 to 7)
and increases greatly from 7 to 13, especially for PT and PJ, less
clearly for PB.  On the contrary, the participation of skeletal
length parameters is almost constant and non negligible (0.50 for
TA;  0.40 for SS).

    Thus, the importance of each parameter in morphological changes
evolves during growth from the ages of 4 to 15.  In particular, corpulence, whose evolution is described quite well by the second axis,
would not be a good indicator of individual morphological type before
the age of 7.

### 3.2.2. Trajectories of the individuals

    In the compromise axes, we represent each individual's trajec-
tory which describes the multivariate evolution of this individual
compared to all the others, that is to say, to the mean.  The compromise individuals can help in interpretation.  Particularly, the
squared cosine of the angle between a compromise individual and the
compromise plane measures the reliability of this individual's re-

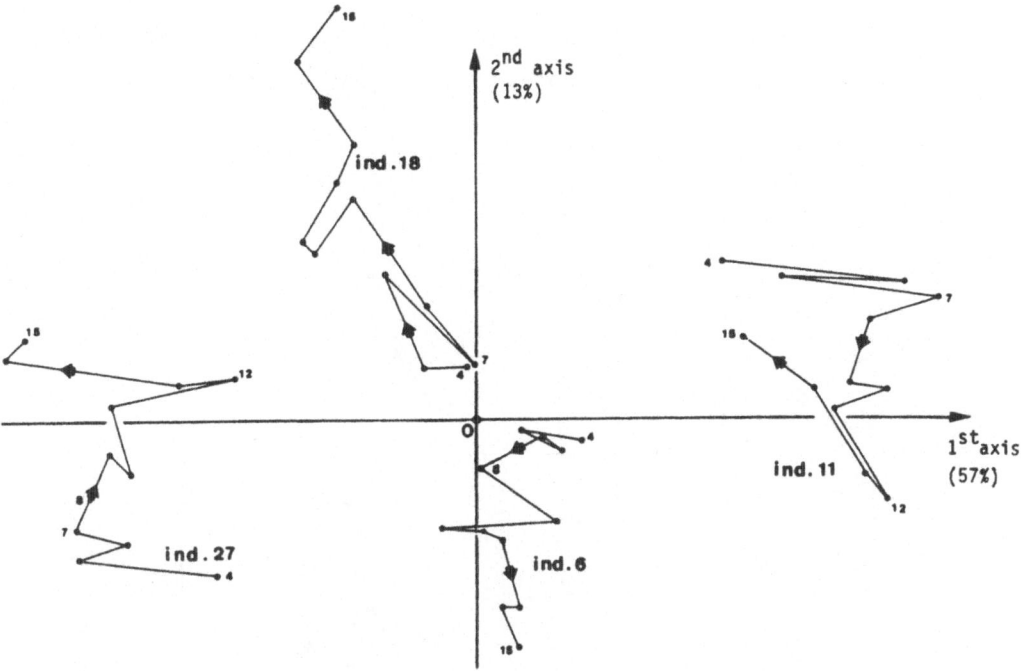

Fig.3. Four characteristic types of individual trajectories on the
       compromise plane :
       "heavy, stout, and short" individual : girl number 18
       "light, thin, and tall" individual : girl number 6
       "light, thin, and short" individual : girl number 11
       "heavy, stout, and tall" individual : girl number 27


presentation (individuals of low squared cosine should not be considered).

       Therefore, for well represented individuals, we can interpret
their trajectories in  terms of position, direction and form (Fig.3) :
       - vertically ascending (descending) trajectories express increase
(respectively decrease) in corpulence parameters (PB, PT, and PJ)
relative to corresponding average growth;
       - horizontal trajectories in direction of negative (positive)
abscissa express increase (respectively decrease) in weight (PD);
       - "turning trajectories" express stability relative to the ave-
rage growth of the following parameters PD, PB, PT, PJ, TA, and SS.

Every combination of these preceding types of trajectory are, of course, possible an may be interpreted.  See for example figures 4 and 5.

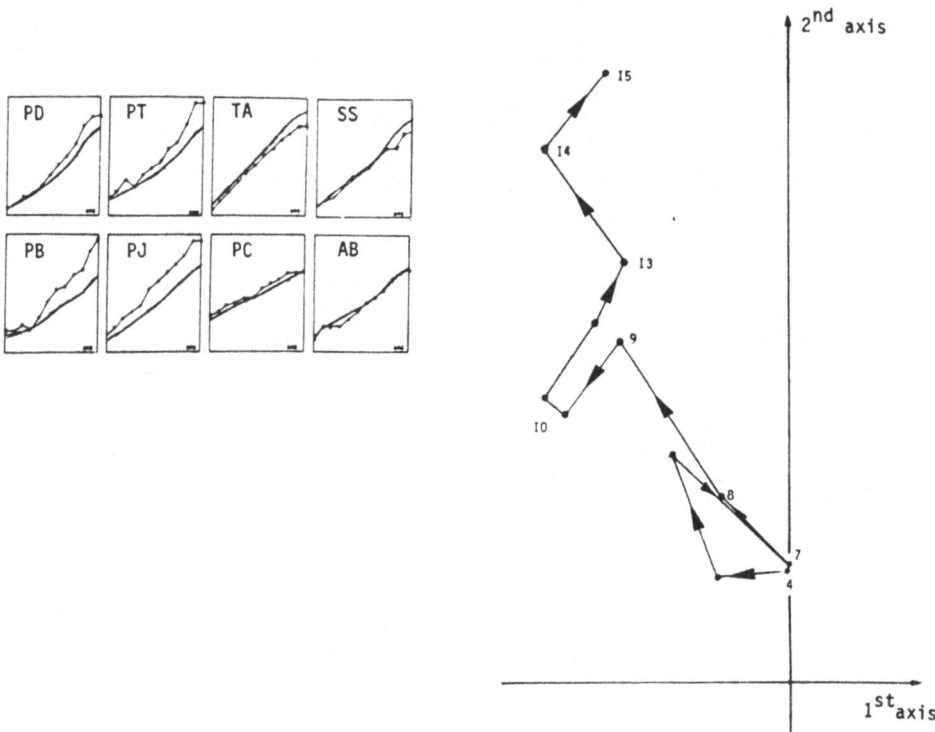

Fig.4. Details of figure 3 :
        Girl number 18, "heavy, stout, and short" individual.
        Comparison of the observed growth curve of each parameters
        (••••) with the corresponding average growth curve (————).
        For each of the parameters PD, PT, PJ, and PB, the observed
        growth curve deviates from the average  curve by higher
        values from 7 or 8 to 15 years old, whereas the growth of
        parameters TA and SS deviates from the average growth by
        lower values.  These observations are revealed on the tra-
        jectory by the vertical ascending start which can be noti-
        ced at 7 years old.

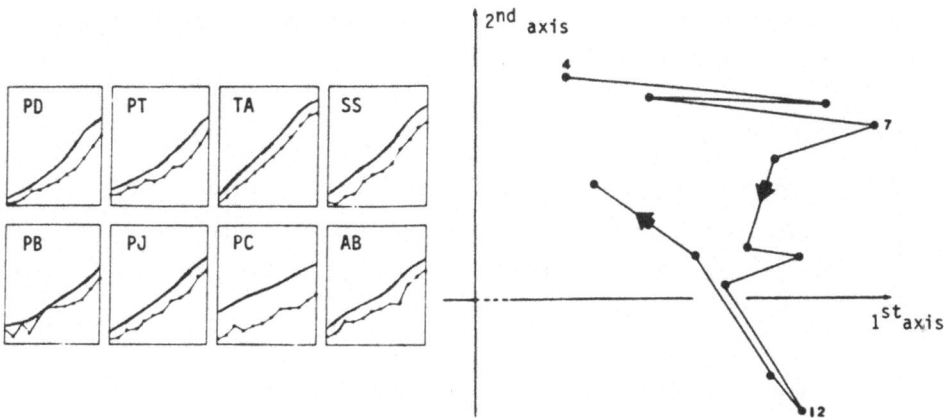

Fig.5. Girl number 11, "light, thin, and short" individual.
        Between 4 and 7 years old, one notices the stability of the
        trajectory on the second axis, which corresponds to the
        "zigzaging" evolution of the arm circumference (PB) and the
        stability of the individual growth relative to the average
        growth, especially concerning weight (PD), arm circumference
        (PB), and chest circumference (PT).
        This trajectory is globally vertical descending from 7 to
        12 years old, then ascending.  This observation corresponds
        to the deviation of the individual growth from the average
        growth by lower values, from 7 to 12 years old, concerning
        stoutness parameters especially PD, PT, and PB;  from 12 to
        15 years old the individual growth tends to catch the average
        growth up.

        The global position of individual trajectories indicates :

        weight characteristics (generally in relation with corpulence
characteristics) relative to average weight : negative (positive)
abscissa correspond to heavy (respectively light) individuals;

        length characteristics relative to average length : negative
(positive) ordinates correspond to tall (respectively short) individuals.  However, it is to be noticed that length characteristics (TA,
SS) are not as well described as weight or corpulence characteristics,
perhaps because of their lack of varability along growth.  Consequent-

ly, length structures may be dominated by corpulence evolution and so
are not always clearly distinguishable.

Conclusion

   In this paper, we use the STATIS multivariate data analysis
method to describe morphological development of normal children.
The data consists of 8 morphological parameters taken on 30 girls at
12 different ages.

   The global analysis shows the chronological growth, and seems
to mark off two particular ages : 7 and 11.  This is not surprising
for the age of 11 which immediatly precedes puberty, but the age of
7 raises a more delicate question;  we may think that the age of 7
would be a first step of the morphological type from which a new
evolution would surface.  However, this hypothesis has yet to be
verified.

   The close-up analysis of the variables shows two biological
types of parameters in terms of type of body dimensions and type of
growth :
     * skeletal parameters which have regular growth;
     * stoutness parameters which are more sensitive to environment.

   The morphological typology in terms of corpulence and length is
obtained by the individuals' close-up analysis, and revealed on graphic representations.  Nevertheless, interpreting only the positions
of each compromise individual is not always sufficient.  In fact,
global morphological type much depends on morphological evolution
during the entire growth span.  Thus, the interpretation in terms of
morphology can be deducted only from individual trajectories.

   In conclusion, we regret that, for this type of data, one problem with the STATIS method is that it does not include time as such.
So, in the global analysis for example, chronological pattern is the
first and most evident to be shown, but may partially hide a finer,

less evident subpattern.  In addition, the absence of an evident link between global and close-up analysis makes the interpretation diffi-cult.

However, the STATIS method, applied to this data, seems to be efficient, particularly in the description of each individual mor-phological evolution relative to the average evolution.

## References

A.Barre, M.Cl.Bernard, A.Carlier, Y.Escoufier, B.Fichet, T.Foucart, J.Y.Lafaye, C.Lavit (1985). Analyse conjointe de plusieurs tableaux de données. Numéro spécial, Statistique et Analyse des Données, 10, 1.

B.Escofier, J.Pages (1985). Le traitement des variables qualitatives et des tableaux mixtes par analyse factorielle multiple. Quatrièmes Journées Internationales Analyse des Données et Informatique, Versailles 9-11 Octobre 1985, France.

Y.Escoufier (1980). Exploratory Data Analysis when Data are Matrices. Recent Development in Statistical Inference and Data Analysis. K.Matusita Editor, p.45-53.

T.Foucart (1983). Une nouvelle approche de la méthode STATIS. Revue de Statistique Appliquée, vol XXXI, n°2, p.61-75.

F.Glaçon (1981). Analyse conjointe de plusieurs matrices de données. Thèse, Université Scientifique et Médicale de Grenoble, France.

C.Lavit, C.Perez Hugalde (1985). The STATIS method applied to econo-mic data:  multivariate evolution of the spanish provinces. Poster aux Quatrièmes Journées Internationales Analyse des Données et Informatique, Versailles 9-11 Octobre 1985, France.

Y.Le Foll (1982). Pondération des distances en Analyse Factorielle. Statistique et Analyse des Données, 7, 1, p.13-31.

P.Robert, Y.Escoufier (1976): A unifying tool for linear multivariate statistical methods : the RV-Coefficient. Appl.Stat., C,25 (3), p.257-265.

M.Sempé (1979). Auxologie : méthodes et Séquences . Théraplix, Paris, 205 p.

SECTION 1.4

SOLUTION USING SPECTRAMAP

P.J. Lewi, G. Galomme and J. Van Hoof

Research Laboratories, Janssen Pharmaceutica NV

Beerse, Belgium

## Introduction

Lavit and Pernin (1985) have submitted for analysis a three-way structure of data on thirty young girls with eight morphometric measurements, recorded over twelve consecutive years.

The solution proposed here is based on spectral map analysis (SMA). Spectral mapping is especially useful when a significant size component is apparent in the data and when an interpretation in terms of characteristic ratios is indicated. Spectramap is the name of a particular program that performs SMA.

Our purpose is to construct simple graphic means for the classification of individual subjects and for the detection of abnormal growth patterns. The result of our analysis is presented stepwise, in the manner one would proceed in practice. First, the global features of the data are surveyed. Once the most characteristic ratios of measurements have been identified, more detailed studies will be undertaken in a simplified framework.

Briefly, SMA involves logarithmic re-expression of the data, row- and column-wise centering, factorization and simultaneous plot-

31

ting of row- and column-markers in the plane of dominant factors. Orthogonal projection in this plane reproduces the (doubly-centered) data. Weights can be assigned to individual rows and columns according to their importance (Lewi, 1982 and 1984a). SMA is related to factorial analysis of correspondences (Benzécri, 1973). The emphasis in SMA, however, is put on ratios rather than on distances in a chi-square metric.

Spectral mapping allows to identify visually those measurements whose ratios differentiate most among the data. In the case of three measurements, a spectral map can be produced in the form of the well-known triangular chart. The use of triangular maps in data analysis can be traced back to Newton and to the early studies of the composition of light in terms of three primary colors (Lichtenberg, 1775).

### i. Method

Spectral map analysis (SMA) has been devised initially for the classification of biological activity spectra in pharmacology (Lewi, 1976 a,b,c; 1978; 1979; 1980a). Later on, the method proved to be applicable to other types of biological data and to data arising from surveys, marketing, finance and in many other fields (Lewi, 1980b; 1984b).

The idea behind SMA is that objects (drugs, respondents, companies, countries, etc.) can be widely different in size or importance, although being similar in shape or other relative characteristics. On the other hand, objects can be close in absolute terms, while possessing completely divergent spectra of differential properties. Spectral mapping attempts to display characteristic ratios. The method can be said to perform an analysis of multiple contrasts or of multiple ratios.

When the objective of the analysis is to study characteristic ratios between measurements, a logarithmic re-expression of the data

is mandatory.  (It is easily understood that an analysis of differences between logarithms amounts to an analysis of logarithms of
ratios).  This poses no special problem in the present longitudinal
data, as all data have been defined on ratio scales (weights, lengths
and circumferences).

The next step in the analysis is to define suitable weights for
the items represented by rows and columns.  In the case of homogeneous
data (i.e. data that can be added row- and column-wise such as in true
contingency tables or cross-tabulations) weight coefficients are derived from the marginal totals of the table.  With heterogeneous data,
weight coefficients for column-items have to be defined identically.
Weight coefficients for row-items can be associated to any one column
of the data which represents a suitable measure of size or importance
(such as body weight or body length).

From the weight coefficients one constructs two diagonal matrices  W  and  $W^*$, which are associated to row- and column-items respectively.  Elements on the principal diagonals of these weighting
matrices are equal to the normalized weight coefficients.  All other
elements are set to zero.

By means of these weighting matrices, we compute the expectations
m  and  $m^*$  of row- and column-items respectively :

$$m_i = \sum_j X_{ij} W^*_{jj}$$

$$m^*_j = \sum_i W_{ii} X_{ij}$$

where  X  represents the data matrix (after logarithmic re-expression).

The global expectation  m  of the data-items is defined by means
of :

$$m = \sum_i \sum_j W_{ii} X_{ij} W^*_{jj}$$

The next step in the procedure results in a doubly-centered data
table  Y :

$$Y_{ij} = X_{ij} - m_i - m_j^* + m$$

Double-centering of the data preserves the symmetry between the
rôles of row- and column-items. It has been practiced in the analysis of similarities between psychological profiles (Cronbach and
Gleser, 1953).

For computational reasons, the doubly-centered data $Y$ are
reduced to a table $Z$ with unit global variance :

$$Z_{ij} = Y_{ij}/v_y^{1/2}$$

where $v_y$ represents the global (weighted) variance of the data
items :

$$v_y = \sum_i \sum_j W_{ii} Y_{ij}^2 W_{jj}^*$$

After these operations, a weighted variance-covariance matrix
$C$ can be defined, from the data table $Z$, using matrix multiplication notation :

$$C = Z^t.W.Z$$

Using standard factorization procedures (e.g. Jacobi's or
Hotelling's algorithms in the case of moderately-sized problems) we
obtain the eigenvector matrix $F$ and the associated diagonal eigenvalue matrix $D$ :

$$D = F^t.W^*.C.W^*.F$$

where $F$ is an orthogonal matrix with respect to the weighted matrix
product :

$$F^t.W^*.F = I$$

with $I$ the unit matrix associated to the space of the factor-items.

Using the eigenvector matrix $F$ and the data table $Z$ we
compute the representation of the row-items in factor space (the
so-called factor scores) $S$ :

$$S = Z.W^*.F$$

Finally, we derive the representation of the column-items in factor space (the so-called factor loadings) L :

$$L = F.D^{1/2}$$

Note that L has been scaled by the square root of the eigenvalue matrix D. This ensures that both factor scores L and factor loadings S possess identical factor variances :

$$S^t.W.S = L^t.W^*.L = D$$

The factor scores S and factor loadings L reproduce the variance-covariances among row- and among column-items respectively. The resulting map, however, may not reproduce the doubly-centered data Z by means of orthogonal projection of the row-items upon a structure of column-items (and vice versa). In order to ensure re-production of the data table itself one may opt for a scheme such as employed in the biplot method (Gabriel, 1971) :

$$S' = Z.W^*.F$$

$$L' = F$$

Here, the factor scores S' reproduce the covariances among column-items. The factor loadings L',however, possess unit variance. In the case of three column-items and with constant column-weights, these can be represented on the spectral map at the vertices of an equilateral triangle. In this case, the symmetry between rôles of rows and columns is destroyed.

Another scheme, of our own, distributes the square root of the factor variance D equally among the factor scores and factor loadings :

$$S'' = Z.W^*.F.D^{-1/4}$$

$$L'' = F.D^{1/4}$$

This way, factor scores S'' and factor loadings L'' possess equal variance. Furthermore, orthogonal projection of S'' upon L'' reproduces the data table  Z :

$$S''.L''^t = Z.W^*.F.F^t = Z$$

A mapping in the plane of the first two dominant factors produces the basic spectral map.  The third most important factor is
coded on the map by varying the thickness of the contours of the
representative symbols.  A symbol appearing with a thick outline is
located above the plane of the map.  A thin outline indicates that
the symbol lies below the plane.

By convention, row-items are identified by circles and column-
items are denoted by squares.  The sizes of circles and squares are
varied according to the size or importance of the corresponding row-
and column-items.  Items that cannot be adequately represented this
way are identified by means of a broken contour of their representative symbol.  SMA allows to position at the same time row- and
column-items with respect to the functional items that are used in
the calculation of the map.  This allows the display of badly-
represented items without affecting the computed result.

Taking into account the variance of the sizes of the symbols
and the variance of the three dominant factors, a considerable part
of the global variance of the data  X  can be reproduced in a single
two-dimensional diagram (i.e. the size component and three factors
of contrast).  The latter global variance  $v_x$  is related to the
global variance after double-centering  $v_y$  and to the variances of
row- and column-expectation  $v_m$, $v_m^*$  (Lewi, 1982) :

$$v_x = v_y + v_m + v_m^*$$

Hence, the degree of reproduction  r  can be derived from :

$$r = (v_f + v_m + v_m^*)/v_x \qquad \text{with} \quad v_f \leqslant v_y$$

where  $v_f$  equals the variance contributed to  $v_y$  by the three most
dominant factors.

SMA differs from correspondence analysis mainly in the initial steps
of the procedure (Lewi, 1984a). Instead of logarithmic re-expression,
correspondence analysis divides each element of a contingency table
by its expectation (i.e. the product of the corresponding row- and
column-marginals). Also, in correspondence analysis weight coefficients are strictly derived from the marginal totals of the table
(Benzécri, 1973).

## 2. Preliminary analysis of age vs eight measurements

A global analysis has been performed on a compressed two-way
table. The average values of eight morphologic measurements computed over thirty individuals at twelve subsequent years are represented in Table 1. The corresponding Spectramap is shown in Figure 1.
On this map, circles represent ages and squares refer to morphologic measurements.

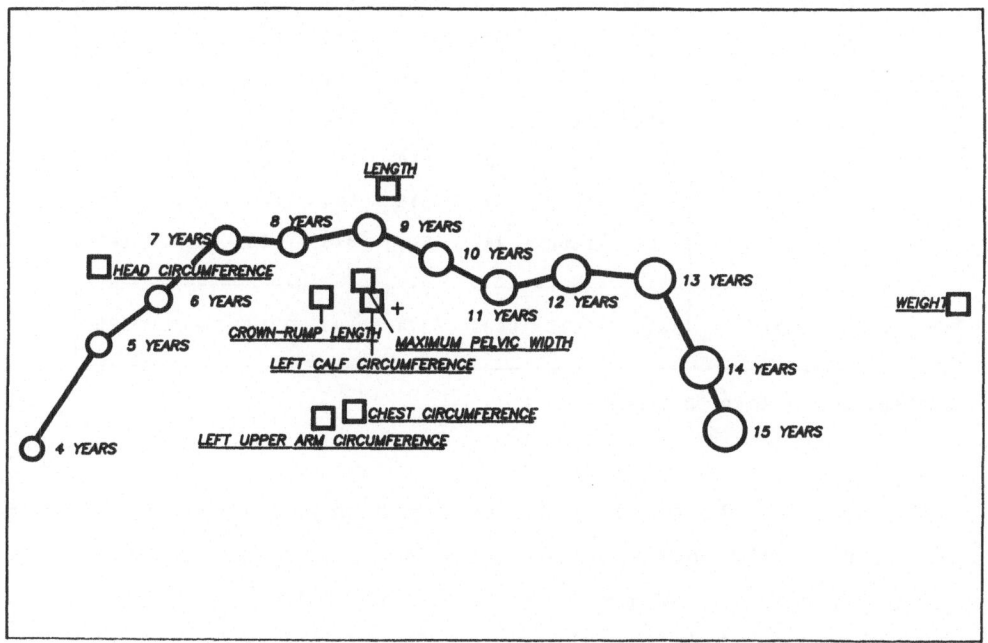

Fig. 1

Table 1

| A : WEIGHT |
| B : LENGTH |
| C : CROWN-RUMP LENGTH |
| D : HEAD CIRCUMFERENCE |
| E : CHEST CIRCUMFERENCE |
| F : LEFT UPPER ARM CIRCUMFERENCE |
| G : LEFT CALF CIRCUMFERENCE |
| H : MAXIMUM PELVIC WIDTH |

|                   | A     | B     | C    | D    | E    | F    | G    | H    |
|-------------------|-------|-------|------|------|------|------|------|------|
| 1 :  4 YEARS      | 1519  | 998   | 576  | 495  | 535  | 163  | 214  | 167  |
| 2 :  5 YEARS      | 1700  | 1060  | 597  | 502  | 549  | 164  | 223  | 178  |
| 3 :  6 YEARS      | 1885  | 1119  | 622  | 507  | 569  | 167  | 230  | 187  |
| 4 :  7 YEARS      | 2106  | 1176  | 641  | 512  | 583  | 170  | 242  | 194  |
| 5 :  8 YEARS      | 2359  | 1234  | 666  | 517  | 605  | 178  | 252  | 200  |
| 6 :  9 YEARS      | 2669  | 1288  | 686  | 520  | 620  | 185  | 262  | 207  |
| 7 : 10 YEARS      | 2986  | 1341  | 705  | 524  | 645  | 193  | 273  | 213  |
| 8 : 11 YEARS      | 3342  | 1401  | 733  | 529  | 679  | 202  | 286  | 222  |
| 9 : 12 YEARS      | 3789  | 1475  | 771  | 535  | 714  | 209  | 300  | 235  |
| 10 : 13 YEARS     | 4355  | 1541  | 806  | 540  | 752  | 216  | 313  | 250  |
| 11 : 14 YEARS     | 4767  | 1581  | 833  | 545  | 800  | 228  | 326  | 258  |
| 12 : 15 YEARS     | 5011  | 1600  | 849  | 549  | 825  | 238  | 336  | 265  |
| TT : TABLE TOTAL  | 36490 | 15814 | 8486 | 6274 | 7877 | 2313 | 3258 | 2578 |

AVERAGE VALUES OVER 30 GIRLS

The areas of the circles have been made proportional to the ave-rage-weights of the corresponding age groups, in order to restore the component of size on the map. The positions of the age groups on the map depend on their affinities for the various measurements.

Note that the rôles of age groups and of measurements are interchangeable.  (The method is invariant under a transposition of rows
and columns in the data table).

Interpretation of the map of Figure 1 can be stated as follows.
Along the horizontal axis Weight increases most rapidly with age.
Head circumference varies most slowly.

In the middle part of the map one finds a cluster formed by the
remaining six measurements that vary more or less in synchrony.
Along the vertical axis one suspects a differential effect between
Length and Chest or Upper arm circumferences.  Length is favored over
circumference during early age.  The ratio of length to circumference
remains constant during the middle period and reverses during the
last years of the series.

In order to study this effect more closely we decided to zoom-in
on the middle section of the map.  This has been achieved by removing Weight from the analysis and by concentrating on the remaining
seven measurements.

## 3. Analysis of age vs seven measurements

In Figure 2 we show the spectramap of twelve age groups with
respect to a reduced morphometry.  Weight has been removed from the
analysis and has been projected on the new result.  Its position
appears at the extreme right border, while in fact it lies far outside the frame of the map.

The previously observed effects find confirmation on Figure 2.
Head circumference is shown to grow least rapidly when compared to
length, width and circumference measurements.  A strong differential
effect is observed between Length and Chest or Upper arm circumference.  The proximity between the latter two measurements is the
result of the strong intercorrelation between the ratios of Length
to Chest circumference and of Length to Upper arm circumference.

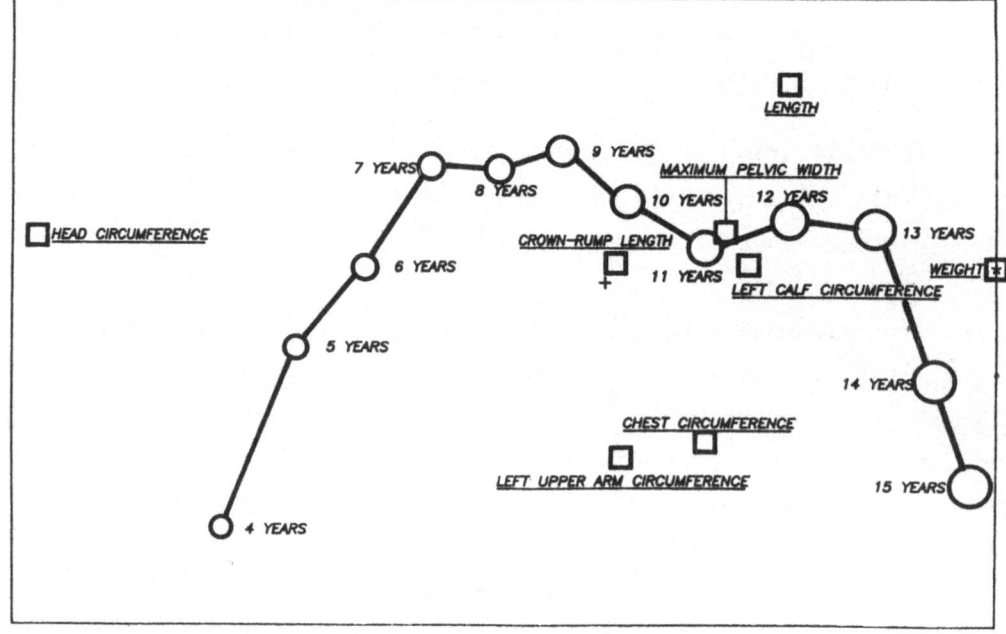

Fig. 2

Length increases more rapidly than Chest circumference between 4 and 7 years. From 7 to 13 years the ratio of Length to Chest circumference appears to be almost constant. Between 13 to 15 years Chest circumference grows more rapidly than Length.

The most relevant characteristic ratios for this morphometry are based on the three outstanding poles of the map : i.e. Head circumference, Length and either Chest of Upper arm circumference. We chose Chest circumference, since this measurement may possess the smallest relative error. (We found, however, that Chest and Upper arm circumferences are almost interchangeable in the present analysis.)

## 4. Analysis of age vs three measurements

Figure 3 shows the pattern of age groups with respect to Head circumference, Length and Chest circumference. The triangular Spectramap differs from the previous ones by the fact that the three

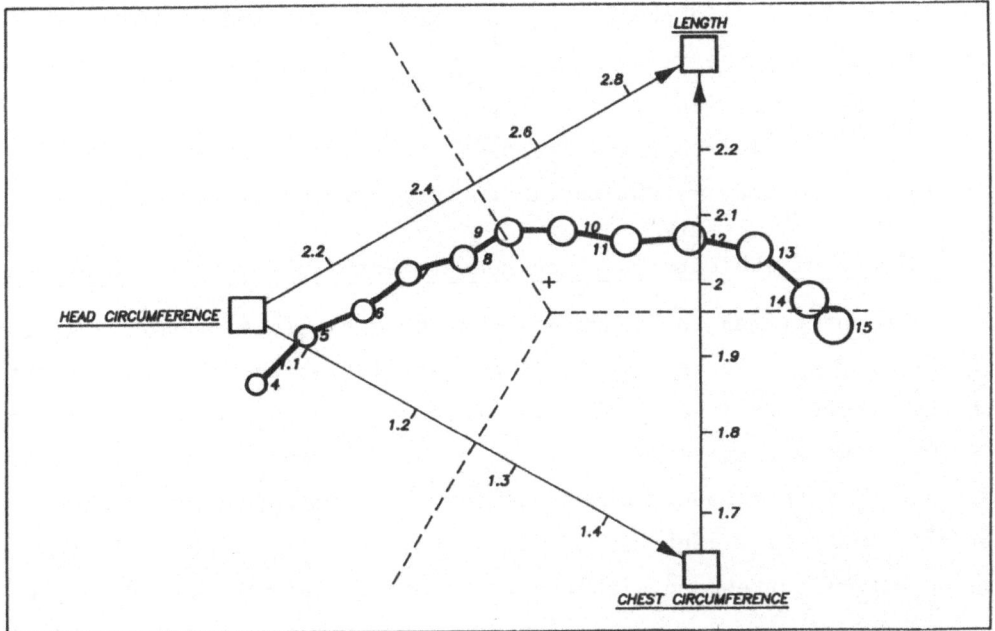

Fig. 3

poles have been fixed at the corners of an equilateral triangle.  As
a result one obtains more regular calibrations of the three ratios
than would be possible on a general type of mapping.




    Note that the coordinates of a point of the triangular map are
obtained by perpendicular projection upon the sides of the triangle.
(This follows from the arguments developed in the method section.
Matrix multiplication of factor scores with the transpose of factor
loadings reproduces the contrasts among the measurements).  For
example, by vertical projection one establishes that the ratio of
Length to Chest circumference in the age group of 10 years amounts
to 2.08.  Likewise, the corresponding ratio of Length to Head circumference equals 2.55.  Any two triangular coordinates suffice to
locate a given point on the triangular map.

When Figure 3 is compared with Figure 2, we find that the general pattern is preserved.  Hence, we will base our subsequent analyses on this simplified triangular Spectramap.

It is recalled that the areas of the circles are proportional to the average body weight in the corresponding age groups.

## 5. Analysis of subjects vs three measurements in selected age groups

The Spectramap in Figure 4 shows three clusters of thirty girls at the ages of 4 years (left), 10 years (middle) and 15 years (right) in the reduced frame of Head circumference, Length and Chest circumference.  Note that the areas of the circles, representing subjects, are scaled with respect to body weight.  The ellipses around the clusters represent contours of 95% probability.  The barycenters of the clusters correspond with the locations of the corresponding age

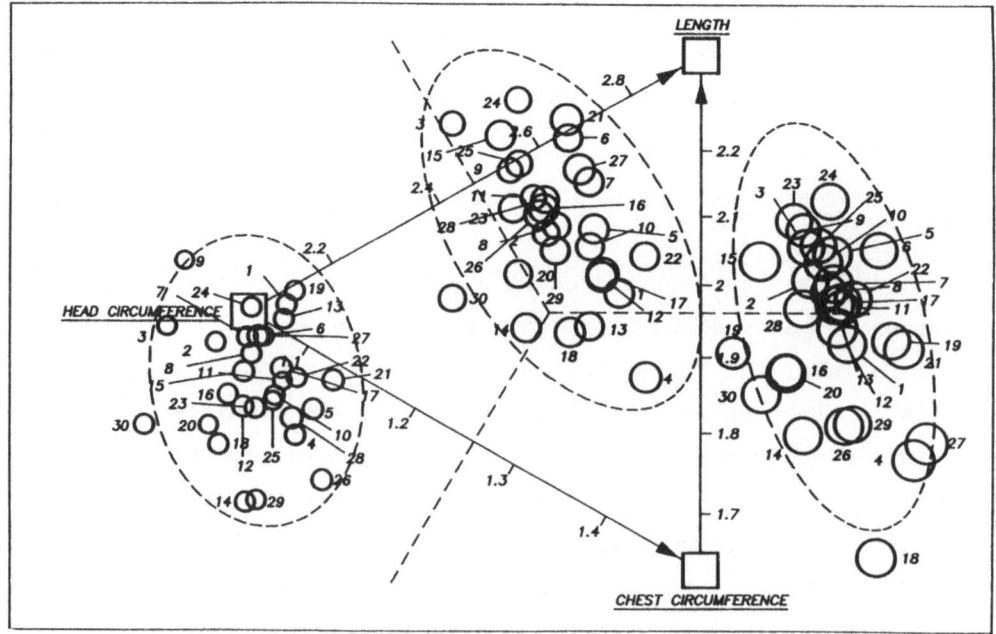

Fig. 4

groups at 4, 10 and 15 years on the previous Figure 3. One observes
a gradual increase in size and excentricity of the ellipses with advancing age. The meaning of this might be that the Length/Chest circumference ratio becomes more variable with age when compared to the
Length/Head circumference ratio.

It is also observed that the major axis of the ellipses remain
more or less orthogonal to the Length/Head circumference axis on the
map. This suggests that this ratio possesses a smaller variability
than the other two ratios of Length/Chest circumference and Head/
Chest circumference.

The midpoint of the map defines the modal point over the combined subjects and age groups.

Of course, clusters and equiprobability ellipses can be constructed for each of the twelve age groups. We only show three age
groups for the purpose of illustration. Figure 4 could be used for
the detection of pathological morphologies, retarded developments
and precocious growth patterns.

## 6. Analysis of subjects vs three measurements at a selected age

Figure 5 represents the Spectramap obtained by zooming-in on the
middle cluster fo Figure 4. It represents an enlarged view of the
10-year group. The three ellipses represent contours of 60, 80 and
95 percent probability, respectively.

Similar diagrams may be constructed for each of the twelve age
groups. They may be used to position an individual within its own
age group in order to detect abnormal morphologic ratios.

The broken line segments through the mid-point of the map deli-
mit three areas of diagnostic interest. In the upper right corner,
Length dominates over both Head and Chest circumferences. In the
lower right corner Chest circumference is dominant (e.g. subject
19). Finally, in the leftmost area both Length and Chest circumferences are retarded with respect to Head circumference (e.g. subject 30).

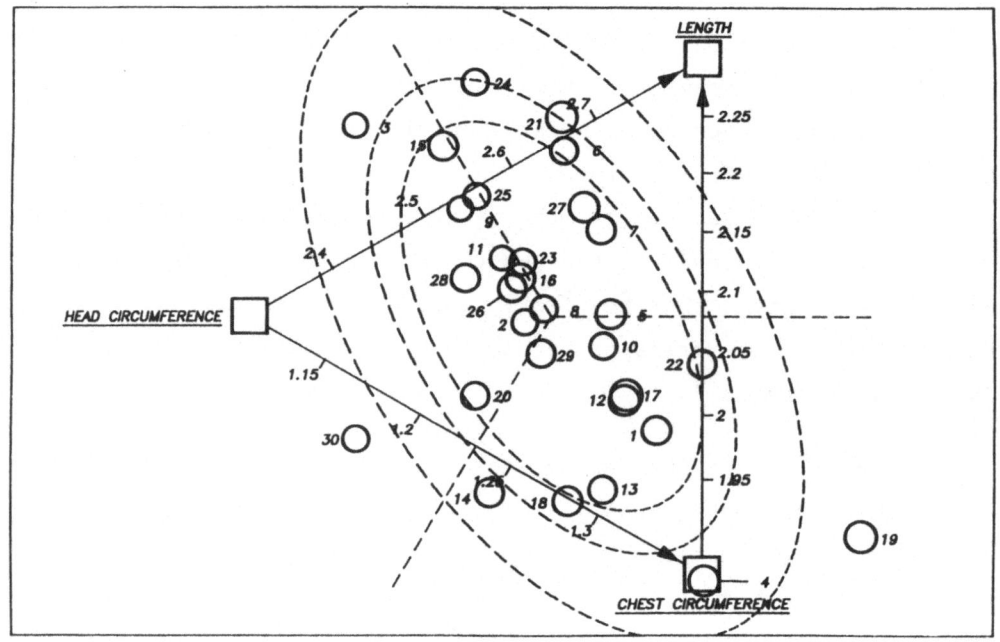

Fig. 5

## 7. Analysis of age vs three measurements for selected subjects

Figures 6, 7 and 8 show simplified morphologic growth patterns. Each of the three points on the broken lines represent the position of a subject at 4, 10 and 15 years respectively. Figure 6 shows two apparently normal growth patterns for a leptosomatic type (subject 3) and for a pyknic type (subject 14). In Figure 7 we have grouped the growth curves of three individuals which fail to show a reversal of the length/circumference ratio. The growth patterns of subjects 1 and 13 are almost flat.

The other extreme is shown in Figure 8, where a very sharp change in the growth pattern is observed in subjects 21 and 26.

It appears that growth patterns can be classified into three categories. These three categories are differentiated by either normally inflected, abnormally flat or abnormally inflected growth patterns.

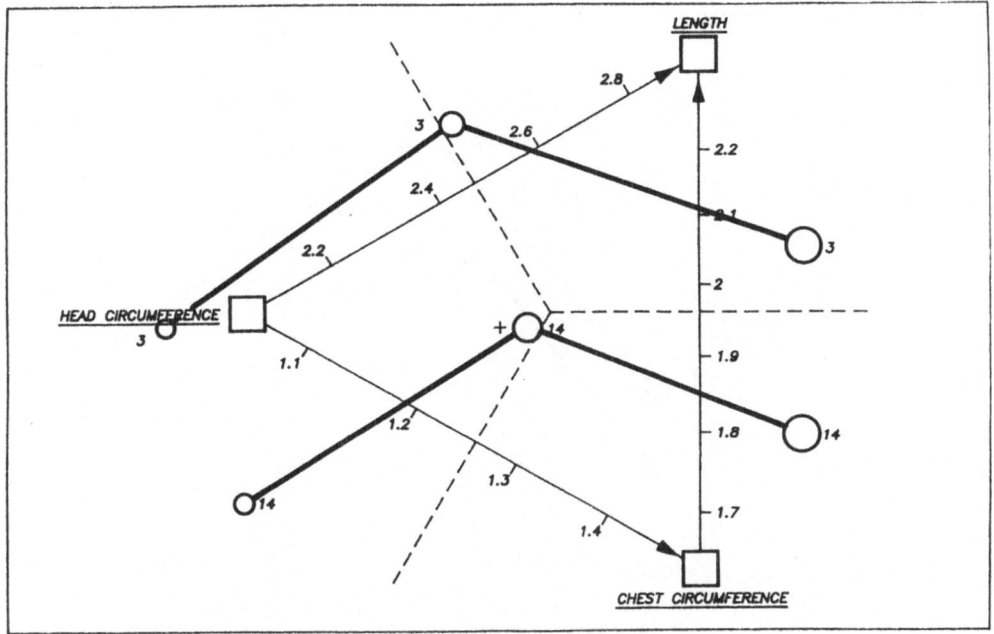

Fig. 6

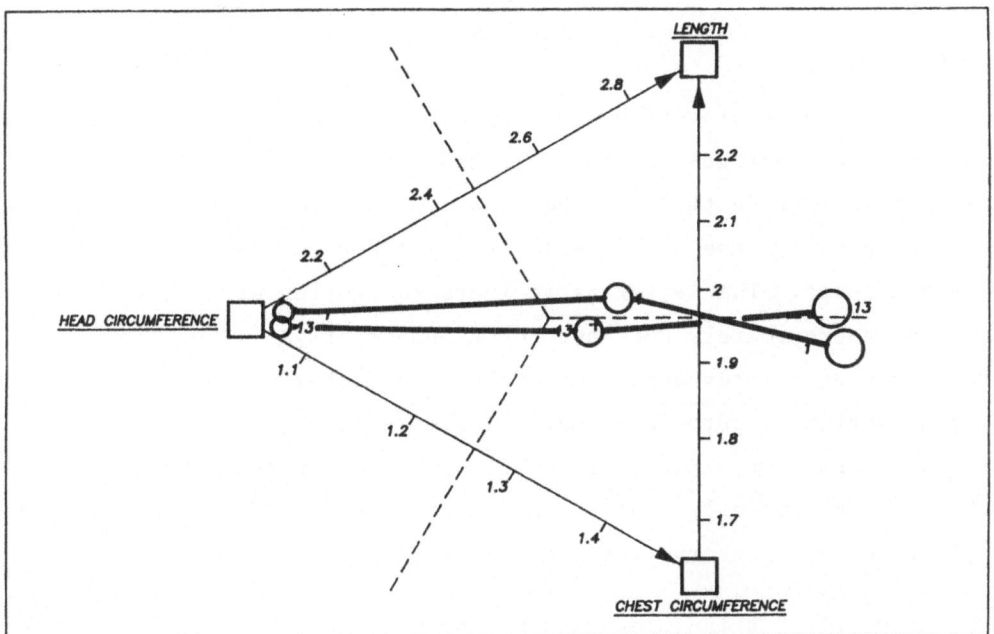

Fig. 7

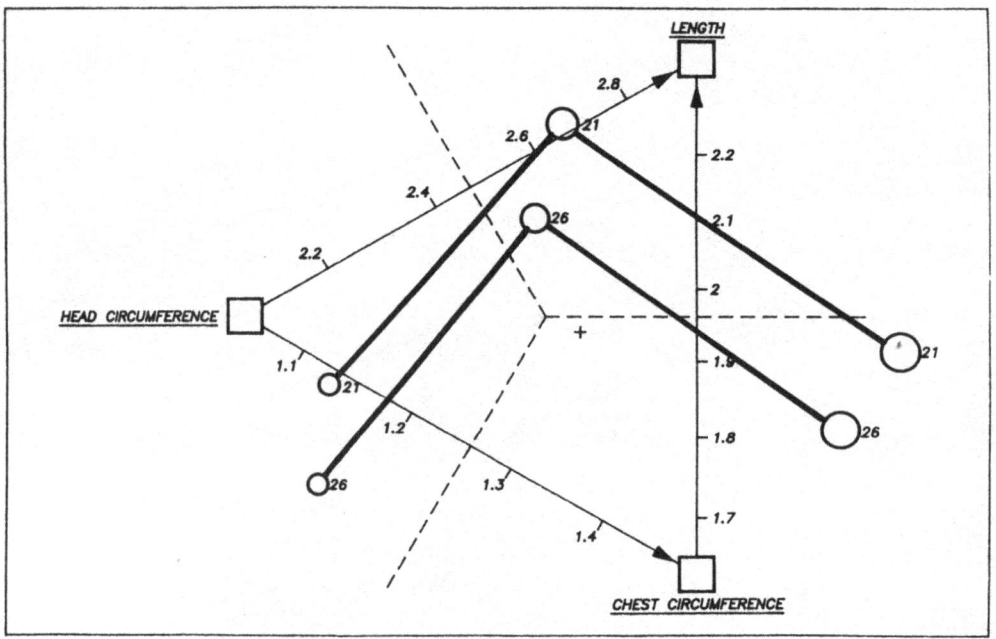

Fig. 8

Conclusion

In the analysis of the problem of longitudinal morphometric measurements by spectral mapping we found two important ratios that
explain salient features in the data. One ratio explains the overall
rate of growth; the other reflects the change of the length/circum-
ference ratio. The latter ratio increases during early childhood,
remains approximately constant for a number of years and finally
decreases at a later age. Our analysis produces a graphic means for
the detection of pathologic morphologies at any given age, which
requires only three measurements (Length, Head circumference and
Chest circumference). The graphical representation may help in classifying growth patterns into broad categories.

Simplified triangular maps based on these three key measurements
have been derived for specific age groups (Figure 5). Once these
three measurements have been obtained for a particular subject, its

corresponding location on the map can be established manually.  It
suffices to compute two characteristic ratios (e.g. Length to Chest
circumference and Length to Head circumference) and to determine the
point of intersection at the corresponding coordinates on the triangular map.  These three measurements should be readily accessible in
medical practice.

Note that the calibrated diagram is used in the same way as a
Cartesian diagram, except for the fact that the coordinate axes are
at an angle of sixty rather than ninety degrees.

SMA can be shown to be a robust method of analysis.  Small per-
turbations or inaccuracies in the data do not greatly alter the
appearance of the map.

The prime objective of SMA in this study has been to visualize
contrasts among subjects and between measurements made upon them.
Since these contrasts are expressed in the form of ratios, they can
be readily verified from the original data.  Spectral maps add value
to data that often have been acquired at great cost of time and money.

Summarizing, the main advantages of using SMA in longitudinal
measurements on growing children is the direct link between the tabu-
lation and graphicatior (by means of ratios).  As a collateral bene-
fit of SMA, one may obtain simplified spectral maps in the form of
an equilateral triangle with a perpendicular projection rule.

References

J.-P.Benzécri (1973). L'analyse des données, Vol.II, L'analyse des
        correspondances, Dunod, Paris.
L.J.Cronbach and G.C.Gleser (1953). Assessing similarities between
        profiles, Psychol.Bull., 50, 456-473.
K.R.Gabriel (1971). The biplot-graphic display of matrices with
        application to principal components analysis, Biometrika, 58,
        453-467.

C.Lavit and M.-O.Pernin (1985). Multivariate and longitudinal data on growing children. Problem submitted to the 4th Symposium on Data Analysis, Brussels.

P.J.Lewi (1976a). Spectral mapping, a technique for classifying activity profiles of chemical compounds. Drug.Res. (Arzneim.Forsch.), 26, 1295-1300.

P.J.Lewi, W.F.M.Van Bever and P.A.J.Janssen (1976b). Classification and discrimination for data analysis in pharmacology. Eur.J. Pharmacol., 35, 403-407.

P.J.Lewi and F.C.Colpaert (1976c). On the classification of antidepressant drugs. Psychopharmacology, 49, 219-224.

P.J.Lewi (1978). Classification of dopamine antagonists from pharmacological data, Life Sc., 23, 519-522.

P.J.Lewi (1979). The use of multivariate statistics in industrial pharmacology. In : Encyclopaedia of pharmacology and therapeutics. Section 7, A.C.Delaunois, Ed., Vol.3, Pergamon Press, Oxford, pp.1149-1228.

P.J.Lewi (1980a). Multidimensional analysis of pharmacological data. Rev.pure appl.pharmacol., 2, 229-290.

P.J.Lewi (1980b). Multivariate data analysis in structure-activity relationships. In : Drug Design, Vol.X, E.J.Ariëns, Ed., Academic Press, New York, pp.307-342.

P.J.Lewi (1982). Multivariate analysis in industrial practice. Research Studies Press (J.Wiley), Chicester, Engl.

P.J.Lewi (1984a). Multivariate data representation in medicinal chemistry. In : Chemometrics. Mathematics and statistics in chemistry. B.R.Kowalski, Ed., Reidel Publ., Dordrecht, The Neth., pp.351-376.

P.J.Lewi (1984b). Picturing the pharmaceutical environment by Spectramap, Pharmacy Intl., 5, 44-46.

G.Chr.Lichtenberg (1775). Opera inedita Tobiae Mayeri I, Goettingen (Translated in English by E.G.Forbes, McMillan, London, 1971).

SECTION 1.5


SOLUTION USING "LONGI"

Jacques Pontier and Marie-Odile Pernin

Laboratoire de Biométrie, Université Claude Bernard

Villeurbanne, France

## 1. Introduction

Auxology, which is "la science métrique de la croissance"
(the metric science of growth) according to Godin (1903), is, in
human biometrics, a very important example of a field which requires
a methodology adapted to the analysis of multivariate longitudinal
data.

Auxological data are multivariate, as generally several characters are concerned by each individual observation. These characters are either quantitative (measurements) or qualitative (skeletal or sexual maturation). Auxological data are longitudinal,
for a correct study of growth needs to be supported by repeated
observations of the same individuals, throughout their growth period.

In practical auxology, the difficulties are numerous, essentially due to the fact that the gathering of data continues over a
period of about twenty years. For example, during this time, a
number of the chosen individuals do not participate sufficiently;
some even disappear altogether. This inevitable and gradual phenomenon of disappearance leads to the problem of missing data, and

any method of analysis of this type of data should take such problems into account.  The set of auxological data, based on the auxological survey carried out in France from 1953 to 1975, under the
successive responsibilities of Dr Marie-Paule Roy-Pernot and Dr
Michel Sempé, does not elude the phenomenon of gradual disappearance
of individuals under study.  Out of the 497 children present at the
beginning of the survey, 30 (18 boys and 12 girls) remained 22 years
later, at the end of the survey.  Interested readers may, for historical and technical details of this survey, refer to Sempé & al.
(1976; 1979).

Here our purpose is to produce a method for building multivariate auxological indices, each of which possesses any previously
defined properties pertinent to the phenomenon of growth.

Let us examine one given individual, from year to year;  we
observe an evolution of every character under study (generally, this
evolution goes towards increase).  The evolutions of two different
characters are not necessarily parallel.  Figure 1 shows the case
(girl n°6) of length, which has a regular evolution, and chest
circumference, evolution of which is quite irregular.  So, one of
our aims will be the search for a global evolution index, a function of the observed characters, which summarizes at any given age
the stage of growth in terms of the entire set of these characters.
We wish to bring out the growth phenomenon, eliminating if possible
the variability among individual subjects.  So this global index
should depend as much as possible on the age of the individual, and
as little as possible on the set of characters of the individuals.

Let us now examine, for one given character, how the individuals from the same cohort are situated among themselves.  We
observe that an individual situation is possibly modified from year
to year.  Figure 2 shows the evolutions of relative situations of
individuals numbers 6, 11, 18, 27, 28, 30, for the character "weight".

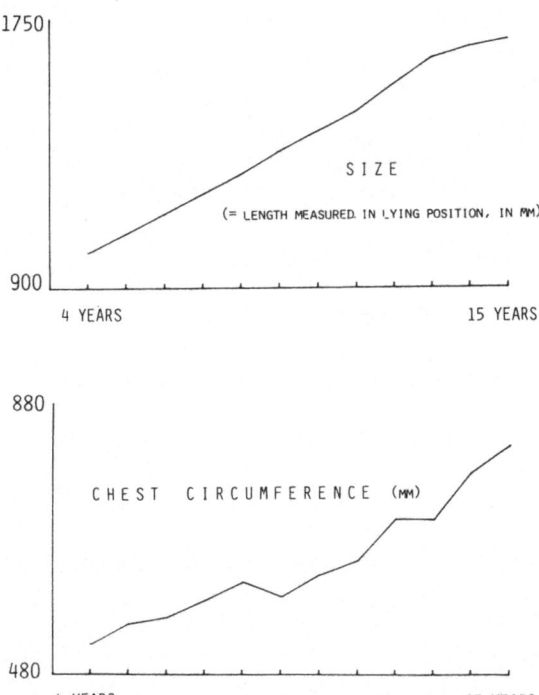

Fig.1. An example of the evolution of two somatic characters (length,
and chest circumference) for the same individual (number 6)


This situation is characterized by the date-to-date standardized
weight (at each date, the mean and standard deviation of the sample
are calculated, and at this date, data are transformed by subtracting
this particular mean and dividing the result by this particular standard deviation). This artificial "variable" seems to us better adapted than the simple "individual rank" variable because of the possibility of a continuous variation, a better knowledge of the gap
between two individuals, and the possibility of introducing a new
individual without any modification of the values determined for the
others. In Figure 2, for example, we observe that individual number
11 is always the lightest of the group; that number 28, the heaviest
when 4 years old, is progressively overtaken by numbers 27, 18 and

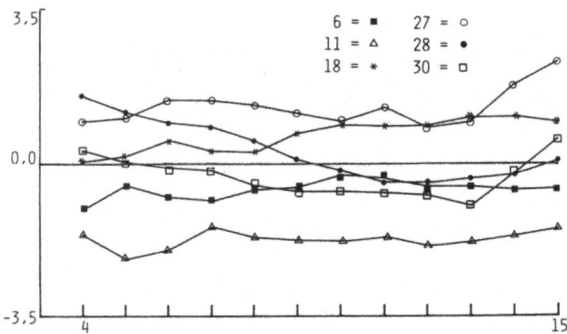

Fig.2. Relative evolutions of weight, for 6 individuals (numbers 6, 11, 18, 27, 28, 30).

30; etc.  So, our second aim will be the construction of a global situation index, combining the characters thus standardized date by date.  This index will give each individual, at any time, a general "score" which situates this individual "as exactly as possible" among the entire group of individuals, depending on the observed values for every character.  We wish to eliminate from this index, the aspects due to the growth phenomenon (which is common to all individuals), and to bring out the aspects due to individuality of each subject.  So this global situation index should depend as much possible on the set of characters of individuals and as little as possible on the age of the individuals.

## 2. Principles and practical use of "LONGI"

Our purpose is to build some multivariate functions of the observed characters, so as to reach the above-mentioned goals.  These functions will be linear combinations of variables.  In order to facilitate comparison of groups of individuals, these functions will be standardized (zero mean and unit variance).  The method we have used to build these functions is the "LONGI" method, the principles and practical use of which we shall describe in §§ 2.2 and 2.3. Beforehand, we present in § 2.1 the so-called "Euclidian model", which will serve to describe both the problems posed by the study, and their solution by means of LONGI.

## 2.1. The Euclidian model

We call an observation unit, the fact of observing for one individual, at a given date (age), states of the p characters (variables) $X_1, X_2, \ldots, X_p$. Therefore, the observation of the individual $I_i$, at the date $D_j$, leads to the observation unit $\omega_{ij}$, to which are associated :

$$\begin{cases} i, \text{ the number of the observed individual} \\ j, \text{ the number of the date of this observation} \\ x_{1ij}, x_{2ij}, \ldots, x_{pij}, \text{ the respective observed numerical values} \\ \qquad \text{of the } p \text{ characters} \end{cases}$$

Let us designate $\Omega$ the finite, weighted set of all observation units, and $[\Omega]$ the set of the applications $\Omega \to \mathbb{R}$. The total number of observation units is m, less than or equal to $n \times k$ (n = size of the set of individuals; k = the number of all observation dates; there is a possibility of missing data). The weight of the observation unit $\omega$ is designated $\pi(\omega)$. So, $[\Omega]$ is an m-dimensional Euclidian space, with the scalar product $\varphi$ defined as follows :

$$X, Y \in [\Omega] \quad \to \quad \varphi(X,Y) = \sum_{\omega \in \Omega} \pi(\omega) X(\omega) Y(\omega)$$

The set of the m indicative functions of the observation units (i.e., the indicative function $f_{ij}$ of the unit $\omega_{ij}$ is such that $f_{ij}(\omega) = 1$ if $\omega = \omega_{ij}$, = 0 if $\omega \neq \omega_{ij}$), is a $\varphi$-orthogonal basis of $[\Omega]$; this basis will be called the indicative basis of $[\Omega]$.

All available information is presented in the form of a table such as that shown in Table 1. Each row of the table is associated with an observation unit, each column has an interpretation as the matrix of an element of $[\Omega]$ (matrix relative to the indicative basis).

Table 1. Presentation of data

| o. u. | $I_1$ | $I_2$ | ... | $I_n$ | $D_1$ | $D_2$ | ... | $D_k$ | $X_1$ | $X_2$ | ... | $X_p$ |
|---|---|---|---|---|---|---|---|---|---|---|---|---|
| $\omega_{11}$ | 1 | 0 | ... | 0 | 1 | 0 | ... | 0 | $x_{111}$ | $x_{211}$ | ... | $x_{p11}$ |
| $\omega_{12}$ | 1 | 0 | ... | 0 | 0 | 1 | ... | 0 | $x_{112}$ | $x_{212}$ | ... | $x_{p12}$ |
| ⋮ | ⋮ | ⋮ | | ⋮ | ⋮ | ⋮ | | ⋮ | ⋮ | ⋮ | | ⋮ |
| $\omega_{1k}$ | 1 | 0 | ... | 0 | 0 | 0 | ... | 1 | $x_{11k}$ | $x_{21k}$ | ... | $x_{p1k}$ |
| ⋮ | ⋮ | ⋮ | | ⋮ | ⋮ | ⋮ | | ⋮ | ⋮ | ⋮ | | ⋮ |
| $\omega_{n1}$ | 0 | 0 | ... | 1 | 1 | 0 | ... | 0 | $x_{1n1}$ | $x_{2n1}$ | ... | $x_{pn1}$ |
| $\omega_{n2}$ | 0 | 0 | ... | 1 | 0 | 1 | ... | 0 | $x_{1n2}$ | $x_{2n2}$ | ... | $x_{pn2}$ |
| ⋮ | ⋮ | ⋮ | | ⋮ | ⋮ | ⋮ | | ⋮ | ⋮ | ⋮ | | ⋮ |
| $\omega_{nk}$ | 0 | 0 | ... | 1 | 0 | 0 | ... | 1 | $x_{1nk}$ | $x_{2nk}$ | ... | $x_{pnk}$ |

Each one of the $I_i$ elements (indicative functions of the individuals) is $\phi$-orthogonal to each of the others. The $I_i$'s generate
in $[\Omega]$ the n-dimensional subspace $[I]$, of which their set is a
$\phi$-orthogonal basis. Each of the $D_j$ elements (indicative functions
of observation dates) is $\phi$-orthogonal to each of the others. The
$D_j$'s generate in $[\Omega]$ the k-dimensional subspace $[D]$, of which
their set is a $\phi$-orthogonal basis.

Any set of elements $Y_1, Y_2, \ldots, Y_q$ in $[\Omega]$ generates a subspace
$[Y]$, the dimension of which is less than or equal to q. This set
is not necessarily a basis of $[Y]$.

Let us consider X, a character observed for the individuals
under study. We associate two artificial variables X' and X"
with this X. These variables X' and X" have the following
definitions :

* X' is obtained from X by a global standardization, that

is :

$$X'(\omega_{ij}) = [X(\omega_{ij}) - m_X]/s_X$$

where $m_X$ and $s_X$ are respectively the overall mean and standard deviation of X; this transformation eliminates artefacts due to differences of measurement units or magnitudes.

* X" is obtained from X, by a standardization at any date, on the whole set of the observations made on this date :

$$X''(\omega_{ij}) = [X(\omega_{ij}) - m_{Xj}]/s_{Xj}$$

where $m_{Xj}$ and $s_{Xj}$ are respectively the mean and standard deviation of X at the date $D_j$; this transformation eliminates the trend of mean and the trend of variance with the passage of time.

## 2.2. Principles of the LONGI method

In both situations (search for a global evolution index, and search for a global situation index), the problem may be expressed as follows : search for a linear combination of variables $Y_1, Y_2, \ldots, Y_q$, with an additive constant, so that this combination will be as near as possible to the subspace [A], and as far as possible from the subspace [B]. According to the classical properties, in [$\Omega$], of the least squares proximity, $\phi$-orthogonality, and canonical analysis between two subspaces, we choose the following solution to the problem :

The linear combination we are looking for is, in the subspace [$\tilde{Y}$] generated by $\{1, Y_1, Y_2, \ldots, Y_q\}$, the first non-trivial canonical component obtained by canonical analysis between [$\tilde{Y}$] and $[A] \cap [B]^\perp$, where $[B]^\perp$ denotes the $\phi$-orthogonal supplement of [B] in [$\Omega$].

So, the search for a global evolution index (a multivariate analogue of an evolutive character), leads us to search for a linear combination of variables $1, X_1', X_2', \ldots, X_p'$, depending the most on observation dates and the least on individuals. The solution we choose is the element of $[\tilde{X}']$ which is nearest to the subspace $[D] \cap [I]^{\perp}$. This element is, in $[X']$, the first non-trivial canonical component in the canonical analysis between $[\tilde{X}']$ and $[D] \cap [I]^{\perp}$.

In the same way, the search for a global situation index (a multivariate analogue of a variable such as $X_s''$) leads us to search for a linear combination of variables $1, X_1'', X_2'', \ldots, X_p''$, depending the most on individuals and the least on observation dates. So the solution is, in $[\tilde{X}'']$, the first non-trivial canonical component in the canonical analysis between $[\tilde{X}'']$ and $[I] \cap [D]^{\perp}$.

## 2.3. Practical use of the LONGI method

The two above-mentioned canonical analyses are easily made if we know a basis (if possible a $\phi$-orthogonal, or even better a $\phi$-orthonormal basis) in each of the four subspaces $[\tilde{X}']$, $[\tilde{X}'']$, $[D] \cap [I]^{\perp}$, $[I] \cap [D]^{\perp}$.

In $[\tilde{X}']$, we choose the principal basis formed by the unit vector $1$ and the principal components (associated with nonzero eigenvalues) of the set of variables $\{X_1', X_2', \ldots, X_p'\}$. In $[\tilde{X}'']$, we choose the principal basis formed in a similar manner.

As for the bases in $[D] \cap [I]^{\perp}$ and $[I] \cap [D]^{\perp}$ respectively, we build them using a method which proceeds from the properties of canonical analysis between two subspaces $[A]$ and $[B]$ of $[\Omega]$ as described below.

Canonical analysis between $[A]$ (a a-dimensional subspace), and $[B]$ (a b-dimensional subspace), consists in the search for eigenvalues and eigenvectors of the two matrices :

$$\Phi_{AA}^{-1} \Phi_{AB} \Phi_{BB}^{-1} \Phi_{BA} \quad (\text{dim. } a \div a) \quad \text{and} \quad \Phi_{BB}^{-1} \Phi_{BA} \Phi_{AA}^{-1} \Phi_{AB} \quad (\text{dim. } b \div b)$$

$[\Phi_{UV}$ designates the matrix the elements of which are scalar products between vectors from a set $\{U\}$ and vectors from a set $\{V\}$; here $[A]$ and $[B]$ designate respectively the bases of each subspace $[A]$ and $[B]$ of $[\Omega]]$

These two matrices have $\nu + \mu$ common eigenvalues :
- $\nu$ eigenvalues which are equal to 1 ($\nu$ may be zero);
- $\mu$ eigenvalues which are strictly between 0 and 1 ($\mu$ may be zero).

The $a-\nu-\mu$ (resp. $b-\nu-\mu$) other eigenvalues are all equal to zero.

In any of the two subspaces $[A]$ and $[B]$, each of the eigen-vectors (canonical components) are $\phi$-orthogonal to each of the others.

The subspace $[A]$ (resp. $[B]$) is a direct sum of the three following subspaces
- $[A_1]$ (resp. $[B_1]$), $\nu$-dimensional subspace generated by those eigenvectors associated with unit eigenvalues;
- $[A_2]$ (resp. $[B_2]$), $\mu$-dimensional subspace generated by those eigenvectors associated with eigenvalues strictly between 0 and 1;
- $[A_3]$ (resp. $[B_3]$), $(a-\nu-\mu)$-dimensional (resp. $(b-\nu-\mu)$-dimensional) subspace generated by those eigenvectors asso-ciated with null eigenvalues.

We have the following :

- $[A_1] = [B_1] = [A] \cap [B]$;
- Any non-null element from $[A_2]$ (resp. $[B_2]$) forms an angle with any non-null element from $[B]$ (resp. $[A]$), this angle is neither zero nor right;
- Any non-null element from $[A_3]$ is $\phi$-orthogonal to any non-null element from $[B]$ : $[A_3] = [A] \cap [B]^{\perp}$;

•   Any non-null element from  $[B_3]$  is $\phi$-orthogonal to any non-null element from  $[A]$ : $[B_3] = [B] \cap [A]^\perp$.

Obviously, it is the very last part of this long proposition which leads to the solution of our problem. A $\phi$-orthogonal basis of  $[D] \cap [I]^\perp$, and a $\phi$-orthogonal basis of  $[I] \cap [D]^\perp$, are obtained from a complete canonical analysis between  $[D]$  and  $[I]$. These two bases are each formed by the respective sets of eigenvectors associated with null eigenvalues. Note that this sort of eigenvalue necessarily exists, at least for one of the two subspaces, if dimensions  k  and  n  are different. Also note the possibility that for any space  $[D]$  or  $[I]$, none of the eigenvalues is null (for example when the proportion of missing data is high). In this case, lacking a better method, we use the eigenvector associated to the least (non-null) eigenvalue, to define the basis we seek. Then the final phase of the resolution of the problem simply consists, in this case, of a multivariate linear regression, $\phi$-orthogonal projection of this eigenvector on  $[\tilde{X}']$  or on  $[\tilde{X}'']$, depending on the situation.

## 3. Application of the LONGI method to the case of 8 somatic characters

We applied the LONGI method to the multivariate longitudinal data from 30 girls, each of them examined 12 times (at the ages of 4 years, 5 years, ..., 15 years), for 8 somatic characters (weight, length, crown-coccyx length, head circumference, chest circumference, left upper arm circumference, left calf circumference, maximum pelvic width). No data is missing. This is a résumé of our results.

### 3.1. First global evolution index

This index is the first canonical function (component) from the canonical analysis between  $[\tilde{X}']$  and  $[D] \cap [I]^\perp$. This index, designated as IGE1, globally expresses the simultaneous development of characters  $X_1, X_2, \ldots, X_8$, for a given individual.

Table 2. Line 1 : correlations between chronological age and variables
$X_1', X_2', \ldots, X_8'$;
Line 2 : correlations between the function IGE1 and the same
variables.

|      | $X_1'$ | $X_2'$ | $X_3'$ | $X_4'$ | $X_5'$ | $X_6'$ | $X_7'$ | $X_8'$ |
|------|--------|--------|--------|--------|--------|--------|--------|--------|
| age  | .931   | .971   | .953   | .798   | .911   | .839   | .926   | .915   |
| IGE1 | .943   | .994   | .969   | .813   | .921   | .848   | .942   | .930   |

We observe :

● a high linear correlation (+ .976) between IGE1 and chronological
age of individuals;

● high correlations between IGE1 and each of the 8 variables;  these
correlations are systematically higher than those between chronological age and these variables (see Table 2).

Figure 3 shows, for 6 individuals (as an example), the variation
of IGE1 versus chronological age.  It can be compared to any univariate case (refer to Fig.1).

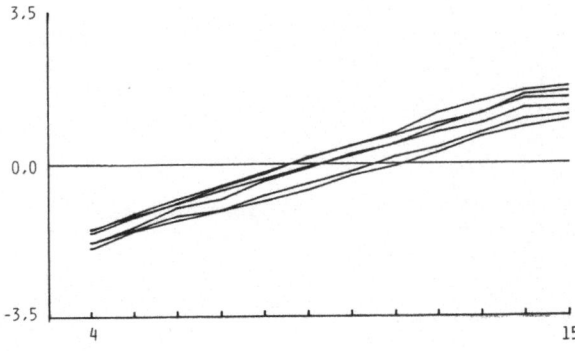

Fig.3. IGE1 versus chronological age (individuals numbers 6, 11, 18,
27, 28, 30)

Table 3. Correlations between IGS1 and variables $X_1'', X_2'', \ldots, X_8''$

|      | $X_1''$ | $X_2''$ | $X_3''$ | $X_4''$ | $X_5''$ | $X_6''$ | $X_7''$ | $X_8''$ |
|------|------|------|------|------|------|------|------|------|
| IGS1 | .762 | .794 | .831 | .878 | .504 | .384 | .475 | .785 |

## 3.2. First global situation index

This index is the first canonical function from the canonical analysis between $[\widetilde{X}'']$ and $[I] \cap [D]^{\perp}$. This index, designated as IGS1, gives each individual a global (multivariate) score which situates this individual, at any age, among the whole set of individuals.

The linear correlation between IGS1 and chronological age is null, as expected. Correlations between IGS1 and the date-to-date standardized variables $X_1'', X_2'', \ldots, X_8''$ are miscellaneous (see Table 3).

As an example, Figure 4 shows variations of IGS1 versus chronological age, for the same six individuals as those in Figure 3.

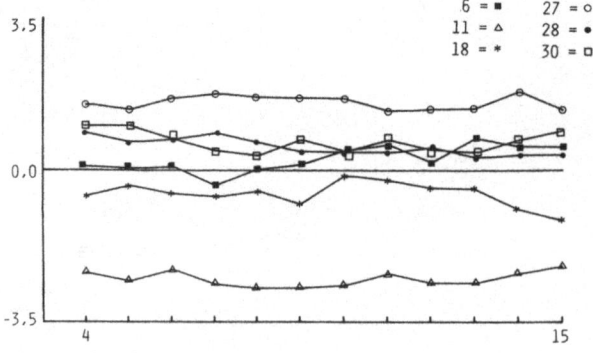

Fig.4. IGS1 versus chronological age (same six individuals as in Fig.3)

Table 4. Values of the coefficients of determination for each of the
8 IGE indices, and for each of the 8 IGS indices.  The percentages are relative to the sum of the coefficients.

|     | IGE   |      | IGS   |      |
| --- | ----- | ---- | ----- | ---- |
| 1.  | .9574 | 58%  | .9282 | 18%  |
| 2.  | .5071 | 88%  | .8784 | 35%  |
| 3.  | .1183 | 95%  | .7891 | 50%  |
| 4.  | .0621 | 99%  | .7321 | 64%  |
| 5.  | .0071 | 99%  | .7107 | 78%  |
| 6.  | .0064 | 100% | .4733 | 87%  |
| 7.  | .0053 | 100% | .3667 | 94%  |
| 8.  | .0007 | 100% | .2935 | 100% |

## 3.3. Other global indices obtained from the LONGI method

As for any canonical analysis between multi-dimensional subspaces, there exist other canonical pairs, having classical properties of $\phi$-orthogonality with different pairs.  From there, we can
define other global evolution indices IGE2, IGE3, etc., and other
global situation indices IGS2, IGS3, etc., of which we suppose they
"complete", in some sense, the first indices IGE1 and IGS1.  Presently, we do not know how to interpret these complementary indices.  We
think the simultaneous consideration of several of these indices
could be an aid to the typology of individuals, and to that of forms
of evolutions over a period of time.

The knowledge of the determination coefficient (square of the
canonical correlation) associated with each of these global indices,
allows us to appreciate its degree of relative significance (see
Table 4).  The drawing of simultaneous variations of these indices,
in time, for each individual, leads to a good idea of the diversity
of configurations.

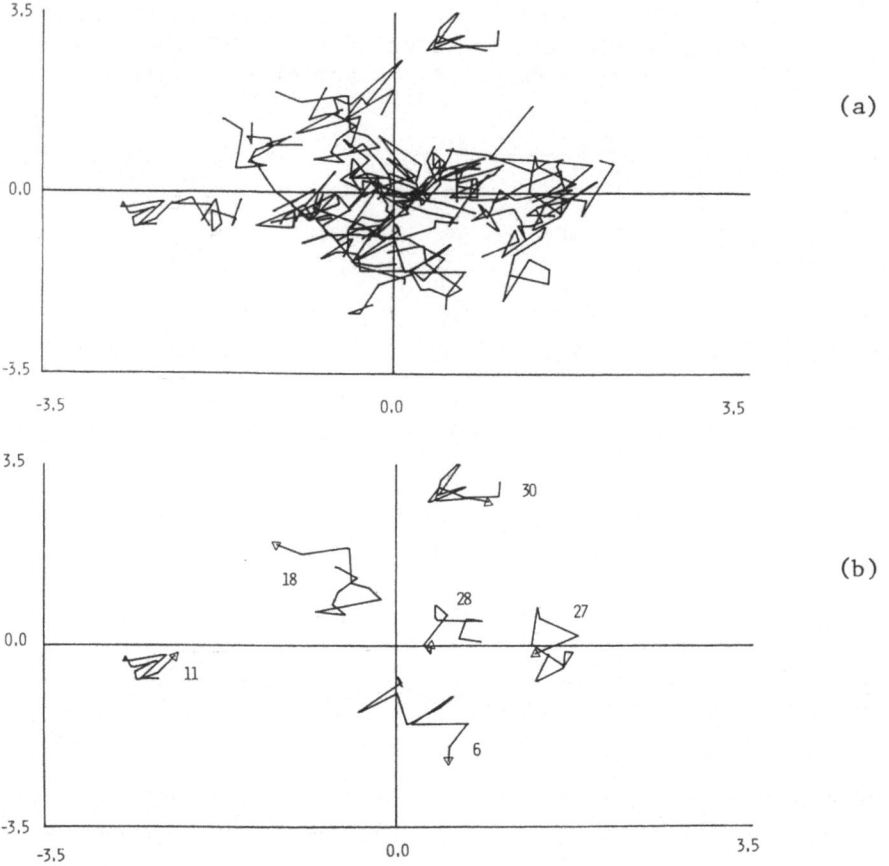

Fig.5. (a) Trajectories of the 30 individuals in the plane IGS1×IGS2.
       (b) Particular trajectories of individuals numbers 6, 11, 18,
           27, 28, 30

So, Figures 5a and 5b show the "trajectories", in the plane
IGS1 × IGS2, in 5a, for the 30 individuals, in 5b, for the particular
individuals numbers 6, 11, 18, 27, 28, 30. Perhaps a typology of
individuals could be supported by the shape and position of these
trajectories in the plane. In the same manner, a typology of individual evolutions could perhaps be based on individual trajectories
in the plane IGE1 × IGE2 (not shown here).

A study of these questions, and others, is in progress; its
results are described in a doctoral theses by M.-O.Pernin (1986).


## 4. Conclusions. Elements for a discussion

From the theory, and from practical use of the LONGI method, we
shall now draw some conclusions both from an auxological, and from
a methodological viewpoint.

### 4.1. Conclusions on the auxological level

The LONGI method not only allows a description of the set of
data, but also produces global indices which are of interest in auxological studies.  We saw a global evolution index, which at any time
evaluates the morphological stage of an individual, and which makes
it possible to follow each individual's evolution.  We also saw a
global situation index, which at any time situates any individual
among his cohort.  Peculiarly, these two indices, when calculated
for a "supplementary individual", provide information of interest
to any person concerned with the normal or pathological growth of a
child.

Let us insist on the relationship between growth and time in
the LONGI method.  Calculations never use chronological time (weeks,
months, years) as measured by a clock independent of the observed
individuals.  The only hypothesis used is that the ages at which observations were made are different;  the chronological order relation
on the set of these ages is not considered : these ages are used as
simple categories in an analysis of categorical data.  Nevertheless,
in results of calculations, we observe a high correlation (+ .976)
between the global evolution index IGE1 and chronological time;
this correlation has the same magnitude as the multiple correlation
coefficient between time and the set of 8 characters.  Therefore,
this leads to the evident fact that the human body is in itself,
by the evolution of its morphology (at least in its growth phase),

a reliable clock.  So the index IGE1 could be used as a means of
measurement or even as a definition of a "biological age" relative
to the set of characters which are included in its formulation.

In the present state of our study, we did not find an evident
typology either among morphologies of individuals, nor among morphological evolutions over a period of time.  We think this lack is
only due to the fact that we did not explore enough possibilities
of the method, especially by a systematic examination of links
between global indices and variables.

### 4.2. Conclusions on the methodological level

LONGI contributes to the methodology of analysis of a particular type of data : multiple multivariate tables indexed by a parameter.  This method is an example of a deliberate use of several
properties of canonical analysis, with a view to building any linear
combinations satisfying non-classical aims.  In particular, eigenvectors associated with null eigenvalues are rehabilitated, whereas
they are generally neglected because of their supposed lack of
interest.

This method was created to solve some problems in the field of
auxology, concerning quantitative characters.  Nevertheless, it
seems that this method, which is based on several canonical analyses,
could for this reason be extended to categorized characters.  Practical problems posed by this possible extension are under study.

As noted above (4.1), the parameter "time", on which the different tables are indexed, has not been included in the formulation
of the method.  In spite of this, any results obtained by this method
are directly and closely linked to this parameter (IGE1 index, for
instance).  Therefore, in this particular case, the method shows its
capacity to point out what in the data is intrinsically linked to
this time parameter.  This fact suggests that this method would show
an equal capacity, when applied to any other case where the indexa-

tion parameter is other than time.  In the case of time, the structure of the parameter is evident, we know it a priori (linear order).
However, for example in the case of a spatial or spatio-temporal,
parameter (pluridimensional localizations), a structure may not be
evident a priori, and could be discovered a posteriori owing to the
LONGI method.

## References

P.Godin (1903). Recherches anthropométriques sur la croissance des
    diverses parties du corps. L'adolescent-type aux différents
    âges pubertaires. (Paris, Maloine éd., 224 p.)

M.-O.Pernin (1986). Contribution à la méthodologie d'analyse de
    données logitudinales. Etude de la croissance chez l'être humain
    (Thèse de Doctorat, Université Claude Bernard Lyon I).

M.Sempé et al. (1976). Rapport d'activité terminal de l'Equipe
    Française. (Centre International de l'Enfance, XIIIème Réunion
    de Coordination des Equipes chargées des Etudes sur la Croissance
    et le Développement de l'Enfant Normal; 60 p. + ann.).

M.Sempé, G.Pédron, M.-P.Roy-Pernot (1979). Auxologie. Méthode et
    séquences. (Paris, Laboratoire Théraplix éd., 165 p.).

SECTION 1.6

SOLUTION USING CLUSTERING METHODS

Antonio Mineo

Institute of Statistics, University of Palermo

Palermo, Italy

Introduction

The main aim of this analysis is to find out typical morphologies from the multivariate and longitudinal data set on growing children and to describe the morphological evolution of the found groups of girls.  The finding out of typical morphologies is, in our opinion, strictly linked to the search of structures in the individuals and in the variables.

Our idea is that if in the evolution process of the girls some typical morphologies exist these morphologies should result in groups or clusters of girls that present, along the whole growing period, characteristics (or values of the biometric parameters) that are similar within and different between groups.

1. The data

The given data set $\{X_{ij}\}_t$ concerns a sample of 30 girls $(j = 1,2,...,30)$ measured each year between their fourth and fifteenth birthdays $(t = 4,5,...,15)$ on the eight morphological parameters $(i = 1,2,...,8)$ :

$X_1$ - Weight (dcg)

$X_2$ - Length (mm)

$X_3$ - Crown-rump length (mm)

$X_4$ - Head circumference (mm)

$X_5$ - Chest circumference (mm)

$X_6$ - Left upper arm circumference (mm)

$X_7$ - Left calf circumference (mm)

$X_8$ - Maximum pelvic width (mm)

## 2. The analysis

Our approach to the problem is iterative and based on the following steps :

1 - At each age group the 30 girls into the best number of
clusters by a suitable clustering method.  As the structures of data
may vary significantly with the age according to the growing process
of every girl, it is important to use a good clustering method to be
sure that the variation of the membership of the classified units in
each group at different ages is really due to the variation in the
structure of the data.

2 - Find out typical morphologies on the basis of the concordance at the different ages of the membership of the units to the
same cluster.

3 - Test the properties of the obtained morphologies to decide
on their acceptance or not.

4 - If the obtained clusters do not constitute typical morphologies, arrange the variables according to the importance of their
contribution to the description of the growing process of the girls
and choose those variables which have the highest explicative power.

5 - Repeat the previous steps 1, 2 and 3 to find out, if possible, better morphologies.

To describe the characteristics of the obtained morphologies
we fit to the series of the mean values of the 8 anthropometric
measures at each age the generalized logistic function :

$$\bar{X} = \frac{L}{1 + e^{f(t)}} \tag{1}$$

where $f(t)$ is a function of the age $t$ that may be developed in Taylor's series

$$f(t) = a_0 + a_1 t + a_2 t^2 + \dots + a_n t^n + \dots \tag{2}$$

and $L$ is an asymptote.


## 2.1. The clustering method

To group the 30 units at each age we use a non-hierarchical clustering method based on a Rational Choice of Seed Points around which the data units of the given data set are grouped by the Nearest Centroid Sorting Algorithm (Mineo, 1985). To avoid the influence of the different weights of the original variables, the clustering method is applied to each of the 12 matrices of 30 units in the 8 standardized variables. Each of these 12 matrices is grouped in k=6,5,4,3,2 clusters to find the best number k of clusters by means of Beale's F test (Beale, 1969).

## 2.2. The choice of the variables

To choose among the 8 variables those that have the highest explicative or discriminant power in describing the growing process of the girls we use a modified version of the "backward elimination method" (Mineo, 1970).

By supposing that each age of the girls may be described by the values of the 8 anthropometric measures, we assume the multiple linear regression model

$$t = b_0 + \sum_{i=1}^{8} b_i (X_i - M_x) \tag{3}$$

and proceed to order the 8 variables according to their explicative power. The chosen algorithm allows us (1) to calculate for each value of

$k_1 = 8(1)1$   the   $\binom{8}{k_1}$   determination coefficients *

$$R^2 = 1 - \frac{\Delta}{\Delta_{tt}} \qquad\qquad (4)$$

between the dependent variable  t  and the set of the  $k_1$  independent variables  $X_i$ , corresponding to all the possible combinations
of the 8 variables taken $k_1$ at the time and (2) to choose the determination coefficient with the highest value.  The combination of the
$8 - k_1$ variables that are absent in the chosen determination coefficient represents the less important variables.

## 2.3. The fitting of the generalized logistic function

The generalized logistic function (1) is a very general model
that may apply to very different situations.  So, before using it
in describing the evolution of the 8 morphological parameters of
the found morphologies, it is important to choose the best degree
of the polynomial (2).  To do this we use a suitable method (Mineo,
1971) that gives the best complete or incomplete polynomial function
to describe the parameters of the morphologies.  The choosing method
is applied to the transformed series

$$f_i^j(t) = z_i^j = \log \frac{\hat{L}_j - \bar{X}_i}{\bar{X}_i} \qquad\qquad (5)$$

in correspondence to each variable  $X_i$  and to some different tentative values  $\hat{L}_j$  of the asymptote  L.  The estimates of the parameters  $a_i$  of the chosen function  $f_i(t)$  and of the asymptote  L
are obtained by an iterative fitting of the chosen model to the
empirical series based on the research of the least squares solution
that minimizes at the same time the goodness of fit index

---

* In (4)  $\Delta$  is the determinant of the correlation matrix of the
age  t  and of the independent variables  $X_i$  at each step of the
choosing process and  $\Delta_{tt}$  the corresponding algebraic complement
of the first element  $r_{tt} = 1$.

$$x^2 = \sum_{t=4}^{15} (z_{it} - f_i(t))^2/f_i(t).\tag{6}$$

## 3. Results and interpretation

The clustering process applied to the 12 matrices (30 × 8) corresponding to each age indicate the best number of clusters k=3.

In Table 1 are shown the memberships of each of the 30 girls to the three clusters during the complete growing period from 4 to 15 years. On the basis of a full concordance of membership of each of the 30 girls to the same cluster we found (indicated by arrows) the elements strictly belonging to each cluster. The points on the bottom of the table indicate the units that present a high concordance of membership (ten or eleven equal values on twelve).

Unfortunately the examination of the characteristics of the clusters did not show the presence of clearly distinguished typical morphologies.

By supposing that this was due to the fact that someone of the 8 variables was not discriminant as far as the growing process of girls was concerned, we proceeded to the choice of the variables.

The resulting order of the eight variables according to their discriminant power was the following:
    1  Length
    2  Chest circumference
    3  Crown-rump length
    4  Head circumference
    5  Left calf circumference
    6  Weight
    7  Left upper arm circumference
    8  Maximum pelvic width

Table 1. Classification of the 30 girls at each age into three clusters by means of the 8 standardized variables

| AGE | | | | | | | | | | | | | Membership of the units to the clusters | | | | | | | | | | | | | | | | | |
|---|1|2|3|4|5|6|7|8|9|10|11|12|13|14|15|16|17|18|19|20|21|22|23|24|25|26|27|28|29|30|
| 4 | 2|2|1|3|2|2|3|1|1|1|1|2|1|2|3|2|3|2|2|1|3|1|3|2|1|3|3|3|2|2 |
| 5 | 2|2|1|3|3|2|3|2|1|2|1|3|2|2|3|3|3|2|2|2|3|2|3|2|1|3|3|3|2|2 |
| 6 | 2|2|1|2|2|2|3|1|1|2|1|3|2|2|3|3|3|2|2|1|3|1|2|2|1|2|3|3|2|2 |
| 7 | 2|2|1|3|2|2|3|2|1|2|1|3|2|2|3|3|3|2|3|2|3|2|2|2|1|2|3|3|2|2 |
| 8 | 2|2|1|2|2|2|3|1|1|2|1|3|2|2|2|2|3|2|3|2|3|2|2|2|1|2|3|2|2|2 |
| 9 | 2|2|1|3|2|2|3|2|1|2|1|3|2|2|3|3|3|2|3|2|3|2|2|2|1|2|3|2|2|2 |
| 10 | 2|2|1|3|2|2|2|2|1|2|1|3|2|2|2|2|3|2|3|2|3|2|2|1|1|2|3|2|2|2 |
| 11 | 2|1|1|3|2|2|3|2|1|2|1|3|2|2|3|3|3|2|3|2|3|2|1|1|1|2|3|2|2|2 |
| 12 | 2|1|1|3|2|2|3|1|1|2|1|3|2|2|3|3|3|2|3|2|3|3|1|1|1|2|3|2|2|2 |
| 13 | 2|1|2|3|2|2|3|1|1|2|1|3|3|2|3|3|2|2|3|1|3|2|2|2|1|2|3|2|2|2 |
| 14 | 2|1|2|3|2|2|2|1|1|1|1|2|2|2|2|2|2|3|3|1|3|2|2|2|1|2|3|2|2|2 |
| 15 | 2|2|1|3|2|2|2|1|1|1|1|2|2|2|2|2|2|3|2|1|2|1|2|2|1|2|3|2|2|2 |

Table 2. Classification of the 30 girls at each age into three clusters by means of the selected standardized variables $X_2$, $X_5$

| AGE | | Membership of the units to the clusters | | | | | | | | | | | | | | | | | | | | | | | | | | | | |
|---|---|---|---|---|---|---|---|---|---|---|---|---|---|---|---|---|---|---|---|---|---|---|---|---|---|---|---|---|---|---|
| | 1 | 2 | 3 | 4 | 5 | 6 | 7 | 8 | 9 | 10 | 11 | 12 | 13 | 14 | 15 | 16 | 17 | 18 | 19 | 20 | 21 | 22 | 23 | 24 | 25 | 26 | 27 | 28 | 29 | 30 |
| 4 | 2 | 2 | 1 | 3 | 3 | 2 | 3 | 2 | 1 | 2 | 2 | 2 | 2 | 2 | 2 | 2 | 3 | 2 | 2 | 1 | 3 | 2 | 3 | 2 | 2 | 3 | 3 | 3 | 2 | 2 |
| 5 | 2 | 2 | 1 | 3 | 2 | 2 | 3 | 2 | 1 | 2 | 1 | 2 | 2 | 2 | 3 | 2 | 3 | 2 | 2 | 2 | 3 | 2 | 3 | 2 | 1 | 3 | 3 | 3 | 2 | 2 |
| 6 | 2 | 2 | 1 | 3 | 2 | 2 | 3 | 2 | 1 | 2 | 1 | 3 | 2 | 2 | 2 | 2 | 3 | 3 | 3 | 2 | 3 | 2 | 3 | 2 | 2 | 3 | 3 | 3 | 2 | 2 |
| 7 | 3 | 2 | 1 | 3 | 2 | 3 | 3 | 2 | 1 | 2 | 1 | 3 | 2 | 2 | 3 | 3 | 3 | 2 | 3 | 2 | 3 | 3 | 3 | 3 | 2 | 3 | 3 | 3 | 2 | 2 |
| 8 | 2 | 2 | 1 | 3 | 2 | 2 | 3 | 1 | 1 | 2 | 1 | 3 | 2 | 2 | 2 | 2 | 3 | 2 | 3 | 2 | 3 | 2 | 2 | 2 | 1 | 2 | 3 | 2 | 2 | 1 |
| 9 | 2 | 2 | 1 | 3 | 2 | 2 | 3 | 2 | 1 | 2 | 1 | 3 | 2 | 2 | 2 | 2 | 3 | 2 | 3 | 1 | 3 | 3 | 2 | 2 | 1 | 2 | 3 | 2 | 2 | 1 |
| 10 | 2 | 2 | 1 | 3 | 2 | 3 | 3 | 2 | 1 | 2 | 1 | 3 | 2 | 2 | 3 | 3 | 3 | 2 | 3 | 1 | 3 | 3 | 2 | 2 | 1 | 2 | 3 | 2 | 2 | 1 |
| 11 | 2 | 2 | 1 | 3 | 2 | 2 | 3 | 2 | 1 | 2 | 1 | 3 | 2 | 1 | 3 | 3 | 3 | 2 | 3 | 1 | 3 | 3 | 1 | 1 | 1 | 2 | 3 | 2 | 2 | 1 |
| 12 | 2 | 1 | 1 | 2 | 2 | 3 | 3 | 1 | 1 | 2 | 1 | 2 | 2 | 1 | 3 | 3 | 2 | 2 | 3 | 1 | 3 | 3 | 1 | 1 | 1 | 2 | 2 | 2 | 2 | 1 |
| 13 | 2 | 1 | 2 | 3 | 2 | 2 | 2 | 1 | 1 | 2 | 1 | 3 | 2 | 1 | 3 | 3 | 2 | 3 | 2 | 2 | 3 | 3 | 1 | 2 | 1 | 3 | 3 | 3 | 3 | 1 |
| 14 | 2 | 1 | 2 | 3 | 2 | 2 | 2 | 1 | 1 | 1 | 1 | 2 | 2 | 1 | 2 | 2 | 2 | 3 | 2 | 1 | 3 | 2 | 2 | 2 | 1 | 3 | 3 | 2 | 3 | 1 |
| 15 | 2 | 2 | 1 | 3 | 2 | 2 | 2 | 1 | 1 | 1 | 1 | 2 | 2 | 1 | 2 | 3 | 2 | 3 | 2 | 1 | 3 | 2 | 2 | 1 | 2 | 3 | 3 | 2 | 3 | 1 |

As the value of the multiple determination coefficient based on the 8 variables resulted 0.95319, while that based on the first two variables, i.e. the length and the chest circumference, had the value 0.94318, with a loss in the value of $R^2$ of only 0.01, we repeated the clustering process on the basis of these two variables *.

In Table 2 are shown the memberships of each of the 30 girls to the three clusters. On the basis of the full concordances (arrows) and of the high concordances (points) of membership we chose the following three clusters **

first cluster     (3,9,11)
second cluster    (1,5,10,13)
third cluster     (4,7,21,27).

In Table 3 are shown the series of mean values $\bar{X}_i$ of the 8 variables at each age and the corresponding series of the per year growing speeds calculated on the whole sample of 30 girls. In Tables 4, 5, and 6 are shown the same series of mean values computed, respectively, on the members of the first, of the second and of the third cluster. As we can see the three cluster may constitute typical morphologies specially so far as the evolution of the mean values and of the average growth speeds are concerned.

Further interesting descriptions of the three groups may be obtained by the fitting of the generalized logistic model (1) to the series of mean values of the preceding tables.

---

*   We made other attempts to classify the girls on the basis of the first six variables and of the first four variables of the ranking. However these attempts did not give good concordances of membership to the clusters.

**  Other clusters may be made by the elements that present similar evolution in the growing process even if they pass from one cluster to another. We will analyse these groups later.

Table 3. Mean values of the 8 variables computed on the 30 elements
of the whole sample

| AGE | MEAN VALUES FOR THE 8 VARIABLES | | | | | | | |
|---|---|---|---|---|---|---|---|---|
|  | $x_1$ | $x_2$ | $x_3$ | $x_4$ | $x_5$ | $x_6$ | $x_7$ | $x_8$ |
| 4 | 1521.5 | 997.9 | 576.5 | 495.0 | 534.8 | 162.9 | 214.0 | 167.1 |
| 5 | 1700.4 | 1060.0 | 597.3 | 501.7 | 549.1 | 163.7 | 222.9 | 178.5 |
| 6 | 1885.3 | 1118.9 | 622.0 | 506.9 | 569.3 | 167.3 | 230.3 | 187.0 |
| 7 | 2105.8 | 1176.3 | 640.7 | 511.8 | 582.7 | 170.2 | 241.8 | 193.5 |
| 8 | 2359.4 | 1233.6 | 666.2 | 515.9 | 605.2 | 177.5 | 252.1 | 200.2 |
| 9 | 2669.4 | 1288.4 | 685.7 | 520.1 | 620.5 | 184.5 | 261.6 | 207.5 |
| 10 | 2987.0 | 1340.9 | 705.3 | 323.9 | 645.5 | 193.3 | 273.4 | 213.0 |
| 11 | 3441.7 | 1401.1 | 733.2 | 528.5 | 679.4 | 202.2 | 285.8 | 222.4 |
| 12 | 3788.6 | 1475.8 | 770.6 | 534.7 | 713.9 | 209.0 | 300.0 | 235.5 |
| 13 | 4355.5 | 1540.6 | 806.1 | 539.6 | 752.1 | 216.4 | 313.5 | 249.8 |
| 14 | 4767.3 | 1581.0 | 833.4 | 545.4 | 799.7 | 228.3 | 326.4 | 258.4 |
| 15 | 5010.9 | 1600.5 | 847.2 | 549.2 | 824.6 | 237.8 | 336.4 | 265.0 |

MEAN VALUES OF THE PER YEAR GROWING SPEEDS

| | | | | | | | | |
|---|---|---|---|---|---|---|---|---|
| 4– 5 | 178.9 | 62.0 | 20.8 | 6.7 | 14.3 | 0.8 | 8.9 | 11.4 |
| 5– 6 | 184.9 | 58.9 | 24.7 | 5.2 | 20.2 | 3.6 | 7.4 | 8.5 |
| 6– 7 | 220.5 | 57.4 | 18.7 | 4.9 | 13.4 | 2.9 | 11.5 | 6.5 |
| 7– 8 | 253.6 | 57.3 | 25.5 | 4.1 | 22.5 | 7.3 | 10.3 | 6.7 |
| 8– 9 | 310.0 | 54.8 | 19.5 | 4.2 | 15.3 | 7.0 | 9.5 | 7.3 |
| 9–10 | 317.6 | 52.5 | 19.6 | 3.8 | 25.0 | 8.8 | 11.8 | 5.5 |
| 10–11 | 454.7 | 60.2 | 27.9 | 4.6 | 33.9 | 8.9 | 12.4 | 9.4 |
| 11–12 | 346.9 | 74.7 | 37.4 | 6.2 | 34.5 | 6.8 | 14.2 | 13.1 |
| 12–13 | 566.9 | 64.8 | 35.5 | 4.9 | 38.2 | 7.4 | 13.5 | 14.3 |
| 13–14 | 411.8 | 40.4 | 27.3 | 5.8 | 47.6 | 11.9 | 12.9 | 8.8 |
| 14–15 | 243.6 | 19.5 | 13.8 | 3.9 | 24.9 | 9.5 | 10.0 | 6.6 |

In order to find the best complete or incomplete polynomial of
the family (2) describing the 8 series of mean values of Table 3 we
used the quoted choosing algorithm in correspondence to four tentative values of the asymptote L. The four values of the parameters
L were chosen one near to the maximum mean value of each series
and the other farther and farther from this value. For instance,

A. MINEO

Table 4. Mean values of the 8 variables computed on the elements of
         the first cluster (3,9,11). Classification based on the
         standardized variables $z_2$, $z_5$

| AGE | MEAN VALUES FOR THE 8 VARIABLES | | | | | | | |
|-----|-----|-----|-----|-----|-----|-----|-----|-----|
|     | $x_1$ | $x_2$ | $x_3$ | $x_4$ | $x_5$ | $x_6$ | $x_7$ | $x_8$ |
| 4 | 1297.7 | 960.3 | 561.0 | 480.0 | 493.7 | 149.3 | 205.0 | 154.3 |
| 5 | 1366.3 | 1015.3 | 567.0 | 484.7 | 510.7 | 149.0 | 209.7 | 160.3 |
| 6 | 1552.3 | 1067.3 | 589.7 | 490.7 | 524.0 | 155.0 | 217.7 | 171.0 |
| 7 | 1729.0 | 1109.7 | 599.3 | 490.0 | 532.0 | 152.3 | 221.0 | 176.3 |
| 8 | 1866.7 | 1169.3 | 636.7 | 494.3 | 557.3 | 160.3 | 230.3 | 183.7 |
| 9 | 2056.7 | 1220.7 | 651.0 | 497.7 | 579.7 | 168.3 | 241.0 | 188.7 |
| 10 | 2263.3 | 1265.3 | 668.3 | 501.7 | 580.7 | 172.0 | 247.7 | 195.7 |
| 11 | 2562.7 | 1327.0 | 692.3 | 506.0 | 614.3 | 174.0 | 260.7 | 203.3 |
| 12 | 2953.3 | 1398.3 | 733.7 | 512.3 | 628.3 | 188.0 | 278.0 | 208.0 |
| 13 | 3463.3 | 1470.3 | 764.7 | 514.3 | 671.7 | 195.0 | 289.7 | 225.7 |
| 14 | 4050.0 | 1524.3 | 804.7 | 523.0 | 742.0 | 215.3 | 311.0 | 239.3 |
| 15 | 4346.7 | 1554.0 | 825.0 | 526.7 | 763.7 | 225.3 | 323.7 | 250.3 |

MEAN VALUES OF THE PER YEAR GROWING SPEEDS

| | | | | | | | | |
|-----|-----|-----|-----|-----|-----|-----|-----|-----|
| 4- 5 | 68.6 | 55.0 | 6.0 | 4.7 | 17.0 | -.3 | 4.7 | 6.0 |
| 5- 6 | 186.0 | 52.0 | 22.7 | 6.0 | 13.3 | 6.0 | 8.0 | 10.7 |
| 6- 7 | 176.7 | 42.4 | 9.6 | -.7 | 8.0 | -2.7 | 3.3 | 5.3 |
| 7- 8 | 137.7 | 59.6 | 37.4 | 4.3 | 25.3 | 8.0 | 9.3 | 7.4 |
| 8- 9 | 190.0 | 51.4 | 14.3 | 3.4 | 22.0 | 8.0 | 10.7 | 5.0 |
| 9-10 | 206.6 | 44.6 | 17.3 | 4.0 | 1.4 | 3.7 | 6.7 | 7.0 |
| 10-11 | 299.4 | 61.7 | 24.0 | 4.3 | 33.6 | 2.0 | 13.0 | 7.6 |
| 11-12 | 390.6 | 71.3 | 41.4 | 6.3 | 14.0 | 14.0 | 17.3 | 4.7 |
| 12-13 | 510.0 | 72.0 | 31.0 | 2.0 | 43.4 | 7.0 | 11.7 | 17.7 |
| 13-14 | 586.7 | 54.0 | 40.0 | 8.7 | 70.3 | 20.3 | 21.3 | 13.6 |
| 14-15 | 296.7 | 29.7 | 20.3 | 3.7 | 21.7 | 10.0 | 12.7 | 11.0 |

for the series of the length we use the four values 1620, 1650, 1700
and 1800 and for the series of the chest circumference the four
values 850, 880, 900 and 950.

For the eight series and all the values of L, we found that
the incomplete polynomial

Table 5. Mean values of the 8 variables computed on the elements of the second cluster (1,5,10,13). Classification based on the standardized variables $z_2$, $z_5$

| AGE | MEAN VALUES FOR THE 8 VARIABLES | | | | | | | |
|---|---|---|---|---|---|---|---|---|
| | $x_1$ | $x_2$ | $x_3$ | $x_4$ | $x_5$ | $x_6$ | $x_7$ | $x_8$ |
| 4 | 1506.75 | 1603.75 | 574.00 | 484.00 | 528.75 | 160.50 | 212.00 | 164.75 |
| 5 | 1691.75 | 1063.50 | 594.75 | 492.50 | 542.50 | 165.00 | 219.50 | 178.00 |
| 6 | 1870.00 | 1119.50 | 620.50 | 498.25 | 569.25 | 169.75 | 228.00 | 190.00 |
| 7 | 2098.00 | 1172.25 | 636.25 | 506.75 | 583.50 | 171.50 | 239.50 | 195.00 |
| 8 | 2309.00 | 1228.00 | 655.00 | 513.00 | 607.75 | 177.50 | 251.50 | 201.25 |
| 9 | 2675.00 | 1285.00 | 680.50 | 515.50 | 631.25 | 188.00 | 261.75 | 208.75 |
| 10 | 3061.75 | 1335.00 | 704.50 | 518.00 | 662.25 | 201.00 | 275.75 | 215.25 |
| 11 | 3387.25 | 1390.50 | 723.25 | 521.00 | 683.75 | 209.00 | 283.75 | 221.00 |
| 12 | 3950.00 | 1467.50 | 764.25 | 529.00 | 753.00 | 267.25 | 306.25 | 231.25 |
| 13 | 4432.50 | 1544.00 | 801.50 | 537.75 | 753.25 | 223.25 | 318.75 | 256.25 |
| 14 | 4778.75 | 1578.50 | 830.50 | 537.75 | 789.00 | 229.25 | 329.75 | 258.00 |
| 15 | 4973.75 | 1597.00 | 842.00 | 541.25 | 802.00 | 238.75 | 337.50 | 265.00 |

MEAN VALUES OF THE PER YEAR GROWING SPEEDS

| | | | | | | | | |
|---|---|---|---|---|---|---|---|---|
| 4- 5 | 185.00 | 59.75 | 20.75 | 8.50 | 13.75 | 4.50 | 7.50 | 13.25 |
| 5- 6 | 178.25 | 56.00 | 25.75 | 5.75 | 26.75 | 4.75 | 8.50 | 12.00 |
| 6- 7 | 228.00 | 52.75 | 15.75 | 8.50 | 14.25 | 1.75 | 11.50 | 5.00 |
| 7- 8 | 211.00 | 55.75 | 18.75 | 6.25 | 24.25 | 6.00 | 12.00 | 6.25 |
| 8- 9 | 366.00 | 57.00 | 25.50 | 2.50 | 23.50 | 10.50 | 10.25 | 7.50 |
| 9-10 | 386.75 | 50.00 | 24.00 | 2.50 | 31.00 | 13.00 | 14.00 | 6.50 |
| 10-11 | 325.50 | 55.50 | 18.75 | 3.00 | 21.50 | 8.00 | 8.00 | 5.75 |
| 11-12 | 562.75 | 77.00 | 41.00 | 8.00 | 69.25 | 8.25 | 22.50 | 10.25 |
| 12-13 | 482.50 | 76.50 | 37.25 | 6.75 | 0.25 | 6.00 | 12.50 | 20.00 |
| 13-14 | 346.25 | 34.50 | 29.00 | 2.00 | 35.75 | 6.00 | 11.00 | 6.75 |
| 14-15 | 195.00 | 18.50 | 11.50 | 3.50 | 13.00 | 9.50 | 7.75 | 7.00 |

$$f(t) = a_0 + a_1 t^2 + a_2 t^3 + a_3 t^4 \qquad\qquad (7)$$

described at least the 99 percent of the variability of the series of Z values and that the loss from the description of a complete polynomial of fifth degree was less than 0.005.

A. MINEO

Table 6. Mean values of the 8 variables computed on the elements of the third cluster (4,7,21,27). Classification based on the standardized variables $z_2$, $z_5$

| AGE | MEAN VALUES FOR THE 8 VARIABLES | | | | | | | |
|-----|--------|--------|--------|--------|--------|--------|--------|--------|
|     | $x_1$  | $x_2$  | $x_3$  | $x_4$  | $x_5$  | $x_6$  | $x_7$  | $x_8$  |
| 4   | 1669.75 | 1037.00 | 602.50 | 504.50 | 552.00 | 166.65 | 226.00 | 179.00 |
| 5   | 1883.75 | 1106.25 | 626.50 | 510.00 | 564.00 | 169.00 | 237.25 | 191.00 |
| 6   | 2110.00 | 1165.50 | 654.75 | 515.50 | 583.50 | 169.25 | 241.25 | 202.00 |
| 7   | 2365.00 | 1228.25 | 676.25 | 519.75 | 601.50 | 175.75 | 255.50 | 209.76 |
| 8   | 2671.25 | 1287.25 | 696.75 | 522.25 | 625.00 | 183.75 | 266.75 | 216.00 |
| 9   | 2991.75 | 1340.75 | 720.25 | 527.75 | 635.50 | 191.75 | 279.25 | 224.75 |
| 10  | 3414.75 | 1392.50 | 731.25 | 529.50 | 661.50 | 200.50 | 286.25 | 232.50 |
| 11  | 3914.00 | 1455.75 | 764.75 | 535.75 | 724.50 | 214.50 | 305.25 | 245.75 |
| 12  | 4351.25 | 1533.25 | 800.75 | 542.25 | 750.00 | 217.25 | 317.25 | 258.50 |
| 13  | 5112.50 | 1590.75 | 844.50 | 548.25 | 798.50 | 229.00 | 331.75 | 274.25 |
| 14  | 5661.25 | 1624.00 | 862.75 | 553.75 | 867.50 | 245.25 | 346.25 | 282.25 |
| 15  | 5812.50 | 1639.25 | 874.25 | 553.75 | 879.50 | 252.50 | 354.50 | 284.75 |

MEAN VALUES OF THE PER YEAR GROWING SPEEDS

| | | | | | | | | |
|-----|--------|--------|--------|--------|--------|--------|--------|--------|
| 4– 5  | 214.00 | 69.25 | 24.00 | 5.50 | 12.00 | 2.25 | 11.25 | 12.00 |
| 5– 6  | 226.25 | 59.25 | 28.25 | 5.50 | 19.50 | 0.25 | 4.00 | 11.00 |
| 6– 7  | 255.00 | 62.75 | 21.50 | 4.25 | 18.00 | 6.50 | 14.25 | 7.75 |
| 7– 8  | 306.25 | 59.00 | 20.50 | 2.50 | 23.50 | 8.00 | 11.25 | 6.25 |
| 8– 9  | 320.50 | 53.50 | 23.50 | 5.50 | 10.50 | 8.00 | 12.50 | 8.75 |
| 9–10  | 423.00 | 51.75 | 11.00 | 1.75 | 26.00 | 8.75 | 7.00 | 7.75 |
| 10–11 | 499.25 | 63.25 | 33.25 | 6.25 | 63.00 | 14.00 | 19.00 | 13.25 |
| 11–12 | 437.25 | 77.50 | 36.00 | 6.50 | 25.50 | 2.75 | 12.00 | 12.75 |
| 12–13 | 761.25 | 57.50 | 43.75 | 6.00 | 48.50 | 11.75 | 14.50 | 15.75 |
| 13–14 | 548.75 | 33.25 | 18.25 | 5.50 | 69.00 | 16.25 | 14.50 | 8.00 |
| 14–15 | 151.25 | 15.25 | 11.50 | –    | 12.00 | 7.25 | 8.25 | 2.50 |

The generalized logistic function

$$\bar{X}_i = \frac{L}{1 + e^{a_o + a_1 t^2 + a_2 t^3 + a_3 t^4}} \tag{8}$$

described well all the series of mean values in Tables 3, 4, 5 and 6.

Only one of the series corresponding to the variable "head circumference" and all the series of the variable "left upper arm circumference" did not give a finite estimate of $L *$.

It is easy to verify that when $L$ varies from a value $L_o$,
near to $\bar{X}_{i(15)}$, $L_o > \bar{X}_{i(15)}$, to larger values the corresponding
values of the $X^2$ index decrease up to a minimum and then always
increase.

To find the value of $L$ corresponding to the minimum values of
$X^2$ we used an iterative procedure that developes in two stages. In
the first stage we explore the course of the $X^2$ values corresponding to some increasing values of $L$ up to finding two adjacent
values of $X^2$ that are not decreasing. In the second stage we
determine the minimum of $X^2$ by searching for it between the two
found values with a method like the "Regular Falsi".

The complete iterative procedure is based on the iteration of
the following steps :
    1) choose a convenient value of $L$;
    2) estimate the parameter of (7) by means of the least squares
method applied to the transformed values

$$Z_{it} = \log \frac{L - X_{i,t}}{\bar{X}_{i,t}} ;$$

    3) calculate the $X^2$ goodness of fit index.

These three steps are repeated until the minimum value of $X^2$

---

* This was due to the fact that the $X^2$ index did not converge to
a minimum. For these series the $X^2$ values decreased quickly before
reaching some reliable values of L after which they decreased
slower and slower. This lack of convergence might be due to fluctuations of the per year growing speeds in such as small group.

A. MINEO

Table 7. Values of the parameters of the logistic model (7) and of
$X^2$ goodness of fit index computed on the series of mean
values corresponding to the stated variables and groups.

| GROUPS | PARAMETERS | VARIABLES | | | | | | | |
|---|---|---|---|---|---|---|---|---|---|
| | | $X_1$ | $X_2$ | $X_3$ | $X_2-X_3$ | $X_4$ | $X_5$ | $X_7$ | $X_8$ |
| 1st CLUSTER | $a_0$ | .86909 | - .50992 | - .70905 | - .27520 | - .23167 | - .56864 | - .47156 | - .45467 |
| | $a_1$ | - .08316 | - .08573 | - .05685 | - .11953 | - .04707 | - .05634 | - .03878 | - .08185 |
| | $a_2$ | .01636 | .01667 | .01094 | .02381 | .00894 | .01222 | .00656 | .01689 |
| | $a_3$ | - .00112 | - .00114 | - .00078 | - .00161 | - .00062 | - .00088 | - .00047 | - .00110 |
| | L | 4439.50 | 1561.22 | 835.15 | 729.86 | 529.80 | 779.53 | 335.40 | 254.80 |
| | $X^2$ | 2.0667 | .6616 | .3201 | 1.7041 | .0556 | .1078 | .1320 | .1479 |
| 2nd CLUSTER | $a_0$ | .79362 | - .56048 | - .78318 | - .32196 | - .21546 | - .66529 | - .52015 | - .55828 |
| | $a_1$ | - .08364 | - .09988 | - .07760 | - .12456 | - .11915 | - .05614 | - .06811 | - .11374 |
| | $a_2$ | .01407 | .02030 | .01647 | .02489 | .02522 | .00959 | .01217 | .02515 |
| | $a_3$ | - .00095 | - .00143 | - .00120 | - .00173 | - .00165 | - .00071 | - .00084 | - .00170 |
| | L | 5024.70 | 1599.28 | 844.47 | 755.33 | 541.62 | 806.62 | 339.96 | 265.75 |
| | $X^2$ | 7.7507 | .6375 | .2558 | .6619 | .0379 | 1.2543 | .1298 | .4293 |
| 3rd CLUSTER | $a_0$ | .86424 | - .58534 | - .81895 | - .33546 | (*) - | - .52670 | - .56493 | - .57972 |
| | $a_1$ | - .09965 | - .10827 | - .09535 | - .12264 | - | - .04836 | - .06308 | - .10531 |
| | $a_2$ | .01925 | .02162 | .02095 | .02262 | - | .01066 | .01086 | .02365 |
| | $a_3$ | - .00135 | - .00149 | - .00150 | - .00153 | - | - .00088 | - .00073 | - .00173 |
| | L | 5843.90 | 1641.26 | 875.47 | 765.69 | - | 885.21 | 358.77 | 284.97 |
| | $X^2$ | 9.7872 | .8594 | .2509 | 1.0653 | - | 1.2507 | .1951 | .2205 |
| WHOLE SAMPLE | $a_0$ | .80291 | - .54141 | - .77114 | - .29426 | - .22208 | - .58993 | - .54480 | - .58286 |
| | $a_1$ | - .08480 | - .10093 | - .07932 | - .12565 | - .07441 | - .04930 | - .06244 | - .09132 |
| | $a_2$ | .01438 | .01994 | .01638 | .02448 | .01474 | .00950 | .01067 | .01937 |
| | $a_3$ | - .00094 | - .00137 | - .00116 | - .00166 | - .00097 | -.00070 | - .00070 | - .00134 |
| | L | 5096.13 | 1603.48 | 850.24 | 753.79 | 550.94 | 838.30 | 341.77 | 266.10 |
| | $X^2$ | 6.2002 | .6864 | .1380 | .7629 | .0228 | .1834 | .0719 | .2312 |

(*) For this group the value of $X^2$, even if very slowly, diverges.

and of the joint values of the parameters $a_i$ and of the asymptote
L are found.

   In Table 7 are shown the values of the parameters of the logistic model (8) of the $X^2$ goodness of fit index computed on the

Table 8. Observed and computed mean values and per year growing
speeds for the stated variables

| AGE | LENGHT ($x_2$) | | | Grow-Rump Lenght ($x_3$) | | | $x_2 - x_3$ | | |
|---|---|---|---|---|---|---|---|---|---|
| | Observed values | Computed values | Deviations | Observed values | Computed values | Deviations | Observed values | Computed values | Deviations |
| 4 | 997.9 | 1013.6 | -15.7 | 576.5 | 581.4 | - 4.9 | 421.4 | 432.0 | - 10.6 |
| 5 | 1060.0 | 1044.0 | 16.0 | 597.3 | 593.0 | 4.3 | 462.7 | 450.8 | 15.9 |
| 6 | 1118.9 | 1108.9 | 10.0 | 622.0 | 617.6 | 4.4 | 496.9 | 491.4 | 5.5 |
| 7 | 1176.3 | 1179.2 | - 2.9 | 640.7 | 643.8 | - 3.1 | 535.6 | 535.7 | - 0.1 |
| 8 | 1233.6 | 1240.8 | - 7.2 | 666.2 | 666.7 | - 0.5 | 567.4 | 574.4 | - 7.0 |
| 9 | 1288.4 | 1293.0 | - 4.6 | 685.7 | 686.6 | - 0.9 | 602.7 | 606.4 | - 3.7 |
| 10 | 1340.9 | 1343.0 | - 2.1 | 705.3 | 707.1 | - 1.8 | 635.6 | 635.9 | - 0.3 |
| 11 | 1401.1 | 1400.4 | 0.7 | 733.2 | 733.1 | 0.1 | 667.9 | 667.5 | 0.4 |
| 12 | 1475.8 | 1469.3 | 6.5 | 770.6 | 767.6 | 3.0 | 705.2 | 702.3 | 2.9 |
| 13 | 1540.6 | 1537.8 | 2.8 | 806.1 | 805.5 | 0.6 | 734.5 | 732.4 | 2.1 |
| 14 | 1581.0 | 1583.5 | - 2.5 | 833.4 | 834.5 | - 0.9 | 747.6 | 748.8 | - 1.2 |
| 15 | 1600.5 | 1600.4 | 0.1 | 847.2 | 847.1 | 0.1 | 753.3 | 753.3 | - |
| 4- 5 | 62.1 | 30.4 | 31.7 | 20.8 | 11.6 | 9,2 | 41.3 | 18.8 | 22.5 |
| 5- 6 | 58.9 | 64.9 | - 6.0 | 24.7 | 24.5 | 0.2 | 34.2 | 40.7 | - 6.5 |
| 6- 7 | 57.4 | 70.3 | -12.9 | 18.7 | 26.2 | -7.5 | 38.7 | 44.3 | - 5.6 |
| 7- 8 | 57.3 | 62.7 | - 5.4 | 25.5 | 23.0 | 2.5 | 31.8 | 38.8 | - 7.0 |
| 8- 9 | 54.8 | 52.2 | 2.6 | 19.5 | 19.9 | -0.4 | 35.3 | 32.1 | 3.1 |
| 9-10 | 52.5 | 50.0 | 2.5 | 19.6 | 20.5 | -0.9 | 32.9 | 29.4 | 3.5 |
| 10-11 | 60.2 | 57.4 | 2.8 | 27.9 | 26.0 | 1,9 | 32.3 | 31.7 | 0.6 |
| 11-12 | 74.7 | 68.9 | 5.8 | 37.4 | 34.4 | 3,0 | 37.3 | 34.7 | 2.6 |
| 12-13 | 64.8 | 68.5 | - 3.7 | 35.5 | 37.9 | -2.4 | 29.3 | 30.2 | - 0.9 |
| 13-14 | 40.4 | 45.7 | - 5.3 | 27.3 | 28.8 | -1.5 | 13.1 | 16.4 | - 3.3 |
| 14-15 | 19.5 | 16.9 | 2.6 | 13.8 | 12.8 | 1.0 | 5.7 | 4.5 | 1.2 |

series of mean values corresponding to the stated variables and
groups. In the sixth column of the table we find also the values of
the parameters of model (8) and of the $X^2$ index computed on a new
series obtained by the differences between the values of "length"
and those of "Crown-rump length" (see table 8). The $X^2$ and L

values in table 7 show the goodness of the fittings and of the estimates of the asymptote  L *.

In table 8 are shown the observed series of the mean values and
of the per year growing speeds computed on the series of length, of
crown-rump length and of the differences between these two characteristics, the corresponding series obtained by the fitted model (8)
and their deviations.  As we can see, the logistic model (8) fits
well at all ages.

From the computed values of the per-year growing speeds we see
that the growing speed of the length measurements of the girls is
characterized by series that present two maxima;  one at the age of
7 and the other at the age of 12.  Moreover, while the values of the
growing speeds of the crown-rump length result greater in correspondence of the age 12 ($\sim$ 35 mm per year) and lesser in correspondence
of the age 7 ($\sim$ 25 mm per year), those of the length of the remaining
part of the body (i.e. legs, head, neck) result greater at the age 7
($\sim$ 40 mm per year) and lesser at the age 12 ($\sim$ 30 mm per year).

These results may be important to understand the evolution
process of the different part of the human body.

In Tables 9, 10, 11 and 12 are shown the series of the mean
values and of the per year growing speeds computed by the model (8)
in correspondence of the stated variables and groups.  The analysis
of the values in these tables may help to understand the evolution
process of the anthropometric characters in the three clusters and
in the whole sample of 30 girls.

The explanation of the differences between the three clusters
is a subject for auxologists.

------

* The asymptote  L  in model (8) would represent the expected
maximum mean value of the character  $X_i$.

Table 9. Computed mean values and per year growing speeds for the
         stated variables corresponding to the three morphologies
         and the whole sample of girls

| AGE | WEIGHT | | | | LENGHT | | | |
|---|---|---|---|---|---|---|---|---|
| | 1st cluster | 2nd cluster | 3rd cluster | Whole sample | 1st cluster | 2nd cluster | 3rd cluster | Whole sample |
| MEAN VALUES | | | | | | | | |
| 4 | 1311.6 | 1564.7 | 1732.5 | 1576.8 | 975.4 | 1018.8 | 1054.2 | 1013.6 |
| 5 | 1375.3 | 1641.6 | 1833.7 | 1655.5 | 1000.9 | 1047.7 | 1087.0 | 1044.0 |
| 6 | 1523.3 | 1831.0 | 2073.3 | 1848.3 | 1056.1 | 1110.4 | 1156.4 | 1108.9 |
| 7 | 1705.9 | 2085.1 | 2375.9 | 2104.4 | 1117.4 | 1177.5 | 1230.4 | 1179.2 |
| 8 | 1891.1 | 2372.2 | 2692.3 | 2388.0 | 1173.0 | 1235.9 | 1294.2 | 1240.8 |
| 9 | 2071.7 | 2681.3 | 3011.8 | 2684.1 | 1221.6 | 1285.6 | 1347.3 | 1293.0 |
| 10 | 2270.3 | 3025.3 | 3369.2 | 3003.6 | 1269.1 | 1334.6 | 1397.8 | 1343.0 |
| 11 | 2538.7 | 3432.4 | 3834.0 | 3377.4 | 1324.7 | 1393.6 | 1455.6 | 1400.4 |
| 12 | 2941.4 | 3915.0 | 4457.3 | 3832.3 | 1393.8 | 1465.8 | 1523.4 | 1469.3 |
| 13 | 3492.9 | 4414.7 | 5147.7 | 4338.9 | 1468.5 | 1536.8 | 1587.7 | 1537.8 |
| 14 | 4039.6 | 4793.9 | 5636.8 | 4772.1 | 1526.6 | 1581.7 | 1626.8 | 1583.5 |
| 15 | 4347.4 | 4972.5 | 5814.0 | 5010.5 | 1553.8 | 1596.9 | 1639.4 | 1600.4 |
| PER YEAR GROWING SPEEDS | | | | | | | | |
| 4- 5 | 63.7 | 76.9 | 101.2 | 78.7 | 25.5 | 29.7 | 32.8 | 30.4 |
| 5- 6 | 148.0 | 189.4 | 239.6 | 192.8 | 55.2 | 62.7 | 69.4 | 64.9 |
| 6- 7 | 182.6 | 254.1 | 302.6 | 256.1 | 61.3 | 67.3 | 74.0 | 70.3 |
| 7- 8 | 185.2 | 287.1 | 316.4 | 283.6 | 55.6 | 58.4 | 63.8 | 62.6 |
| 8- 9 | 180.6 | 309.1 | 319.5 | 296.1 | 48.6 | 49.7 | 53.1 | 52.2 |
| 9-10 | 198.6 | 344.0 | 357.4 | 319.5 | 47.5 | 49.0 | 50.5 | 50.0 |
| 10-11 | 268.4 | 407.1 | 464.8 | 373.8 | 55.6 | 59.0 | 57.8 | 57.4 |
| 11-12 | 402.7 | 482.6 | 623.3 | 454.9 | 69.1 | 72.2 | 67.8 | 68.9 |
| 12-13 | 551.5 | 499.7 | 690.4 | 506.6 | 74.7 | 71.0 | 64.3 | 68.5 |
| 13-14 | 546.7 | 369.2 | 489.1 | 433.2 | 58.1 | 44.9 | 39.1 | 45.7 |
| 14-15 | 307.8 | 178.6 | 177.2 | 238.4 | 27.2 | 15.2 | 12.6 | 16.9 |

Synthetically we can say that the three clusters may constitute
good estimations of three typical morphologies so characterized :
    the first : long-limbed girls that are of short stature;
    the second : thickset girls that are of average stature;
    the third : long-limbed girls that are of tall stature.

Table 10. Computed mean values and per year growing speeds for the
          stated variables corresponding to the three morphologies
          and the whole sample of girls

| AGE | CROWN–RUMP LENGHT | | | | (LENGHT)–(CROWN–RUMP LENGHT) | | | |
|---|---|---|---|---|---|---|---|---|
|  | 1st cluster | 2nd cluster | 3rd cluster | Whole sample | 1st cluster | 2nd cluster | 3rd cluster | Whole sample |
| | MEAN VALUES | | | | | | | |
| 4 | 559.7 | 579.6 | 607.6 | 581.4 | 414.8 | 437.9 | 446.5 | 431.9 |
| 5 | 568.3 | 590.8 | 621.5 | 593.0 | 432.1 | 456.4 | 463.2 | 450.8 |
| 6 | 587.1 | 614.2 | 649.7 | 617.5 | 469.3 | 496.0 | 506.4 | 491.4 |
| 7 | 608,6 | 638.9 | 677.9 | 643.8 | 509.6 | 530.7 | 552.5 | 535.7 |
| 8 | 629.3 | 660.5 | 700.7 | 666.7 | 544.3 | 575.6 | 593.9 | 574.4 |
| 9 | 649.0 | 679.4 | 718.9 | 686.6 | 572.9 | 606.5 | 629.1 | 606.4 |
| 10 | 670.0 | 699.7 | 737.4 | 707.1 | 599.2 | 635.6 | 660.9 | 635.8 |
| 11 | 695.9 | 726.6 | 763.2 | 733.1 | 629.3 | 668.1 | 692.6 | 667.5 |
| 12 | 729.4 | 762.9 | 799.7 | 767.6 | 665.4 | 704.4 | 723.9 | 702.3 |
| 13 | 768.2 | 802.1 | 839.1 | 805.5 | 700.5 | 735.3 | 748.7 | 732.4 |
| 14 | 803.4 | 830.4 | 865.3 | 834.3 | 722.1 | 751.0 | 761.6 | 748.8 |
| 15 | 825.1 | 842.0 | 874.1 | 847.1 | 728.9 | 754.9 | 765.2 | 753.3 |
| | PER YEAR GROWING SPEEDS | | | | | | | |
| 4– 5 | 8.6 | 11.2 | 13.9 | 11.6 | 17.3 | 18.5 | 18.7 | 18.8 |
| 5– 6 | 18.8 | 23.4 | 28.2 | 24.5 | 37.2 | 39.6 | 41.2 | 40.6 |
| 6– 7 | 21.5 | 24.7 | 28.2 | 26.3 | 40.3 | 42.7 | 46.1 | 44.3 |
| 7– 8 | 20.7 | 21.6 | 22.8 | 22.9 | 34.7 | 36.9 | 41.4 | 38.7 |
| 8– 9 | 19.7 | 18.9 | 18.2 | 19.9 | 28.6 | 30.9 | 35.2 | 32.0 |
| 9–10 | 21.0 | 20.3 | 18.5 | 20.5 | 26.3 | 29.1 | 31.8 | 29.4 |
| 10–11 | 25.9 | 26.9 | 25.8 | 26.0 | 30.1 | 32.5 | 31.7 | 31.7 |
| 11–12 | 33.5 | 36.3 | 36.5 | 34.5 | 36.1 | 36.3 | 31.3 | 34.8 |
| 12–13 | 38.8 | 39.2 | 39.4 | 37.9 | 35.1 | 30.9 | 24.8 | 30.1 |
| 13–14 | 35.2 | 28.3 | 26.2 | 28.8 | 21.6 | 15.7 | 12.9 | 16.4 |
| 14–15 | 21.7 | 11.6 | 8.8 | 12.8 | 6.8 | 3.9 | 3.6 | 4.5 |

Other differences between the three clusters and the whole
sample may be obtained by the investigation of the values in the
previous tables. Very interesting, in our opinion, are the comparisons of the values of the chest circumference between the second and

Table 11. Computed mean values and per year growing speeds of the
stated variables corresponding to the three morphologies
and the whole sample of girls

| AGE | HEAD CIRCUMFERENCE | | | | CHEST CIRCUMFERENCE | | | |
|---|---|---|---|---|---|---|---|---|
| | 1st cluster | 2nd cluster | 3rd cluster | Whole sample | 1st cluster | 2nd cluster | 3rd cluster | Whole sample |
| | | | | MEAN | VALUES | | | |
| 4 | 482.2 | 485.3 | – | 497.0 | 497.7 | 532.7 | 556.5 | 539.3 |
| 5 | 483.9 | 490.0 | – | 499.9 | 505.7 | 541.2 | 564.5 | 547.1 |
| 6 | 487.4 | 498.8 | – | 505.8 | 522.6 | 560.7 | 581.4 | 564.3 |
| 7 | 491.4 | 506.9 | – | 511.8 | 540.6 | 584.8 | 600.4 | 584.4 |
| 8 | 495.0 | 512.5 | – | 516.8 | 556.2 | 609.9 | 619.2 | 604.3 |
| 9 | 498.3 | 516.2 | – | 520.7 | 569.5 | 635.9 | 639.5 | 624.2 |
| 10 | 501.6 | 519.0 | – | 524.2 | 584.0 | 663.9 | 666.2 | 646.5 |
| 11 | 505.6 | 522.4 | – | 528.2 | 605.0 | 695.7 | 705.2 | 675.0 |
| 12 | 510.6 | 527.5 | – | 533.5 | 638.4 | 730.7 | 758.5 | 712.5 |
| 13 | 516.5 | 533.9 | – | 539.8 | 684.7 | 764.3 | 816.8 | 756.7 |
| 14 | 522.4 | 539.0 | – | 545.7 | 732.7 | 789.2 | 860.6 | 798.0 |
| 15 | 526.7 | 541.2 | – | 549.3 | 764.9 | 801.8 | 880.1 | 824.7 |
| | | | | PER YEAR | GROWING SPEEDS | | | |
| 4 – 5 | 1.7 | 4.7 | – | 2.9 | 8.0 | 8.5 | 8.0 | 7.8 |
| 5 – 6 | 3.5 | 8.0 | – | 5.9 | 16.9 | 19.5 | 16.9 | 17.2 |
| 6 – 7 | 4.0 | 8.1 | – | 6.0 | 18.0 | 24.1 | 19.0 | 20.1 |
| 7 – 8 | 3.6 | 5.6 | – | 5.0 | 15.6 | 25.1 | 18.8 | 19.9 |
| 8 – 9 | 3.3 | 3.5 | – | 3.9 | 13.3 | 26.0 | 20.3 | 19.9 |
| 9 – 10 | 3.3 | 2.8 | – | 3.5 | 14.5 | 28.0 | 26.7 | 22.3 |
| 10 – 11 | 4.0 | 3.4 | – | 4.0 | 21.0 | 31.8 | 39.0 | 28.5 |
| 11 – 12 | 5.0 | 5.1 | – | 5.3 | 33.4 | 35.0 | 53.3 | 37.5 |
| 12 – 13 | 5.9 | 6.4 | – | 6.3 | 46.3 | 33.6 | 58.3 | 44.2 |
| 13 – 14 | 5.9 | 5.1 | – | 5.9 | 48.0 | 24.9 | 43.8 | 41.3 |
| 14 – 15 | 4.3 | 2.0 | – | 3.6 | 32.2 | 12.6 | 19.5 | 26.7 |

the third cluster. Clearly larger samples may enable us to obtain
better estimates of typical morphologies in the growing process of
the girls.

Table 12. Computed mean values and per year growing speeds of the
stated variables corresponding to the three morphologies
and the whole sample of girls

| AGE | LEFT CALF CIRCUMFERENCE | | | | MAXIMUM PELVIC WIDTH | | | |
|---|---|---|---|---|---|---|---|---|
| | 1st cluster | 2nd cluster | 3rd cluster | Whole sample | 1st cluster | 2nd cluster | 3rd cluster | Whole sample |
| | MEAN VALUES | | | | | | | |
| 4 | 206.5 | 213.2 | 228.7 | 216.3 | 155.9 | 169.0 | 182.7 | 170.8 |
| 5 | 209.1 | 217.7 | 233.1 | 220.4 | 159.8 | 174.5 | 188.1 | 175.2 |
| 6 | 215.1 | 227.8 | 243.0 | 229.9 | 168.2 | 185.6 | 199.0 | 184.4 |
| 7 | 222.7 | 239.8 | 254.9 | 241.2 | 177.1 | 196.4 | 209.9 | 194.0 |
| 8 | 230.9 | 251.6 | 266.8 | 252.5 | 184.5 | 204.4 | 218.7 | 201.9 |
| 9 | 239.5 | 262.9 | 278.2 | 263.2 | 190.1 | 209.8 | 225.6 | 208.2 |
| 10 | 249.2 | 274.5 | 289.8 | 273.8 | 195.0 | 214.3 | 233.0 | 214.2 |
| 11 | 260.9 | 287.6 | 302.5 | 285.3 | 201.0 | 220.8 | 243.7 | 222.2 |
| 12 | 275.4 | 302.7 | 317.1 | 298.7 | 210.5 | 232.2 | 258.6 | 233.9 |
| 13 | 292.6 | 318.3 | 332.6 | 313.4 | 224.4 | 247.5 | 273.8 | 248.1 |
| 14 | 310.0 | 330.7 | 346.1 | 327.0 | 239.7 | 260.0 | 282.5 | 259.7 |
| 15 | 323.8 | 337.4 | 354.5 | 336.3 | 250.3 | 264.9 | 284.7 | 264.9 |
| | PER YEAR GROWING SPEEDS | | | | | | | |
| 4 – 5 | 2.6 | 4.5 | 4.4 | 4.1 | 3.9 | 5.5 | 5.4 | 4.4 |
| 5 – 6 | 6.0 | 10.1 | 9.9 | 9.5 | 8.4 | 11.1 | 10.9 | 9.2 |
| 6 – 7 | 7.6 | 12.0 | 11.9 | 11.3 | 8.9 | 10.8 | 10.9 | 9.6 |
| 7 – 8 | 8.2 | 11.8 | 11.9 | 11.3 | 7.4 | 8.0 | 8.8 | 7.9 |
| 8 – 9 | 8.6 | 11.3 | 11.4 | 10.7 | 5.6 | 5.4 | 6.9 | 6.3 |
| 9 – 10 | 9.7 | 11.6 | 11.6 | 10.6 | 4.9 | 4.5 | 7.4 | 6.0 |
| 10 – 11 | 11.7 | 13.1 | 12.7 | 11.5 | 6.0 | 6.5 | 10.7 | 8.0 |
| 11 – 12 | 14.5 | 15.1 | 14.6 | 13.4 | 9.5 | 11.4 | 14.9 | 11.7 |
| 12 – 13 | 17.2 | 15.6 | 15.5 | 14.7 | 13.9 | 15.3 | 15.2 | 14.2 |
| 13 – 14 | 17.4 | 12.4 | 13.5 | 13.6 | 15.3 | 12.5 | 8.7 | 11.6 |
| 14 – 15 | 13.8 | 6.7 | 8.4 | 9.3 | 10.6 | 4.9 | 2.2 | 5.2 |

References

E.M.Beale (1969). Euclidean cluster analysis. Bull.of the Int.Stat.
    Inst.

A.Mineo (1970). La scelta delle variabili nell'analisi della regres-
    sione multipla e delle funzioni discriminanti. Statistica, n.2

A.Mineo (1971). Sulla scelta del grado di una funzione di regres-
    sione polinomiale nell'analisi delle serie storiche. Statistica,
    n.3, 529-554.

A.Mineo (1985). A new criterion for the choice of seed points for a
    nearest centroid cluster analysis.   Rivista di Statistica
    Applicata, vol.18, n.4, 191-198.

R.Pearl and L.J.Reed (1923). On the mathematical theory of popula-
    tion. Metron, Vol.VIII, n.1.

Section 1.7

# SOLUTIONS USING A THREE-MODE PRINCIPAL COMPONENT ANALYSIS AND SOME COMPARISON RESULTS WITH THE OTHER APPROACHES

Pieter M.Kroonenberg

Department of Education, University of Leiden

Multivariate longitudinal data on the morphological development of young girls were analysed using three-mode principal component analysis. In particular, the deviations from the average growth curves were examined to investigate differential growth patterns of (groups of) girls. Results show girls to differ with respect to their growth in stoutness, skeletal length, and skeletal width. Most girls' growth patterns can succinctly be described as linear combinations of the average growth curves and three 'Girl Types'. The present analysis is compared with several other analyses with different techniques.

The present paper has two aims : the first and primary aim is to present an analysis of the multivariate longitudinal growth data collected by Sempé (1979, and this volume) employing three-mode principal component analysis. The secondary aim is to compare our analyses with the other analyses of the same data presented in this volume. In particular, comparisons are made with Lavit & Pernin's* work, and to a lesser extent with that of Lewi & Calomme, Pontier

---

* References without a date refer to papers in this volume.

89

& Pernin, and Mineo. In the discussion of the method employed, a
rough theoretical comparison between our method and STATIS (for references see Lavit & Pernin) is worked out. It should be stated at
the outset that the present paper, in contrast with the oral presentation, was written with all other papers available. This implies
that many of the ideas of other authors implicitly or explicitly have
gone into shaping this paper, so that we are very much in debt to
them. To bring some unity in the presentations it has been attempted
to adopt the notation and terminology of other authors, especially
that of Lavit & Pernin.

The growth data, yearly scores of 30 children on 8 morphological
variables from their fourth to their fifteenth year, may be treated
in two parts (1) a set of average growth curves for each of the variables, i.e. a 12 (years) by 8 (variables) matrix M with the averages of each of the variables at each point in time, and (2) the
deviations from these average growth curves, i.e. a $30 \times 8 \times 12$
data block of girls by variables by years. Even though in an independent paper of these data one should provide a proper discussion
of the trends and patterns in M, such an analysis is not presented
here, as Lewi & Calomme have already done so (their Figures 1 and 2,
and Table I), and to a lesser extent Mineo as well. The average
growth curves themselves are portrayed as parts of Figures 4 and 5
in Lavit & Pernin. It should, however, be noted that in both Figures
1 and 2 of Lewi & Calomme not the averages themselves are displayed
but a double-centred version of them (see their Introduction).

With respect to the second part, we will not analyse the deviations from the average growth curves per variable directly, but first
scale these values such that the deviations across variables may be
compared. The deviations themselves are not comparable without
scaling because of different measurement scales. In practice there
seem to be two options for this kind of scaling : either removing the
scale (here defined as the sum of squares, and called variation) per

variable per occasion, or removing the scale per variable over all
occasions together.   In connection with their global indices Pontier
& Pernin (section 1.3) remark that the first kind of scaling eliminates the variation of the variables across time, and the second kind
of scaling eliminates artifacts due to differences of unit of measurement or order of magnitude.   In the present case we considered it
undesirable to remove the increase or decrease in variation from the
deviation scores as such variation is an essential feature of the
transition to maturity, namely children start to diverge in their
physical characteristics.   To eliminate this process from the analysis by equating the sums of squares per variable per year seems
rather artificial.

More formally, the data analysed with three-mode principal component analysis have the following form :

$$\hat{x}_{ijk} = (x_{ijk} - \bar{x}_{.jk})/s_{.j.} \qquad i = 1...n; \ j = 1...p; \ k = 1...K \quad (1)$$

with $\bar{x}_{.jk} = (1/n) \Sigma_i x_{ijk}$, and $s^2_{.j.} = (1/nK) \Sigma_i \Sigma_k (x_{ijk} - \bar{x}_{.jk})^2$.
Table 1 of Lewi & Calomme and Table 3 of Mineo give the $\bar{x}_{.jk}$, and
the $s_{.j.}$ are 20.32 (Weight), 6.86 (Length), 5.18 (Crown-Rump Length),
3.53 (Head), 6.16 (Chest), 3.84 (Arm), 4.06 (Calf), and 3.16 (Pelvis).
The $\hat{x}_{ijk}$ to be analysed represent the deviations from the average
growth curves, or the curves representing the growth of an average
girl.   In other words, an average girl would have $\hat{x}_{ijk} = 0$ for all
i, j, and k.   Considering the way the sample was constituted (Sempé),
i.e. of normal French girls, it is not unreasonable to accept the
existence of such an average girl, and thus averaging over several
growth curves can be accepted as a meaninful procedure.   In fact,
girl 5 is practically this average girl.

## 1. Three-mode principal component analysis

The present description of the technique owes much to Tucker
(1963; 1966), Kroonenberg (1983), and Kroonenberg, Lammers, and

Stoop (1985). In order to make the presentation comparable to that
of STATIS described by Lavit & Pernin, their notation will be used
as much as possible, rather than that of the above references. The
major exception is that we have eliminated the reference to their
metric D (section II-2ff) as in this example it was chosen in such
a way that it is irrelevant for the subtive outcomes of their (and
our) analyses.

## 1.1. (Two-way or two-mode) principal component analysis

Let X be a (n × p) data matrix of p quantitative variables
observed for n individuals. The singular value decomposition of X
is defined as

$$X = T\Lambda^{\frac{1}{2}} U' = YU' = TZ',\qquad\qquad(2)$$

where U is the eigenvector matrix of the sample scalar-product
matrix for variables V = X'X, and T the eigenvector matrix of the
scalar-product matrix for individuals W = XX'. If we define
$Y = T\Lambda^{\frac{1}{2}}$, then the columns of Y contain the coordinates of the individuals on the columns of U, the principal axes of V. Simi-
larly, $Z = U\Lambda^{\frac{1}{2}} = X'Y\Lambda^{-\frac{1}{2}}$ contains the coordinates of the variables
on the columns of T, the principal axes of W.

## 1.2. Three-mode principal component analysis

Before describing the technique a few general remarks about
STATIS are in order as we have placed the present technique within
the STATIS framework, or rather terminology. The STATIS approach
is characterised by the motto "Interstructure – Compromis – Intrastructure", in which the Interstructure describes the structure
between the K matrices $X_k$, the Compromis describes a common component space for the individuals derived from a weighted average of
the scalar-product matrices for individuals $W_k = X_k X'_k$, and finally
the Intrastructure describes the variables loadings for the components of the Compromis, and the scores for individuals at each time

point  k  around their barycentric or common solution.  The three-mode analysis presented here bypasses the Interstructure.  To investigate this structure one should use a different form of three-mode principal component analysis as described in Tucker (1966) or Kroonenberg and De Leeuw (1980), and many other authors.

Suppose the same population of individuals is observed on the same  p  variables at  K  points in time, then  $X = (X_1, \ldots, X_k, \ldots, X_k)$  is the resulting set of  (n × p)  data matrices.  Analogously to the singular value decomposition in the two-way case a generalized singular value decomposition for  X  may be defined, such that

$$X_k = TC_kU' \quad (k = 1 \ldots K) \tag{3}$$

The eigenvector matrices  T  and  U  may be interpreted as in the two-way case, and they are the 'compromise' solutions for the individuals and variables, respectively.  The compromise solutions take the place of separate solutions  $T_k$  and  $U_k$  for each point in time k.  The  $C_k$  contain the generalized singular values for occasion  k, and generally they are not diagonal as they would be for the separate solutions.  Furthermore, they do not necessarily contain non-negative numbers as in the two-way case.  The  $c_{abk}^2$, however, add up to the total sum of squares if a complete solution is obtained, and to the fitted sum of squares if an approximate solution is derived, and each  $c_{abk}^2$  may be interpreted as a variation or sum of squares accounted for as in the two-way case.  The form of three-mode principal component analysis discussed here is based on Tucker (1972), rather than on Tucker (1966) in which a three-mode analysis is proposed which also contains a compromise solution for the third (or remaining) mode.

If we define the symmetric  (n × n)  matrix  $W_k = X_kX_k'$  as the scalar-product matrix for individuals for the k-th point in time, then using (3)  $W_k = TC_kU'UC_k'T' = TC_kC_k'T'$, and  $W = \Sigma_k W_k = T(\Sigma_k C_kC_k')T'$.  As  $W = XX'$,  T  may be chosen to be the  (n × n) eigenvector matrix of $W$, and thus $\Sigma_k \, C_k C_k' = \Lambda$ the diagonal eigen-value matrix with decreasing values on its diagonal. Analogously to the two-way case a $\ddot{Y}$ may be defined as $\ddot{Y} = T\Lambda^{\frac{1}{2}}$, and $Z = (Z_1, \ldots, Z_k)$ with $Z_k = X_k'T = UC_k'T'T = UC_k'$, or equivalently $Z_k = X_k'\ddot{Y}\Lambda^{-\frac{1}{2}}$. The $Z_k$ are the component scores, i.e. the columns of $Z_k$ contain the coordinates of the variables on the compromise axes of the individuals.

Similarly, one may define the symmetric $(p \times p)$ matrix $V_k = X_k'X_k$ as the scalar-product matrix between variables for the k-th point in time. Then $V = \Sigma_k \, V_k = U(\Sigma_k \, C_k'C_k)U'$, and as $V = X'X$, $U$ may be chosen to be the $(p \times p)$ eigenvector matrix of $V$ with $M = \Sigma_k \, C_k'C_k$ the corresponding matrix with descending eigenvalues on the diagonal. When one defines $\ddot{Z} = UM^{\frac{1}{2}}$, and $Y = (Y_1, \ldots, Y_k)$ with $Y_k = X_k U = TC_k U'U = TC_k$, or equivalently $Y_k = T\ddot{Z}M^{-\frac{1}{2}}$ as the component scores, then the columns of $Y_k$ contain the coordinates of the individuals at occasion $k$ on the compromise axes for the variables.

In general, when not all $n$ eigenvectors of $W$, and not all $p$ eigenvectors of $V$ are of interest but in both cases only a limited, not necessarily equal, number of eigenvectors are relevant, one has to solve the eigenvector-eigenvalue problems of $W$ and $V$ simultaneously by iteration in order to find an optimal solution for both compromises at the same time. This solution has the property that the variation accounted for by both compromises is the same and is due to the same part of the data. Such a solution may be found by minimizing the loss function

$$\Sigma_k \, \| X_k - TC_k U' \|^2 \tag{4}$$

over all orthonormal $U$ and $T$, and all arbitrary $C = (C_1, \ldots, C_k)$. Kroonenberg and De Leeuw (1980; Kroonenberg, 1983) show that the solution to this problem is that $\hat{U}$ is the eigenvector matrix of $\hat{W} = \Sigma_k \, X_k \hat{T}\hat{T}'X_k'$, $\hat{T}$ is the eigenvector matrix of $\hat{V} = \Sigma_k \, X_k'\hat{U}\hat{U}'X_k$,

and $\hat{C}_k = \hat{T}'X_k\hat{U}$ $(k = 1...K)$. Note that the solution of each mode
takes into account the reduction over the other mode, and that when
all eigenvectors are computed $\hat{V}$ and $\hat{W}$ reduce to $V$ and $W$, and
thus $\hat{T}$ and $\hat{U}$ to $T$ and $U$. In the approximate solution we will
define, unlike above, the component scores directly as $\hat{Z}_k = \hat{U}\hat{C}_k' =$
$\hat{U}\hat{U}'X_k'\hat{T}$, and $\hat{Y}_k = \hat{T}\hat{C}_k = \hat{T}\hat{T}'X_k'\hat{U}$, again taking into account the reduction in dimension in the other mode. Also here $\hat{Z}_k$ and $\hat{Y}_k$ reduce
to $Z_k$ and $Y_k$ when all components are computed.

What makes three-mode principal component analysis really different from techniques like STATIS is that information is available
on two compromise solutions at the same time, and moreover that the
relationships between the two compromise solutions at each point in
time can be expressed numerically via the so-called 'core slices'
or 'core planes' $C_k$ with the generalized singular values. An
element $c_{abk}$ indicates the weight that is given (or importance
that is attached) to the combination of the a-th column of $T$, and
the b-th column of $U$ in the description of a score $x_{ijk}$ by the
model (3), as can be seen from (3) in sum notation

$$x_{ijk} = \Sigma_a \Sigma_b c_{abk} t_{ia} u_{jb}. \tag{5}$$

Inspecting the vectors $(c_{ab1},...,c_{abK} \mid a = 1...A; b = 1...B;$
$A \leqslant n; B \leqslant p)$ one can judge how the combination of the a-th component for the individuals and the b-th component for the variables
change over time (see Figure 2).

In summary, after the (approximate) decomposition of the k-th
slice of the three-mode data matrix $X_k$ in $TC_kU'$ (dropping the
carets for convenience), one has available a compromise solution for
individuals (T), a compromise solution for variables (U), for each
occasion a matrix $C_k$ with mutual weights for the combinations of
axes from the two compromise solutions, furthermore component scores
of the individuals on the variable axes $(Y_k)$ for each time point,
and similarly for the variables on the individual axes $(Z_k)$.

If a visual display is desired of the relationship between variables and individuals at each occasion within the framework of the compromise solutions, $TC_kU'$ may be decomposed as

$$TC_kU' = TE_k\Gamma_kF_k'U' = (TE_k\Gamma_k^{\frac{1}{2}})(UF_k\Gamma_k^{\frac{1}{2}})' = \tilde{T}_k\tilde{U}'_k \qquad (6)$$

and the $\tilde{T}_k$ and $\tilde{U}_k$ may be simultaneously displayed in a single plot. ($E_k\Gamma_kF_k'$ is the singular value decomposition of $C_k$). Note that after the basic matrices $T$, $U$, and $C$ have been derived all other information is directly based on them without referring back to the original data matrix.

When the special weighting of the $W_k$ (and $V_k$) in STATIS is of relatively minor importance, i.e. $\Sigma_k \alpha_kW_k \simeq \Sigma_k \hat{W}_k$ (and $\Sigma_k \beta_kV_k \simeq \Sigma_k \hat{V}_k$), then the practical differences will be small. In the present data the $\alpha$'s are not very different as can be seen from Figure 1 of Lavit & Pernin. It should be noted that STATIS determines the Interstructure in a way, which has no direct equivalent in the present form of three-mode analysis, nor in Tucker's (1966) version be it that it can be shown that the third compromise solution is a special kind of Interstructure.

## 2. Results and interpretation

The analysis of the data will be discussed in several parts. First, we will present some statistics on the overall solution, then we will discuss the compromise solutions for individuals and variables, next we will portray how the relationships or mutual weights change over time, and give on the basis of these changes a general description how girls may deviate from the average growth partterns, and finally we will portray, and briefly discuss, the development of individual girls over time, both via 'differential growth curves' and via 'trajectories' in the compromise component space for variables.

## 2.1. Fit

In contrast with Lavit & Pernin's two-dimensional solution, we will examine a three-dimensional solution for both individuals and variables;  in other cases different numbers of components may be necessary for the two compromise solutions, but three components for each seems adequate in the present case.  The overall fitted sum of squares accounted for 77% of the total variation in the centred and scaled data.  Due to the simultaneous derivation of the compromise solutions, both of them fit the same part of the data, they differ, however, in the way they divide the fitted sum of squares over their principal axes.  The axes of the subject compromise solution account for 55, 14, and 7% of the variation in the data, respectively, while those of the variable compromise solution accounted for 57, 14, and 6% respectively.  Note that these figures compare very well with the STATIS compromise solution for individuals of 57 and 13% for the first two axes.

As mentioned in the introduction the differences in variability over the years are still contained in the data set, and in Figure 1 this variability is shown in terms of the total and the fitted sums of squares.  The figure shows that differences between girls measured over all variables increase until their thirteenth year, and diminish in the next two years available.  The data allow, however, no extra-

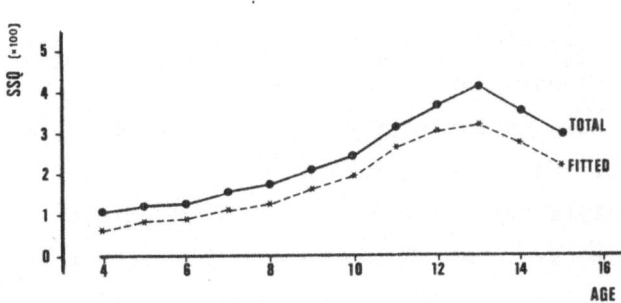

Fig.1. Total and fitted sums of squares per year

polation beyond the fifteenth year (see Mineo who attemps to fit
logistic curves and to estimate upper asymptotes, which asymptotes
can be taken as upper bounds for at least the averages beyond the
fifteenth birthday.  For two variables, head circumference and left
upper arm circumference no asymptotes could be found).  The increasing variability suggests that the onset and speed of growth is different over girls, such that slower growth catches up, at least
partially, with faster growth.  Whether the variability returns to
its original level at the age of five cannot be judged from the
present data, but it seems unlikely that it will.

## 2.2. Compromise solution for girls

In Figure 2 the compromise solution for girls is shown;  the
lengths of the axes are proportional to their explained variation.
As the first axis (I) carries most weight the configuration has
roughly the shape of a flattened elongated ellipsoid.  No specific subgroups can be seen, and the second and third axis are largely dominated by a few girls;  the second axis by 4, 6, 14(E),
18(I), 20(K), and 24(O), and the third axis by 14(E), 19(J), 22(M),
and 30(U).  (The letter between parentheses refers to the label in
this and other figures).  One way of using the 'girl space' is to
consider the axes as (morphological) Types or ideal types, and
describe each girl as a linear combination of such Types.  One could
also refer to average girls in the centre of the figure as a type
(as in Mineo, cluster 2), but this seems not a proper thing to do.
A Type is thus defined as a girl who loads exclusively on one single
axis, and the loading of a real girl on such an axis indicates how
relevant the axis is for that girl.  As no distinct subgroups seem
to be present, the Types defined here should not be taken as clusters with qualitative differences.  In fact, Mineo's cluster analyses define clusters which can largely be recovered by cutting up
the first principal axis into three parts.  His first cluster consists of girls 4, 7, 21(L), and 27(R) on the positive side of the

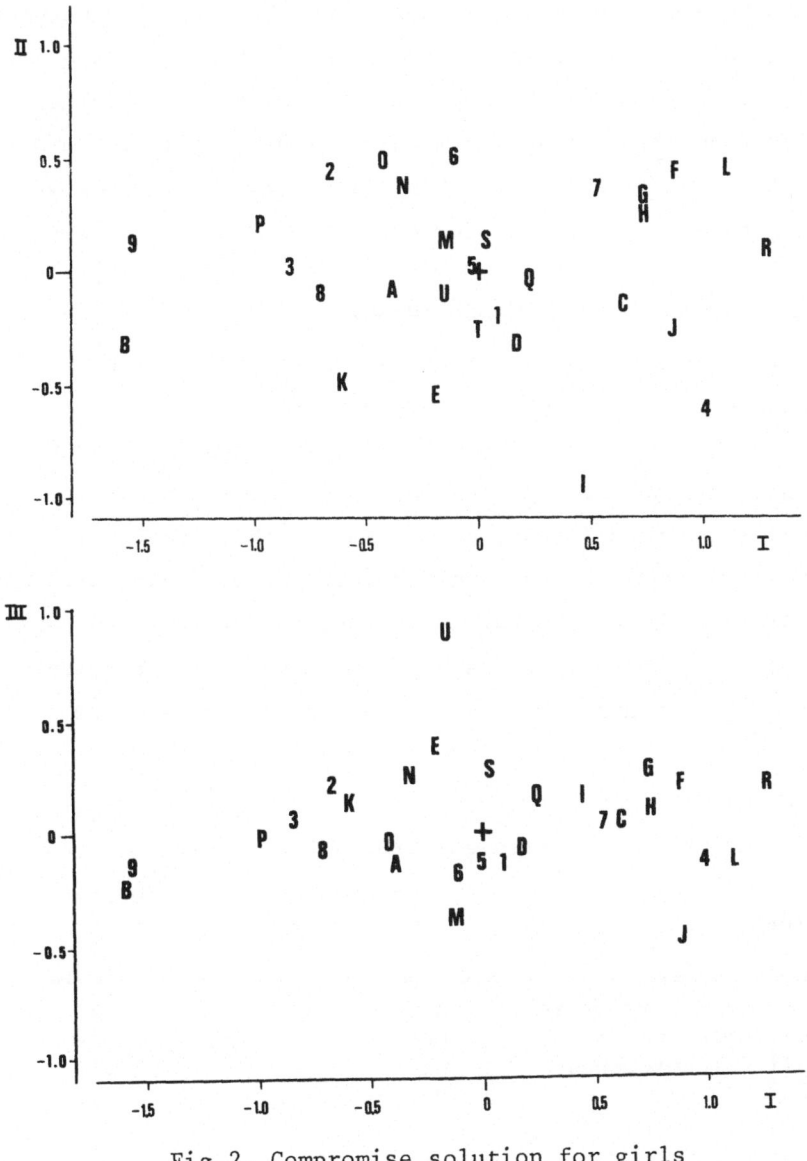

Fig.2. Compromise solution for girls

first axis, his second cluster consists of girls 1, 5, 10(A), and 13(D) located near the centre of Figure 2, and finally the third cluster consists of girls 3, 9, and 11(B) on the negative side of the first axis.

Even though there are three axes, we will, for convenience
define six Girl Types, one for each side of an axis.  Type I+
(Type I-) is defined by the positive (negative) side of the first
axis, and thus 27(R) is approximately a Type I+ girl and 9 a
Type I- girl, while 30(U) is a type III+ girl.  Using the variables
we will give more substantive descriptions of the Girl Types.

## 2.3. Compromise solution for variables

Instead of showing the principal components for the variable
compromise solution, we will direct our attention to a varimax
rotated version of them.  The primary motivation is that the variables can be divided into three groups (see Lavit et Pernin, section 1.3)  Stoutness (chest, calf, arm, weight), Skeletal Length
(length, crown-rump length), and Skeletal Width (head, pelvis),
and as Table 1 shows, the varimax transformed components pass roughly through the centroids of each of the groups of variables.  By
using the varimax axes we can describe the variable compromise solu-

Table 1. Variable compromise solution after varimax

| Variables | Mnemonics[a] | | Component[b] | | |
|---|---|---|---|---|---|
|  | K | L&P | 1 | 2 | 3 |
| Left Upper Arm Circumference | Arm | PB | .56 | -.21 | .04 |
| Chest Circumference | Chest | PT | .52 | .02 | -.05 |
| Left Calf Circumference | Calf | PJ | .44 | .01 | .03 |
| Weight | Weight | TA | .43 | .17 | .08 |
| Length | Length | PD | -.03 | .76 | -.06 |
| Crown-Rump Length | CRLen | SS | .03 | .57 | .13 |
| Head Circumference | Head | PC | -.11 | -.10 | .91 |
| Maximum Pelvic Width | Pelvis | AB | .14 | .17 | .37 |
| Percentage of Sum of Squares | | | 35 | 23 | 19 |

[a]K=this paper; L&P=Lavit & Pernin.  [b]Orthonormal components rotated
with varimax

tion in terms of variable groups, rather than ratios of variable
groups, such as Stoutness/Skeletal Length ratio, as Lewi & Calomme
do explicitly, and Lavit & Pernin implicitly.  The varimax transformation redistributes the variation accounted for per axis is such
a way that Stoutness accounts for 35%, Skeletal Length for 23% and
Skeletal Width for 19% of the deviations from the average growth
curves.

## 2.4. Growth characteristics

Next we will investigate how the girl types defined above
develop, in particular with respect to which variable groups this
growth takes place.  In order to discuss this we will first define
some terms.
Positive (negative) growth curves lie above (below) the average
growth curves, and are (approximately) parallel to them.  Accelerating (decelerating) growth curves move away from (towards) the
average growth curves, and positive (negative) crossing growth
curves cross the average growth curves from above (below).

Figure 3 shows the relationships (or mutual weights) between
the axes of the two compromise solutions for each point in time.
The height of the curves is both influenced by the size of the deviation of girls from the average growth curves and by the number
of girls showing such a pattern.  Therefore, comparisons of absolute
size are best made within girl type.

Girl Type I+.  This girl type has accelerating growth curves
uptil a certain age and decelerating growth curves thereafter.
Stoutness accelerates until the eleventh year, levels off between
11 and 14 years old, and decelerates in the last two years.  Thus
this girl type is stouter to start with at the age of 4, and becomes
increasingly more corpulent until her eleventh year, and relatively
speaking only after fourteen years of age the Average Girl starts
catching up, but probably only partly so.  The acceleration for

Skeletal Length ends somewhat earlier (around 12), and the trend is
reversed more suddenly. Furthermore, this girl type tends to take
up roughly the same position in the end as in the beginning. Skeletal Width shows roughly the same pattern. In other words, Type
I+ girls, like 27(R), have accelerating growth on all variable
groups, indicating earlier growth than average, especially in the
skeletal variables for which they end up about where they started.
In the end they tend to be (overly) hefty compared to the Average
Girl. The mirror image of this type of girl is the Type I- Girl
(e.g. 9 and 11(B)), whose growth curves are the mirror images of
the Type I+ Girl. They have decelerating growth curves on all variables, indicating later growth than average. In the end they
regain their relative position with respect to the skeletal variables, but they remain on the slight side.

Girl Type II+. This girl type has a different set of growth
patterns. From an average Stoutness she decelerates until 14 years
of age, and thus becomes, relatively speaking, slighter all the
time. Her relative Skeletal Width is slightly above average and
remains so, on the other hand she keeps on growing in Skeletal
Length all through the observation period. Thus a Type II+ Girl
becomes a tall and skinny girl, and she gets more so all the time.
Not so a Type II- Girl who becomes a more squat and compact girl
all the time.

Girl Type III+. This girl type is yet again different. She
has a decelerating growth curve for Stoutness until 13 years old,
changing from a slightly above average girl to an under-average one.
Realizing that these scores are deviations from the average curves,
the deceleration implies a delayed growth, i.e. this girl type does
not really start growing as her colleagues in Stoutness until thirteen years of age, but then in two years manages to make up for it
entirely, as she ends up at about the same deviation above average
as she started with. Her skeletal growth reflects the same pattern,

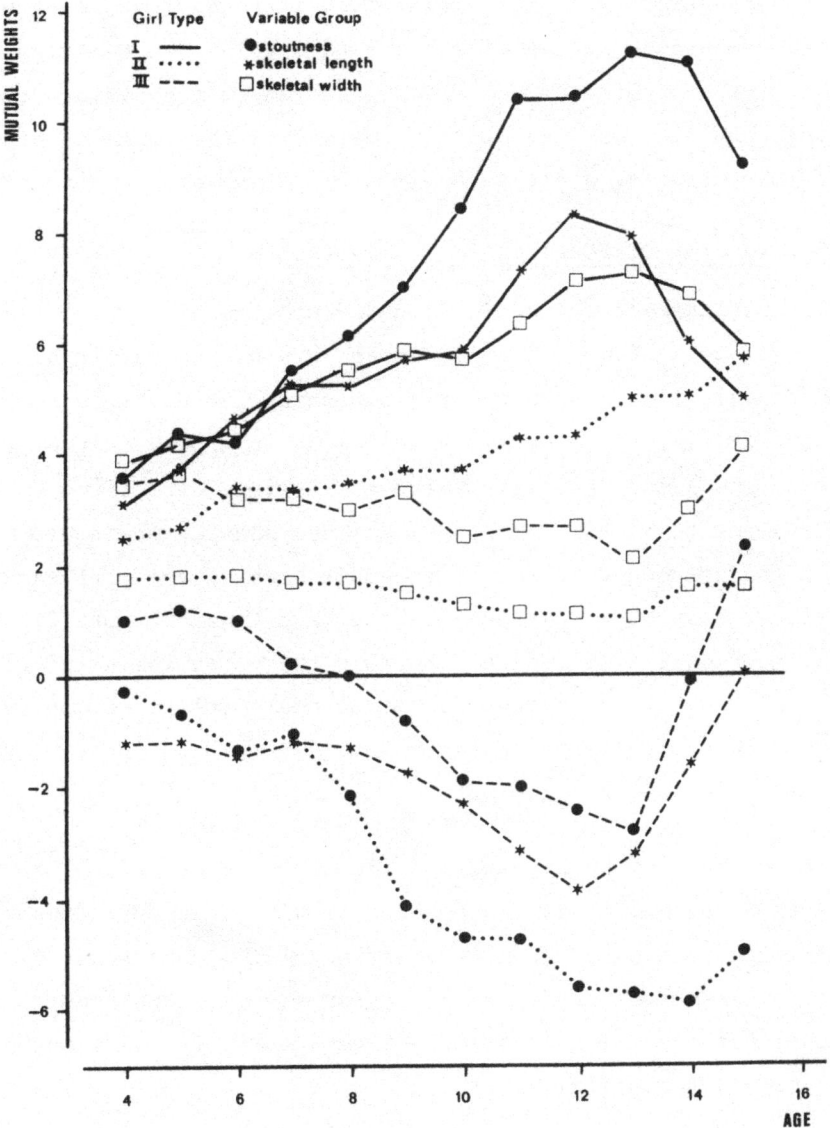

Fig.3. Time trends for mutual wieghts of axes of the variable and
       individual compromise solutions

but note that her Skeletal Width (i.e. her head circumference) ends
up relatively speaking much larger than her Skeletal Length.  Girl
Type III- is, of course, again the mirror image of the Type III+
Girl.

In evaluating different girl types, it seems that it is not so much the appearance of the girls at a particular age that is important in distinguishing between them, but rather the different ways in which they develop. In other words, it is the dynamics of growth that differentiates, and not the static aspects of the situation.

## 2.5. Growth of individual girls : Time trends

If the interest was solely centred on morphological types and their development Figure 3 could serve this purpose admirably, if however one is also interested in inspecting the development of individual girls, then one may describe their growth on the variable groups by comparing them again with the Average Girl using the component scores (i.e. $Y_k = TC_k$) as defined above. One could also inspect the component scores of the variables on the girl compromise axes (i.e. $Z_k = UC'_k$) as Lavit & Pernin do (their Figure 2), but we will not display them here. Because the variables align rather closely with their components, their separate relationships with the girl types are already reflected in our Figure 3 by those of the components or variable groups.

In contrast with the variables, most girls are true linear combinations of girl types, and their growth curves with respect to, for instance, Stoutness are often influenced by more than one of the Stoutness curves in Figure 3. As an example, the growth curve for Stoutness of girl 4 is a linear combination of the Stoutness curves of all three girl types :

$$y_{41k} = t_{41}c_{11k} + t_{42}c_{21k} + t_{43}c_{31k} =$$

$$= .25c_{11k} - .31c_{21k} - .11c_{31k},$$

where $c_{p1k}$ is the Stoutness growth curve (variable axis 1) for girl type $p$ $(p = 1,2,3)$, and the $t_{4p}$ are the loadings of girl 4 on the girl compromise axes. From Figure 3 we see that the signs of the $t_{4p}$ orient all curves with their peaks upwards, so that girl 4

accelerates very rapidly (starts growing early and fast) in Stoutness.  For girl 16(G) the Stoutness curves more or less cancel each
other

$$y_{G1k} = .18c_{11k} + .17c_{21k} + .12c_{31k} \, ,$$

so that her Stoutness curve is nearly flat, and parallel to the average growth curve, just like that of girl 26(Q) for which no Stoutness curve is really important

$$y_{Q1k} = .05c_{11k} - .01c_{21k} + .06c_{31k} \, ,$$

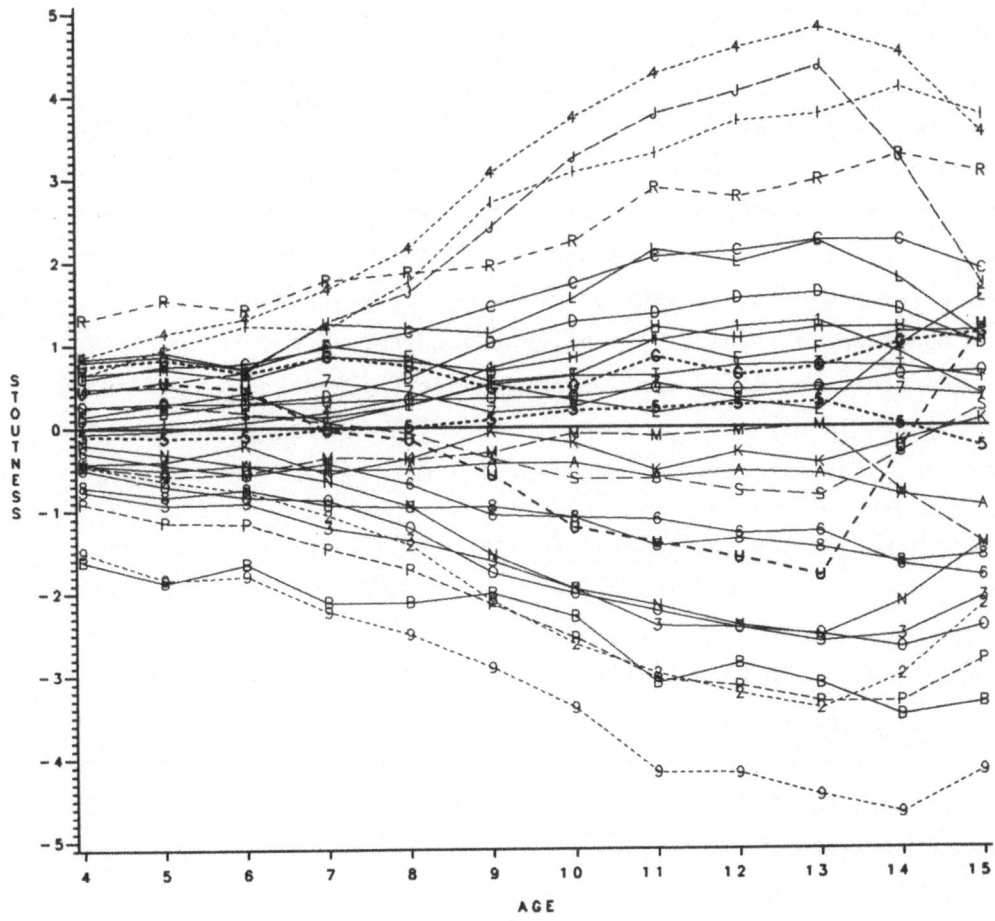

Fig.4. Individual trends for Stoutness

In Figures 4, 5, 6 the individual trends are shown for Stoutness,
Skeletal Length, and Skeletal Width, respectively.  These figures,
of course, confirm the conclusions from Figure 3, but as each girl
is now shown separately, number of girls and size of deviation are
no longer confounded, so that it is easier to make statements about
the absolute size.

   Stoutness.  Differentiation between girls occurs especially in
Stoutness.  As expected from the above discussion, the general trend
is that stout girls above average stay above average, and the stouter
the quicker the acceleration in Stoutness until around 11 to 13 years
of age, while after that the Stoutness decelerates or stabilizes,
with the reverse pattern for slight girls.  There are notable exceptions, for instance, after her thirteenth year girl 19(J) virtually
stopped growing allowing the Average Girl to catch up in Stoutness.
Girl 14(E) suddenly started getting fleshier in her thirteenth year.
The most noticeable deviant pattern is of girl 30(U) who provides the
only example of a substantial crossing curve.  Around her thirteenth
birthday she was relatively slight, but started growing very rapidly
thereafter, to end up considerably above the average growth curve
two years later.  She is probably largely responsible for the form
of the Stoutness curve of the Type III Girl.

   With respect to Skeletal Length, it is clear that the overall
differences are less.  The differences at the age of four between girls
are comparable to those for Stoutness, but Skeletal Length increases
less, and the differentiation in growth curves is not as marked.  The
overall trend is, however, comparable with very few crossing curves,
with taller (shorter) girls tending to become taller (shorter) until
their 12th to 13th birthday.  For some girls Stoutness and Skeletal
Length increase hand in hand (e.g. 19(J)), but this is not true for
all;  for instance, girl 9 makes up for her length deficit but not
her wieght deficit.  Girl 18(I) is very peculiar in that the acceleration of her Stoutness growth is considerable and only seems to be

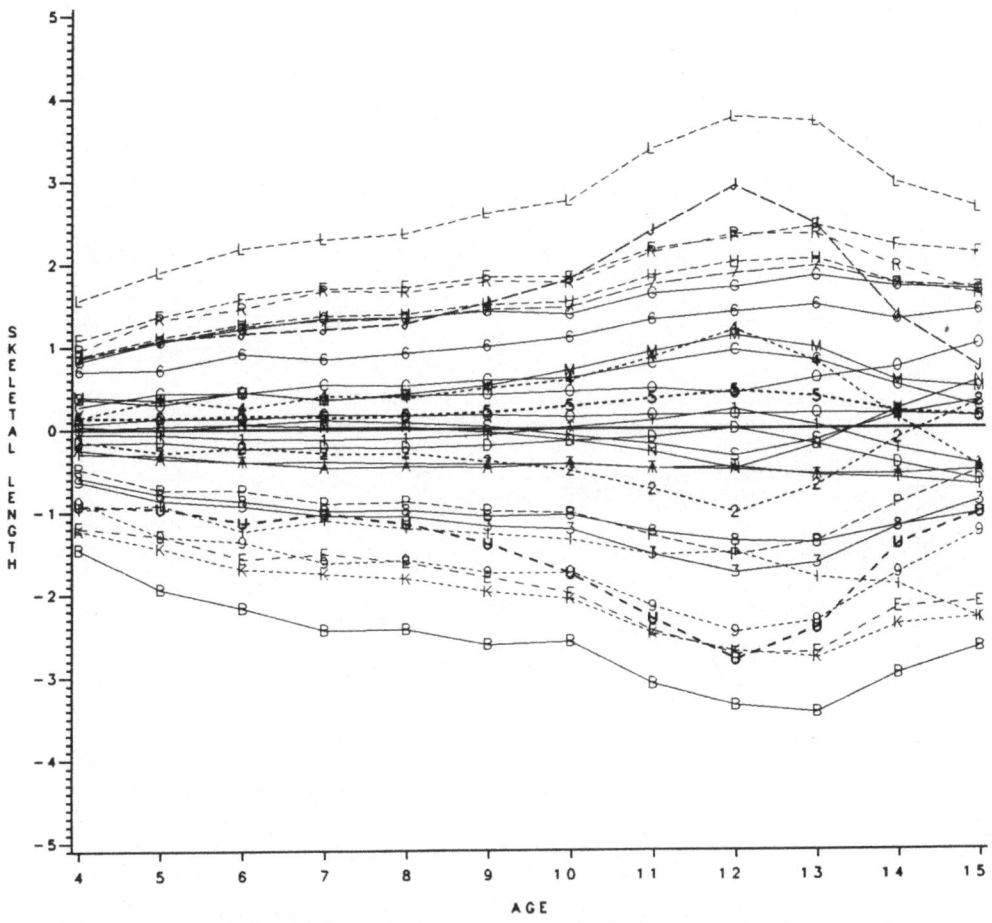

Fig.5. Individual trends for Skeletal Length

levelling off at the age of 15, but she is still decelerating in
Skeletal Length in her 15th year. The relationships between the
various groups for individual girls can however be better examined
via the trajectories shown below.

Skeletal Width (Figure 6) is different in the sense that virtually all girls have curves roughly parallel to the average growth
curves, showing that Skeletal Width, especially head circumference
has a regular and stable growth pattern, and that generally differences between grils are maintained troughout the observation period.

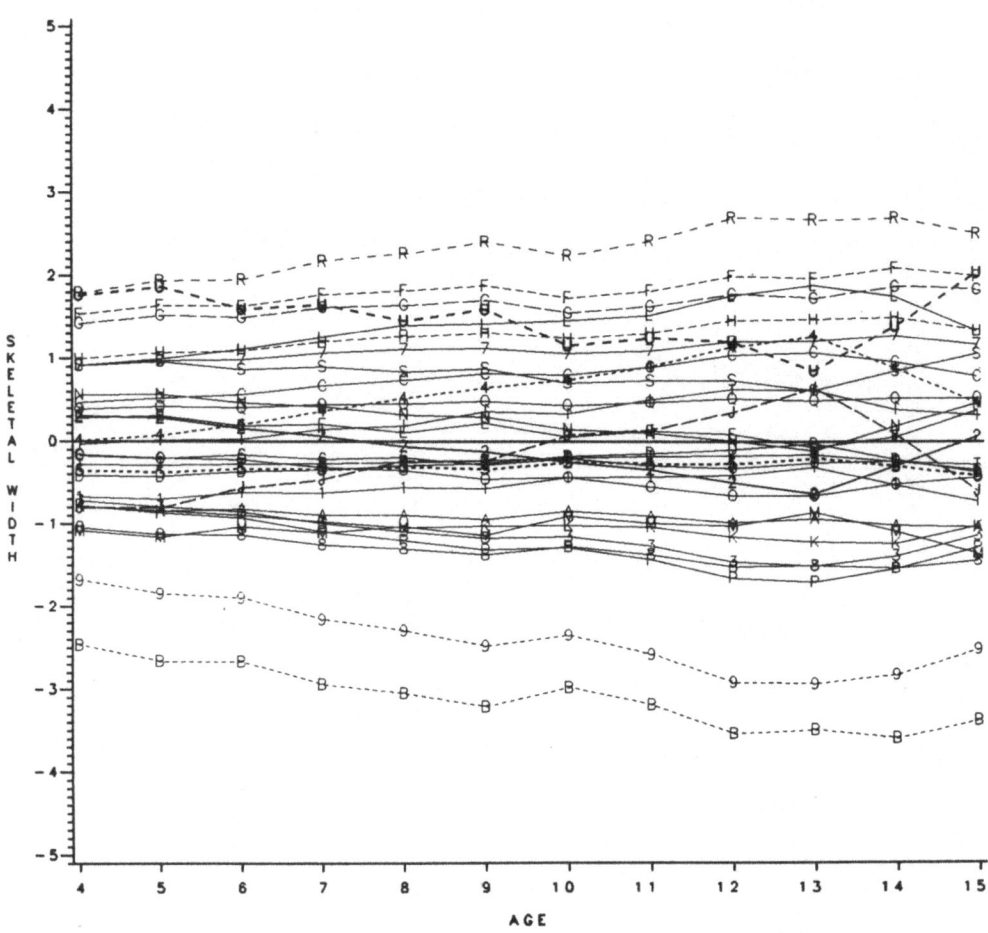

Fig.6. Individual trends for Skeletal Width

Skeletal Width at fifteen years can thus well be predicted from the
information at the age of four. An interesting question that we and
none of the other authors have studied is the predictability for each
of the variables at age 15 from the variables at age four.

## 2.6. Growth of individual girls : Trajectories in variable space

By pairwise plotting component scores on the variable components,
one obtains plots with 'trajectories' for each of the girls showing
how the relative importance of the components change over time (see
also Lavit & Pernin, and Pontiers & Pernin). Note that Lavit & Pernin
in their Figure 3 show the trajectories of the girls in the compro-

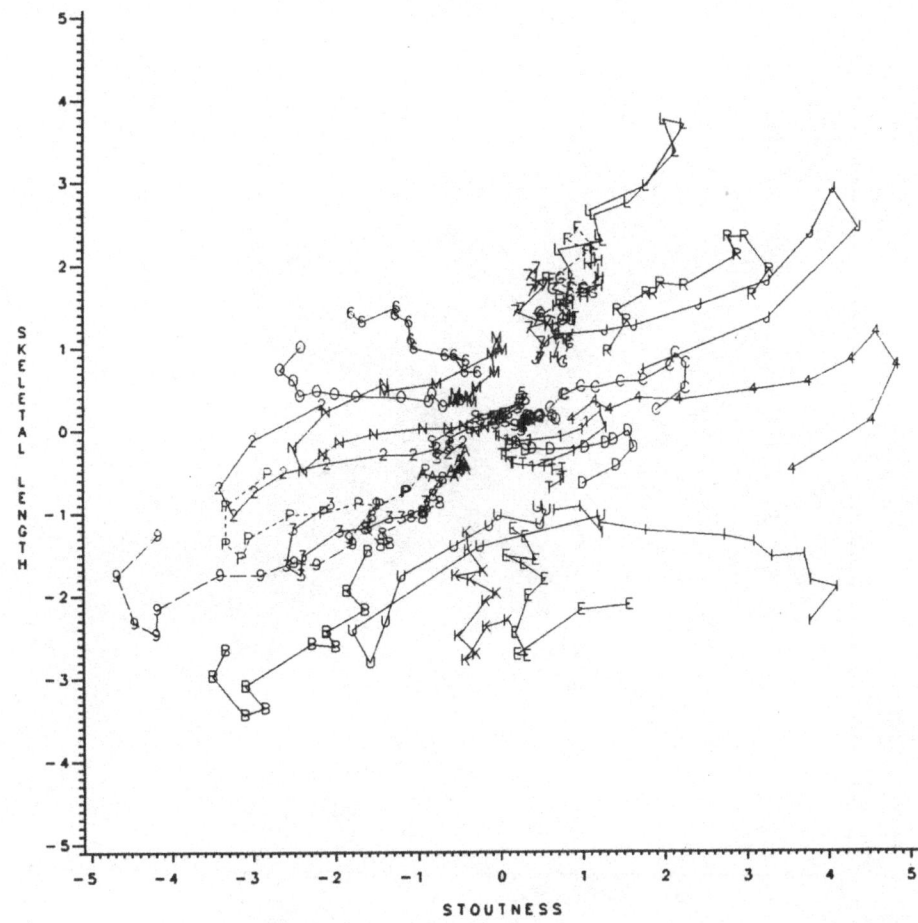

Fig.7. Stoutness versus Skeletal Length

mise solution for individuals, while here in Figures 7 and 8 they
are shown in the compromise solution for variables.  Lavit & Pernin
derived their coordinates per occasion via $\ddot{Y}_k = W_k Y \Lambda^{-1}$ (see their
section 1.3).

In Figure 7 Stoutness versus Skeletal Length is displayed by plotting the vertical coordinates of Figure 4 and 5 against one another.
The elongated pattern shows the correlation between the scores on the
two variable groups, and the longest axis roughly corresponds to the
first principal component.  Trajectories at an angle of the main axis

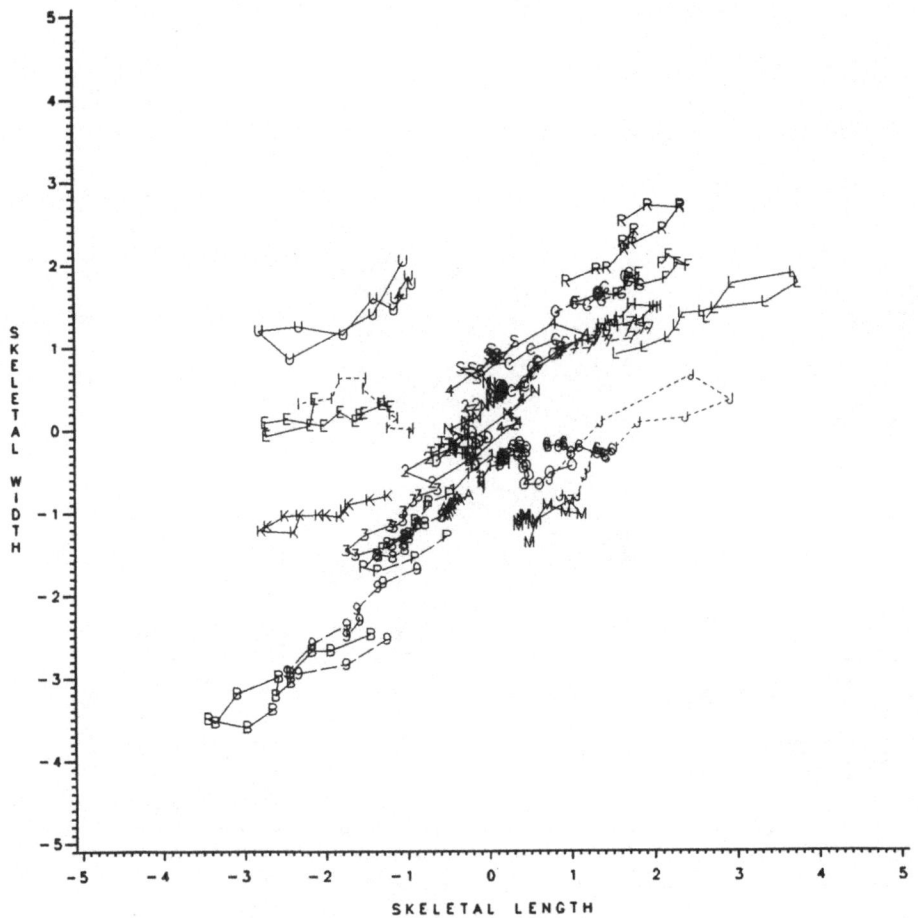

Fig.8. Skeletal Length versus Skeletal Width

of the ellipse show 'imbalances' in growth, i.e. growing stouter but
not taller, and vice versa.  The centre of the plot represents the
Average Girl, or at least her growth pattern.  All clearly visible
trajectories start more or less outwards, indicating an acceleration
away from the average, and most trajectories turn inward again some-
where between the eleventh and thirteenth year, showing that they
tend to become more like the Average Girl later on.  The turning
points of the trajectories indicates changes in the relative rate
of growth between the Variable Groups defining the axes.

Figure 8 gives an impression of differential skeletal growth. Except for nine girls, 6, 14(E), 18(I), 19(J), 20(K), 21(L), 22(M), and 24(0), there is a near perfect correlation between differential growth in Skeletal Length and Skeletal Width.

## 3. Conclusion

In this paper we have tried to give a fairly detailed description of both large and small scale patterns in the morphological growth of girls between 4 and 15 years of age. Different types of girls could be described, be it that some types originate primarily due to a few individuals. The types differ in the way they grow on specific groups of variables. The latter could be grouped into variables indicating Stoutness, Skeletal Length, and Skeletal Width. Several interesting details emerged, for instance, that heavier (lighter) girls tend to grow heavier (lighter) at a faster rate and earlier on, that differences in growth speed between girls are concentrated in the Stoutness variables, and that Skeletal Width (especially head circumference) shows differences in size between the girls but hardly in growth rate, with Skeletal Length taking an intermediate position. Also noteworthy is that with very few exceptions, girls do not crossover to a substantial degree, i.e. as a rule stouter, longer, and/or wider than average girls do not become below average on these variables, but stay above average, with a similar pattern for the slighter, smaller and/or slimmer girls.

Thus the overall impression from the entire analysis is that even though there are differences in growth rates, and the variables in which this becomes manifest, there is little evidence for large scale qualitatively deviating differential growth patterns for the thirty girls. Thirty is, however, not a large sample to base general conclusions on about the deviations from the average growth patterns for all French girls. Some deviating individuals might be part of a larger group of girls with qualitatively different growth patterns, but from the present data alone this cannot be inferred.

References

P.M.Kroonenberg (1983). Three-mode principal component analysis :
    Theory and applications. Leiden, The Netherlands : DSWO Press.

P.M.Kroonenberg & J.De Leeuw (1980). Principal component analysis of
    three-mode data by means of alternating least squares algorithms.
    Psychometrika, 45, 69-97.

P.M.Kroonenberg, C.J.Lammers & I.Stoop (1985). Three-mode principal
    component analysis of multivariate longitudinal organizational
    data. Sociological Methods and Research, 14, 99-136.

M.Sempé (1979). Auxologie : Méthodes et Séquences. Paris : Théraplix.

L.R.Tucker (1963). Implications of factor analysis of three-way
    matrices for measurement of change.  In C.W.Harris (Ed.),
    Problems in measuring change. Madison : University of Wisconsin
    Press.

L.R.Tucker (1966). Some mathematical notes on three-mode factor
    analysis. Psychometrika, 31, 279-311.

L.R.Tucker (1972). Relations between multidimensional scaling and
    three-mode factor analysis. Psychometrika, 37, 3-27.

CHAPTER 2. PREHISTORIC ASSEMBLAGES AND LITHIC ARTIFACTS
FROM A SMALL WEST-EUROPEAN AREA

# PROBLEMS OF CHRONOLOGICAL SERIATION IN THE EPIPALEOLITHIC ("MESOLITHIC")

J.G.Rozoy

Charleville, France

The epipaleolithic ("mesolithic") is a period in prehistory which lasted in our countries around from 9000 to 4000 before our era (uncalibrated conventional radiocarbon years). It was named "the last hunters"period", that is that of the last ones who used to live on hunting (including fishing) only, without coming to produce : neither plant cultivating nor animal breeding. It was also, fundamentally, the period of hunting with bows : the technique of the bow and arrows, through its considerable social consequences, determined the main characteristics of that period.

Most of the remains found in the sites are chipped flints, which are liable to be analysed very deeply, considering the large variety of their shapes. This variety is due as much (and more) to the makers' working habits and to fashions in doing so as to the functions the tools were used to. At a same period in a same area the sites which are rich enough give nearly stable proportions of the various kinds of tools. Both experience and theory show that stability can be obtained only above 100 tools, it is better beyond 200 ones. That situation is quite different from those encountered when studying cemeteries. In that case, all the things usual at

the burial time are not present (far from that) in each tomb.  Here,
on the contrary, usual things are always present in layers corresponding to their ages, and they are there at rates which are nearly
stable at a given time, these proportions evolving with time.
Though, rare specialised sites exist, they have to be taken apart
from chronological study.

Data structure

For the franco-belgian epipaleolithic I could make a list of
119 types (plus a few types of characteristic wastes), distributed
into 14 classes.  There are 7 classes of "common tools" (understand : common to regions and periods) with a rather domestic use :
scrapers (preparing skins), gravers ("burins", working on bone and
antler), borers (making clothes), retouched flakes, retouched or
truncated blades and bladelets, A.S.O. the other 7 classes affect
the arrow "armatures" (points and side cutting edges).  As soon as
the list was established, the classes were arranged in more or less
chronological order, which explains some peculiarities which Mr Lewi
noticed on the graphs he made.  Indeed a rough order had been established for long throug  classical qualitative studies.  It can also
be noticed that some classes of common tools, more particularly retouched flakes (N" 10 to 18), because of their nature not much differentiated, were common to all the evolutive stages, and then very
little significant for chronological evolution.  But they can be
very useful when you differenciate regions.  On the contrary, armatures evolved quickly and they are excellent indicators for chronology.

The flints, which are the only things thoroughly preserved in
all the sites, are now the only countable basis at our disposal for
chronological or interregional comparisons and for the study of the
evolution of industries.  Two problems remain, though, as far as
they are concerned : to which extent do a lot of similar toolkits

prove the existence of a human group with its peculiarities ? Most
French-speaking searchers (but not the American ones) admit to equate
tool-kits with social groups, but it is more a postulate than a demonstration. Besides, probably the toolkits from different times and
places do not correspond to the same actions, some (not constant)
works have probably be done with wooden tools (R.S.O.) which have
since disappeared. That is the reason why it is easier to build a
seriation in the very middle of one period (here, for example) than
to consider the transitions from one period to another (as it is the
case in the first steps of our study).

Some more elements or appreciation : in calcareous sites bones
have been preserved and can give us indications on hunted animals.
Sometimes hearths can be found, with pieces of charcoal which allow
radiocarbon (14-C) datations. Animal bones (burnt or not) give the
same possibility, but a great amount of bones are necessary. These
chronological references are very valuable, with two disadvantages,
though : the indicated dates are affected by statistical variations
(standard deviation) of some 100 to 300 years (and sometimes more
than that), because of strictly physical reasons (the 14-C disintegration is not regular), and, above all, the dated hearth may have
been left on the same ground, which had not been covered in the mean-
while, by other men (prehistorical or not) which occupied the same
place several centuries or several millenia before or after those
who had left the flint tools. For example it is what happened on
the Oirschot VII site (OI 7) polluted by above lying Oirschot VI.


At the present stage of research the problems for which data
analysis may help are mainly :
1. Chronological seriation.
2. Differentiation and more especially comparisons between
   regional human groups.
3. Understanding the modality of evolution.

## 1. Seriation

In other regions putting into chronological order is essentially
the result of observing the superpositions of layers in calcareous
rockshelters, where the desquamation of the roof and walls buries
the remains (at various speeds according to moments and occupations).
Of course the deepest layers are the oldest ones (except in case of
disturbance through erosion and a new deposit, which usually can be
perceived rather well).  An order in the deposit of the layers can
be inferred, but not the speed of evolution.  In the Paris basin,
Belgium and Netherlands, many grounds are siliceous and cannot preserve bone;  besides, rockshelters are rare and consequently there
are no, or very few, possibilities of these stratigraphies, and so
much the less that chronological seriation must be made separately
for each region : indeed the toolkits are different in the various
human groups and evolution in two neighbouring regions may be quite
different from each other.  The diameters occupied by the regional
groups are aroung 100 to 300 km for the studied period, which makes
it more complicated to study than the previous periods when there
were 3 to 5 groups all over France. Actually, we have to find 10,
15 or 20 sites rich enough in each region, with often no more than
3 or 4 among them with liable 14-C datations, and when we happen to
individualize the sub-periods by half-millenia we are very happy
indeed : the present partition is made of 5 stages for some 5000
years, and it is not always possible to subdivide a stage.

Formerly, the chronological classification was made from the
comparisons between the qualitative natures of the "indicatory
fossils" proper to each period.  But that allowed a very rough
seriation only.  The taking account of the proportions of all the
types of tools, and sometimes of the wastes, (bordes method) was a
great progress which allowed a far more accurate approach, particularly with the use of cumulative graphs of tool-kits (Fig.1).  However that method does not allow easily comparisons between more than

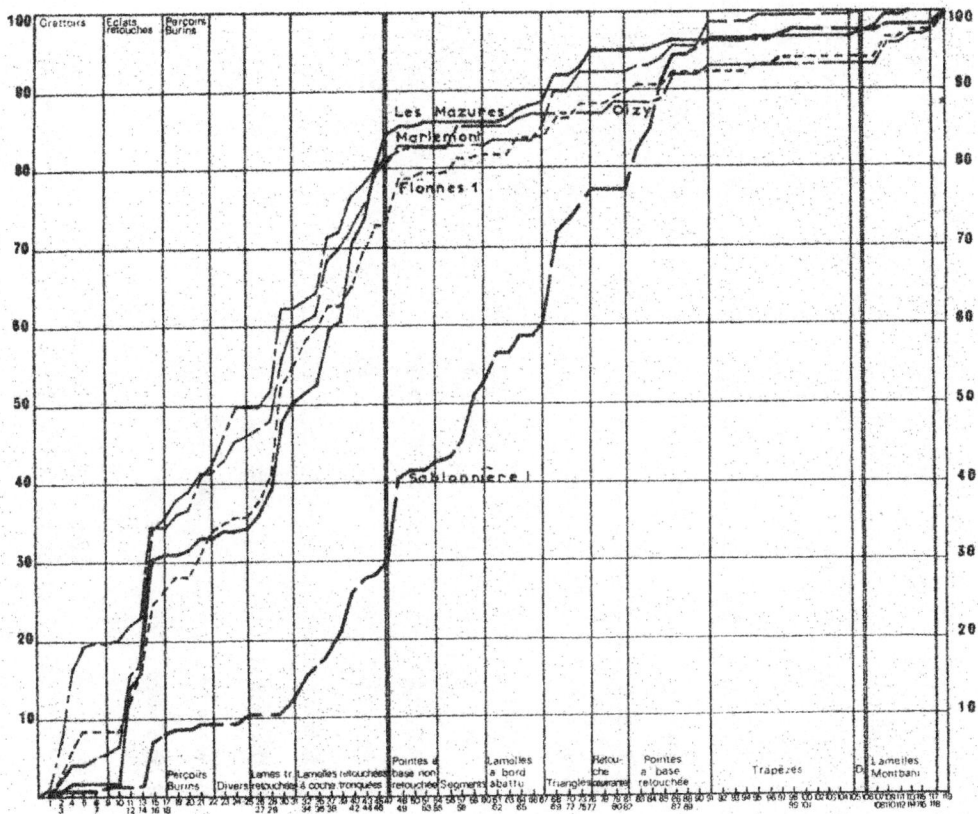

Fig.1.  Cumulative graphs of the Ardennian and Tardenoisian industries.  That type of graphs underlines similarities (between
4 .graphs) and oppositions (with the 5th one) but don't allow
us to perceive easily the evolutions.

J. G. ROZOY

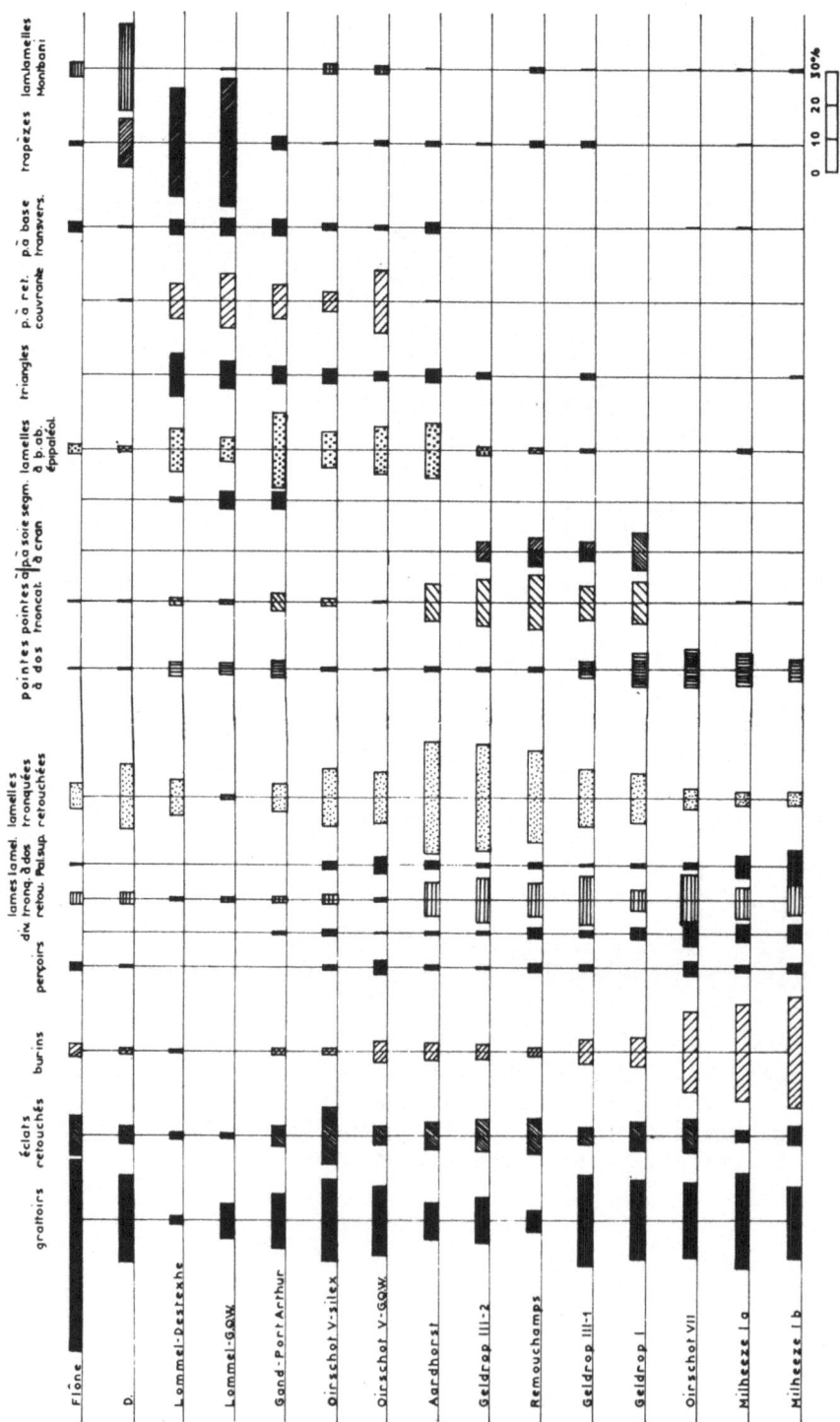

Fig. 2.  Seriation graph for the Belgian-Dutch border all through the Epipalaeolithic.  The 119
types are gathered into 17 classes, some of them are not diagnostic.  No interpolation
is possible.

4 or 5 sites, and leads more easily to analogies or oppositions (what it was conceived for) than on the perception of evolutions. It does not allow to quantify the differences. Quantitative graphical seriations have also been used (Fig.2), but they are restricted to the taking account of classes which gather types considered as near to one another (rightly or wrongly) and using them is neither simple nor easy. Of course only the rank of the sites is given (or proposed), the speed of evolution can be estimated only from the radiocarbon dates, without any possibility of calculating interpolations. Prospects brought by the mathematical methods of data analysis can then be perceived.

2. Regional comparisons

Of course, the differentiation and comparisons between regions can be made only after chronological ordering. Indeed, how useful could it be to compare a region in the 5th millenium to another in the 8th one ? Are the differences between them due to the distance or to passed time ? Classical means, including cumulative graphs are very useful for that, of course, but they give no quantification basis. Now that one is necessary if we wish to judge of the relationships between groups, of the frontiers between groups of different regions a.s.o. The classes and main types of armatures are generally common to several neighboring regions, or even to half Europe at the same period, though, sometimes, with variants in details which could justify the characterization of subtypes. The differences between regions are far more neatly apparent through those which are on common tools, particularly the rate of retouched flakes, the numerical relationships between common tools on blades and bladelets, a.s.o./ Certainly it would be also necessary to add some characteristic wastes ("microburin", which is not a small burin), the taking account of which is not evident from a statistical point of view.

## 3. Human evolution

Also, the study of the modality of evolution can be undertaken only after the resolution of the chronological and geographical problems. We shall have then to distinguish them or put them opposite to one another, and also the way the regional groups part, making the transition from the 5 Magdalenian regions around 10 000 before our era to some 40 epipaleolithic groups around 5000. If we have a number of external criteria large enough (14-C dates) we can also try to quantify the speed of evolution and go further in the discussion about the evolution of human species : some consider there are stable stages separated by short evolutive crisis with quick modification, whereas some others, on the contrary, think they can perceive rather continuous transformation. Of course the variation in the flint tool-kits is only a reflection of their makers' activities; and other criteria (aesthetic, ideologic ones, a.s.o.) should also be taken in account. However it is a major element of appreciation on evolution which is just at hand, for practices of all kinds, using tools, are, of course, an essential acting point for the psychic abilities of our ancestors.

## Reference

J.-G.Rozoy (1978). Les derniers chasseurs. L'épipaléolithique en France et en Belgique. Essai de synthèse. Charleville, 30 CM, 1500 p., 300 pl., drawings of 6500 pieces, tables.

SECTION 2.2

A SERIATION THROUGH CORRESPONDENCE ANALYSIS WITH CONSTRUCTION OF A
CONTINUOUS GRADIENT

J.L. Slachmuylder

CADEPS - Brussels

## 1. Nature of the problem and of its difficulties

The problems of chronological attribution are very common in
prehistoric archaeology, where :

- numerous sites are not stratified and do not provide any radiocarbon date,

- stratigraphy or usual seriation techniques only yield relative
chronologies,

- comparisons with sites dated by radiocarbon are difficult when
these are scarce in the same region, while comparisons with distant
ones are less reliable because these might not belong to the same
cultural entity and thus not show a common and synchronic evolution,

- each assemblage of prehistoric objects forms a unique combination
of typological categories which results from several sources of variation among which time is only one beside others, like seasonality of
activities, functionality of campsites, etc.

- available radiocarbondatings, which are subject to random deviation and pollution, are sometimes contradictory or unrealistic
(Delibrias and Giot, 1970).

   We can summarise the case of the mesolithic sites of the belgo-
dutch border region in terms of data analysis as follows :
given a data table crossing types of archaeological flint objects
with unstratified sites of which a large number are not dated by
radiocarbon, while some of the extant radiocarbondatings might be
very dubious, which is the best way to build a convenient chronology ?

   We have here to deal with two kinds of incertitude :
   - about the data and their categorisation :
Are the descriptive variables (types) sensitive to chronological
variations, and if so, which variables and how sensitive are they ?
Furthermore is it possible to build a unique synthetic variable summarizing a latent chronological gradient which might be based upon
the behaviour of a large number of individual types ?
Are we able to recognize a combination of types useful to the construction of a chronology and to eliminate the other sources variability ?
   - about radiocarbon dates :
Do we have at our disposal, reliable criteria of chronological
control and are we able to recognize the more unlikely dates ?

2.  The data (see Table 1)

   Our data sample have been chosen in a limited area, i.e. Belgian
Limburg and Campine, Dutch Limburg, where it was possible to gather
a large enough number of sites described by the same typology and
belonging to the epipaleolithic (or mesolithic) time.  This period
of about five millenia following the last glaciation has provided
many hunters' campsites showing no agriculture and no husbandry.
The sites belong to a hunting-gathering culture essentially based
upon bow-and-arrow technology (Rozoy, 1978a).

   Almost all sites are unstratified.  More than half of them have
no radiocarbon date.

Table 1

| | AAR | BTH | G31 | G32 | GE1 | GPA | LOD | MIB | MIA | NEE | OI5 | OI7 | OPQ | OVE | SC1 | SC3 | WE1 | WE4 | WE3 |
|---|---|---|---|---|---|---|---|---|---|---|---|---|---|---|---|---|---|---|---|
| 1. | 8 | 0 | 7 | 5 | 3 | 2 | 0 | 2 | 0 | 3 | 15 | 4 | 1 | 0 | 4 | 1 | 7 | 5 | 4 |
| 2. | 9 | 1 | 14 | 7 | 8 | 1 | 1 | 21 | 17 | 1 | 7 | 13 | 3 | 1 | 5 | 1 | 3 | 2 | 1 |
| 3. | 23 | 2 | 30 | 20 | 4 | 0 | 1 | 23 | 41 | 3 | 14 | 13 | 2 | 0 | 5 | 0 | 5 | 6 | 2 |
| 4. | 14 | 6 | 16 | 10 | 4 | 5 | 0 | 12 | 19 | 4 | 34 | 16 | 3 | 15 | 13 | 3 | 12 | 12 | 10 |
| 5. | 5 | 1 | 7 | 2 | 1 | 4 | 0 | 3 | 10 | 1 | 27 | 6 | 0 | 0 | 5 | 1 | 2 | 3 | 2 |
| 6. | 0 | 0 | 0 | 0 | 0 | 1 | 0 | 2 | 3 | 0 | 0 | 0 | 0 | 0 | 0 | 0 | 0 | 0 | 0 |
| 7. | 0 | 2 | 4 | 2 | 0 | 0 | 0 | 13 | 8 | 2 | 17 | 3 | 1 | 4 | 6 | 9 | 17 | 8 | 9 |
| 8. | 1 | 5 | 6 | 1 | 1 | 2 | 0 | 9 | 5 | 2 | 17 | 8 | 1 | 6 | 4 | 4 | 4 | 8 | 5 |
| 9. | 1 | 0 | 0 | 0 | 1 | 0 | 1 | 0 | 9 | 0 | 3 | 0 | 1 | 1 | 0 | 0 | 0 | 1 | 1 |
| 10. | 1 | 1 | 0 | 1 | 0 | 0 | 0 | 2 | 2 | 0 | 7 | 2 | 0 | 1 | 2 | 0 | 1 | 2 | 1 |
| 11. | 1 | 1 | 0 | 0 | 0 | 0 | 0 | 1 | 1 | 0 | 3 | 2 | 0 | 2 | 4 | 0 | 3 | 2 | 1 |
| 12. | 3 | 2 | 1 | 3 | 0 | 0 | 0 | 3 | 1 | 0 | 2 | 4 | 0 | 0 | 3 | 2 | 7 | 10 | 6 |
| 13. | 1 | 3 | 0 | 0 | 0 | 0 | 0 | 0 | 0 | 0 | 2 | 2 | 0 | 0 | 0 | 0 | 0 | 0 | 1 |
| 14. | 3 | 5 | 0 | 3 | 0 | 0 | 0 | 0 | 0 | 2 | 2 | 4 | 0 | 3 | 5 | 2 | 10 | 9 | 9 |
| 15. | 5 | 11 | 2 | 2 | 4 | 5 | 0 | 3 | 1 | 3 | 5 | 6 | 0 | 5 | 3 | 3 | 15 | 12 | 8 |
| 16. | 27 | 15 | 12 | 24 | 3 | 0 | 0 | 12 | 7 | 6 | 42 | 13 | 7 | 2 | 39 | 27 | 61 | 52 | 36 |
| 17. | 4 | 1 | 1 | 0 | 0 | 2 | 0 | 1 | 0 | 0 | 13 | 2 | 0 | 0 | 1 | 2 | 0 | 2 | 0 |
| 18. | 0 | 0 | 0 | 0 | 0 | 0 | 0 | 2 | 0 | 0 | 2 | 0 | 0 | 0 | 0 | 0 | 0 | 0 | 0 |
| 19. | 9 | 0 | 4 | 0 | 0 | 0 | 0 | 9 | 0 | 0 | 8 | 6 | 0 | 1 | 0 | 4 | 6 | 0 | 1 |
| 20. | 0 | 0 | 2 | 0 | 0 | 0 | 0 | 3 | 4 | 0 | 6 | 7 | 0 | 0 | 0 | 0 | 0 | 0 | 0 |
| 21. | 19 | 0 | 7 | 12 | 1 | 2 | 0 | 46 | 41 | 1 | 14 | 23 | 0 | 0 | 13 | 2 | 0 | 0 | 0 |
| 22. | 10 | 0 | 13 | 4 | 6 | 0 | 1 | 86 | 63 | 0 | 4 | 43 | 0 | 0 | 9 | 1 | 0 | 1 | 0 |
| 23. | 3 | 0 | 0 | 0 | 1 | 1 | 0 | 4 | 12 | 0 | 0 | 22 | 1 | 0 | 3 | 0 | 0 | 0 | 1 |
| 24. | 2 | 0 | 0 | 3 | 0 | 0 | 0 | 0 | 1 | 1 | 4 | 0 | 0 | 2 | 0 | 0 | 0 | 1 | 0 |
| 25. | 0 | 0 | 6 | 1 | 2 | 0 | 0 | 17 | 6 | 0 | 5 | 18 | 1 | 0 | 3 | 2 | 0 | 0 | 0 |
| 26. | 0 | 0 | 6 | 3 | 1 | 0 | 0 | 0 | 1 | 0 | 1 | 3 | 0 | 0 | 0 | 0 | 1 | 1 | 0 |
| 27. | 5 | 0 | 11 | 8 | 2 | 0 | 0 | 7 | 7 | 0 | 2 | 4 | 0 | 0 | 2 | 0 | 4 | 4 | 0 |
| 28. | 0 | 2 | 5 | 6 | 0 | 0 | 0 | 6 | 4 | 0 | 4 | 0 | 0 | 1 | 0 | 0 | 11 | 8 | 4 |
| 29. | 6 | 3 | 7 | 7 | 0 | 0 | 0 | 1 | 1 | 1 | 2 | 5 | 0 | 4 | 3 | 5 | 3 | 1 |
| 30. | 34 | 16 | 15 | 19 | 2 | 2 | 1 | 17 | 7 | 5 | 9 | 20 | 7 | 3 | 2 | 15 | 11 | 9 |
| 31. | 11 | 0 | 1 | 3 | 0 | 0 | 0 | 1 | 9 | 0 | 1 | 6 | 0 | 6 | 0 | 0 | 0 | 0 |
| 32. | 3 | 2 | 0 | 2 | 1 | 0 | 0 | 30 | 16 | 0 | 1 | 4 | 0 | 2 | 2 | 5 | 4 | 0 |
| 33. | 11 | 2 | 2 | 2 | 0 | 0 | 0 | 14 | 8 | 4 | 13 | 0 | 0 | 0 | 0 | 8 | 3 | 0 |
| 34. | 0 | 0 | 0 | 0 | 0 | 0 | 0 | 0 | 0 | 0 | 0 | 0 | 0 | 1 | 0 | 0 | 0 | 0 |
| 35. | 0 | 0 | 0 | 0 | 0 | 0 | 0 | 2 | 0 | 0 | 3 | 0 | 0 | 3 | 0 | 0 | 0 | 0 |
| 36. | 0 | 1 | 0 | 0 | 0 | 0 | 0 | 1 | 2 | 0 | 3 | 4 | 0 | 0 | 1 | 0 | 4 | 0 |
| 37. | 41 | 18 | 13 | 22 | 3 | 0 | 0 | 4 | 6 | 0 | 25 | 2 | 0 | 0 | 13 | 4 | 9 | 9 | 5 |
| 38. | 2 | 0 | 0 | 0 | 0 | 0 | 0 | 2 | 2 | 3 | 13 | 1 | 0 | 0 | 0 | 8 | 4 | 7 |
| 39. | 7 | 2 | 0 | 4 | 0 | 0 | 2 | 0 | 0 | 0 | 5 | 0 | 0 | 0 | 3 | 0 | 1 | 2 | 0 |
| 40. | 3 | 4 | 2 | 1 | 1 | 0 | 0 | 0 | 0 | 2 | 3 | 0 | 0 | 0 | 6 | 1 | 15 | 19 | 11 |
| 41. | 6 | 0 | 0 | 0 | 0 | 0 | 0 | 0 | 0 | 0 | 7 | 0 | 0 | 0 | 3 | 0 | 4 | 1 | 0 |
| 42. | 27 | 15 | 1 | 1 | 0 | 0 | 0 | 0 | 0 | 1 | 2 | 0 | 0 | 0 | 19 | 3 | 36 | 37 | 12 |
| 43. | 5 | 0 | 5 | 6 | 0 | 1 | 0 | 0 | 0 | 1 | 2 | 0 | 0 | 0 | 1 | 0 | 3 | 2 | 1 |
| 44. | 19 | 0 | 14 | 20 | 1 | 1 | 0 | 3 | 1 | 0 | 6 | 1 | 2 | 0 | 0 | 9 | 5 | 2 |
| 45. | 7 | 4 | 3 | 8 | 0 | 0 | 4 | 2 | 1 | 1 | 11 | 1 | 1 | 1 | 9 | 5 | 4 | 2 |
| 46. | 9 | 1 | 8 | 14 | 5 | 0 | 0 | 1 | 0 | 2 | 1 | 0 | 3 | 13 | 4 | 24 | 9 |
| 47. | 6 | 9 | 11 | 34 | 5 | 0 | 4 | 1 | 2 | 0 | 11 | 0 | 3 | 10 | 4 | 0 | 13 | 22 | 11 |
| 48. | 58 | 5 | 21 | 30 | 5 | 5 | 1 | 0 | 1 | 21 | 7 | 0 | 1 | 5 | 21 | 3 | 10 | 3 | 3 |
| 49. | 3 | 0 | 10 | 16 | 5 | 0 | 0 | 2 | 0 | 1 | 1 | 0 | 1 | 0 | 0 | 0 | 2 | 1 | 0 |
| 50. | 2 | 0 | 0 | 3 | 0 | 0 | 0 | 0 | 0 | 0 | 0 | 0 | 0 | 2 | 2 | 19 | 17 | 7 |
| 51. | 1 | 0 | 6 | 3 | 5 | 0 | 6 | 13 | 13 | 8 | 3 | 16 | 0 | 0 | 2 | 12 | 0 | 0 | 0 |
| 52. | 4 | 0 | 9 | 1 | 3 | 0 | 14 | 21 | 0 | 0 | 2 | 16 | 0 | 0 | 1 | 0 | 0 | 0 |
| 53. | 2 | 0 | 0 | 0 | 0 | 1 | 0 | 0 | 0 | 4 | 0 | 0 | 0 | 3 | 1 | 0 | 0 | 0 |
| 54. | 1 | 0 | 0 | 0 | 0 | 0 | 0 | 0 | 0 | 2 | 0 | 0 | 0 | 0 | 1 | 0 | 0 | 0 |
| 55. | 0 | 0 | 0 | 0 | 0 | 1 | 0 | 0 | 0 | 1 | 0 | 0 | 0 | 0 | 0 | 0 | 0 | 0 |
| 56. | 0 | 0 | 0 | 0 | 0 | 0 | 0 | 0 | 0 | 0 | 0 | 0 | 0 | 0 | 0 | 0 | 0 | 0 |
| 57. | 0 | 0 | 0 | 0 | 0 | 1 | 0 | 0 | 0 | 0 | 0 | 0 | 5 | 0 | 2 | 0 | 0 | 0 |
| 58. | 0 | 0 | 0 | 0 | 0 | 2 | 0 | 0 | 0 | 0 | 0 | 0 | 0 | 2 | 0 | 0 | 0 |
| 59. | 0 | 0 | 0 | 0 | 0 | 1 | 0 | 0 | 0 | 0 | 0 | 0 | 0 | 0 | 0 | 0 | 0 | 0 |
| 60. | 0 | 0 | 0 | 0 | 0 | 1 | 0 | 0 | 0 | 0 | 0 | 0 | 1 | 0 | 0 | 0 | 0 | 0 |
| 61. | 0 | 0 | 0 | 0 | 0 | 0 | 0 | 0 | 0 | 0 | 20 | 0 | 0 | 0 | 0 | 0 | 0 | 0 |
| 62. | 1 | 0 | 0 | 1 | 0 | 20 | 0 | 0 | 0 | 3 | 0 | 0 | 1 | 0 | 1 | 2 | 14 | 0 | 6 |
| 63. | 0 | 0 | 0 | 0 | 0 | 1 | 2 | 0 | 0 | 0 | 0 | 0 | 0 | 1 | 0 | 1 | 10 | 0 | 7 |
| 64. | 1 | 0 | 2 | 0 | 0 | 0 | 1 | 0 | 0 | 0 | 12 | 0 | 0 | 0 | 1 | 0 | 0 |
| 65. | 26 | 0 | 0 | 2 | 0 | 0 | 3 | 0 | 0 | 0 | 11 | 0 | 0 | 3 | 12 | 4 | 0 | 0 |
| 66. | 38 | 0 | 0 | 2 | 0 | 0 | 3 | 0 | 4 | 0 | 0 | 0 | 0 | 4 | 0 | 1 | 9 | 0 | 1 |
| 67. | 27 | 0 | 0 | 0 | 0 | 1 | 0 | 0 | 1 | 0 | 0 | 0 | 0 | 0 | 1 | 0 | 0 |
| 68. | 15 | 0 | 2 | 4 | 0 | 4 | 9 | 0 | 0 | 3 | 9 | 0 | 0 | 17 | 0 | 1 | 8 | 5 | 1 |
| 69. | 2 | 0 | 0 | 0 | 0 | 0 | 1 | 0 | 0 | 0 | 6 | 0 | 0 | 0 | 4 | 0 | 0 |
| 70. | 0 | 0 | 0 | 0 | 0 | 0 | 0 | 0 | 0 | 0 | 0 | 0 | 0 | 0 | 0 | 0 | 0 |
| 71. | 0 | 0 | 0 | 0 | 0 | 0 | 0 | 0 | 0 | 0 | 0 | 0 | 6 | 0 | 0 | 0 | 0 |
| 72. | 0 | 0 | 0 | 0 | 0 | 0 | 0 | 0 | 0 | 0 | 0 | 0 | 6 | 0 | 0 | 0 | 0 |
| 73. | 4 | 0 | 0 | 0 | 0 | 0 | 2 | 0 | 0 | 0 | 4 | 0 | 5 | 2 | 0 | 2 | 0 |
| 74. | 1 | 0 | 0 | 0 | 0 | 0 | 0 | 0 | 0 | 0 | 0 | 0 | 5 | 0 | 0 | 0 | 0 |
| 75. | 0 | 0 | 0 | 0 | 0 | 0 | 0 | 0 | 0 | 0 | 0 | 0 | 0 | 0 | 0 | 0 | 0 |
| 76. | 0 | 0 | 0 | 0 | 0 | 0 | 0 | 0 | 0 | 0 | 0 | 0 | 0 | 0 | 0 | 0 | 0 |
| 77. | 1 | 0 | 3 | 2 | 0 | 1 | 1 | 0 | 0 | 2 | 0 | 2 | 0 | 0 | 0 | 0 |
| 78. | 0 | 1 | 0 | 0 | 0 | 4 | 3 | 0 | 0 | 1 | 7 | 0 | 1 | 0 | 2 | 1 |
| 79. | 0 | 0 | 0 | 0 | 0 | 3 | 3 | 0 | 0 | 0 | 16 | 0 | 4 | 0 | 3 | 0 |
| 80. | 0 | 1 | 0 | 0 | 0 | 0 | 3 | 0 | 0 | 0 | 2 | 0 | 0 | 1 | 0 | 1 |
| 81. | 2 | 0 | 0 | 0 | 0 | 3 | 0 | 0 | 0 | 11 | 0 | 0 | 0 | 0 | 0 | 1 |
| 82. | 0 | 0 | 0 | 0 | 0 | 0 | 3 | 0 | 0 | 13 | 0 | 1 | 0 | 0 | 7 | 1 | 1 |
| 83. | 0 | 0 | 0 | 0 | 0 | 2 | 0 | 0 | 1 | 0 | 2 | 0 | 0 | 1 | 2 | 0 |
| 84. | 8 | 0 | 0 | 0 | 0 | 0 | 1 | 0 | 0 | 2 | 0 | 0 | 0 | 1 | 0 | 0 |
| 85. | 0 | 0 | 0 | 0 | 0 | 0 | 2 | 0 | 0 | 0 | 0 | 0 | 1 | 0 | 0 |
| 86. | 1 | 0 | 0 | 0 | 0 | 2 | 0 | 0 | 0 | 4 | 0 | 2 | 0 | 0 | 1 | 0 |
| 87. | 6 | 0 | 0 | 0 | 0 | 0 | 2 | 0 | 0 | 4 | 1 | 2 | 0 | 0 | 0 | 1 |
| 88. | 0 | 0 | 0 | 0 | 0 | 0 | 0 | 0 | 0 | 0 | 0 | 0 | 0 | 0 |
| 89. | 0 | 0 | 0 | 0 | 0 | 1 | 1 | 0 | 0 | 0 | 0 | 0 | 1 | 0 |
| 90. | 1 | 0 | 0 | 0 | 0 | 0 | 0 | 0 | 0 | 0 | 0 | 1 | 0 |
| 91. | 0 | 0 | 0 | 0 | 0 | 0 | 0 | 0 | 0 | 1 | 0 | 0 | 0 | 0 |
| 92. | 0 | 12 | 0 | 0 | 1 | 2 | 0 | 0 | 1 | 3 | 0 | 0 | 10 | 8 | 5 |
| 93. | 5 | 13 | 2 | 1 | 0 | 5 | 0 | 0 | 0 | 18 | 0 | 0 | 13 | 16 | 12 |
| 94. | 0 | 0 | 2 | 0 | 10 | 0 | 0 | 0 | 0 | 0 | 44 | 35 | 14 |
| 95. | 1 | 21 | 3 | 0 | 3 | 0 | 0 | 1 | 0 | 0 | 11 | 8 | 8 |
| 96. | 0 | 3 | 1 | 0 | 0 | 0 | 0 | 1 | 0 | 2 | 8 | 8 |
| 97. | 0 | 23 | 0 | 0 | 0 | 0 | 0 | 1 | 0 | 0 | 2 | 4 | 3 |
| 98. | 0 | 1 | 0 | 0 | 8 | 0 | 0 | 0 | 0 | 1 | 0 | 4 | 3 |
| 99. | 0 | 4 | 0 | 0 | 1 | 0 | 0 | 0 | 0 | 0 | 1 | 4 | 1 |
| 100. | 0 | 0 | 0 | 0 | 0 | 0 | 0 | 0 | 0 | 1 | 0 | 2 | 1 |
| 101. | 1 | 0 | 0 | 0 | 0 | 0 | 0 | 0 | 0 | 2 | 0 | 0 | 1 |
| 102. | 0 | 0 | 0 | 0 | 0 | 2 | 0 | 0 | 0 | 0 | 0 | 2 | 0 |
| 103. | 0 | 0 | 0 | 0 | 0 | 0 | 0 | 0 | 0 | 0 | 0 | 0 |
| 104. | 0 | 0 | 0 | 0 | 0 | 2 | 0 | 0 | 2 | 0 | 0 | 0 |
| 105. | 0 | 0 | 0 | 0 | 0 | 0 | 4 | 0 | 0 | 3 | 2 | 2 |
| 106. | 0 | 0 | 0 | 0 | 0 | 0 | 0 | 0 | 11 | 14 | 3 | 2 | 2 |
| 107. | 0 | 5 | 0 | 0 | 0 | 1 | 0 | 0 | 1 | 0 | 0 |
| 108. | 0 | 0 | 0 | 0 | 0 | 0 | 0 | 0 | 23 | 29 | 16 |
| 109. | 0 | 18 | 0 | 0 | 0 | 3 | 7 | 0 | 29 | 29 | 14 |
| 110. | 0 | 19 | 0 | 0 | 1 | 0 | 0 | 0 | 0 | 8 | 10 | 4 |
| 111. | 2 | 1 | 1 | 0 | 0 | 1 | 0 | 0 | 0 | 7 | 7 |
| 112. | 0 | 1 | 0 | 0 | 0 | 1 | 0 | 0 | 2 | 5 | 4 |
| 113. | 0 | 6 | 0 | 0 | 0 | 4 | 0 | 2 | 0 | 3 | 3 |
| 114. | 0 | 4 | 0 | 0 | 0 | 3 | 0 | 1 | 0 | 7 | 3 | 4 |
| 115. | 0 | 3 | 0 | 0 | 0 | 0 | 1 | 0 | 0 | 4 | 3 | 3 |
| 116. | 0 | 6 | 0 | 0 | 0 | 3 | 0 | 1 | 1 | 4 | 3 |
| 117. | 0 | 15 | 0 | 0 | 0 | 2 | 0 | 1 | 1 | 5 | 1 |
| 118. | 0 | 6 | 0 | 0 | 0 | 0 | 0 | 12 | 8 | 0 |
| 119. | 0 | 0 | 0 | 0 | 0 | 2 | 0 | 0 | 1 | 0 |
| 120. | 0 | 0 | 12 | 1 | 0 | 0 | 0 | 0 | 0 | 0 |
| 121. | 0 | 0 | 6 | 19 | 0 | 0 | 0 | 0 | 0 | 0 |
| TOT. | 568 | 302 | 313 | 352 | 81 | 99 | 93 | 402 | 362 | 106 | 579 | 282 | 120 | 149 | 305 | 161 | 639 | 511 | 343 |

The hypothesis of a continuous evolution of these assemblages has been suggested by the French prehistorian Jean-Georges Rozoy, who in 1978 proposed a seriation based on 14 classes of tools (see table 2) but not at the level of individual types (Rozoy, 1978a and 1980).

It seemed interesting to control at this level if the evolution of the toolkit allowed to state more precisely the chronological position of the sites, and to compare the results with extant radio-carbon dates.

The typology used here is the one devised by Docteur Rozoy (1978b).  The sites have been published by him or by Professor Vermeersch of the K.U.L. (Rozoy, 1978a; Vermeersch, 1974, 1982, 1982a, 1982b and 1984).  They are properly mesolithic sites but also ahrens-burgian and tjongerian settlements (see tables 2 and 3).

3. Choice of a method

There is an abundant literature giving various techniques. Especially, many iterative algorithms of seriation on the one hand (Craytor and Johnson, 1968; Goldmann, 1979) and multivariate methods on the other (large bibliography in Marquardt (1978) and Rozoy (1984).

Most iterative techniques of seriation give a mere ranking but Hill has shown that some of these may be translated in terms very close to those of principal components analysis and those of corres-pondence analysis (Hill, 1973 and 1974; Buyse, 1983).

Moreover multivariate analysis is today known to have many uses as a seriation instrument (for instance, Jacobi et al., 1980; Kendall, 1971, Gosden, 1984).  Correspondence analysis offers the advantage of simultaneous representation of individuals and variables in the same mathematical space, which favours optimal selection of variables and the control of a real gradient among the data.

Table 2. List of types (after Rozoy, 1978b; Vermeersch and Lauwers
for the english translation)

Class 1. End-scrapers

1   Long end-scraper on a blade
2   Short end-scraper on a blade
3   Broken end-scraper on a blade
4   Single end-scraper on a flake
5   End-scraper on a retouched flake
6   Cricular scraper
7   Thumbnail scraper
8   Other end-scraper on a flake
9   Core-like scraper
10  Denticulated end-scraper
-------------------------------------------------------------------

Class 2. Retouched flakes

11  Thick denticulated flake
12  Thin denticulated flake
13  Thick truncated flake
14  Thick retouched flake
15  Thin truncated flake
16  Thin retouched flake
17  Side-scraper
18  Raclette
-------------------------------------------------------------------

Class 3. Borers and burins

19  Borer
20  Taraud
21  Dihedral burin
22  Burin on truncation
-------------------------------------------------------------------

Class 4. Divers common tools

23  Pièce émoussée
24  Pièce esquillée
25  Divers
-------------------------------------------------------------------

Class 5. Retouched and truncated blades

26  Blade with a concave truncation
27  Blade with a straight truncation
28  Blade with an oblique truncation
29  Blade with distal retouch
30  Blade with continuous retouch
31  Backed knife
-------------------------------------------------------------------

(continued)

Table 2 (Continued)

Class 6. Backed and truncated bladelets

32  Atypically backed bladelet
33  Partially backed bladelet
34  Backed bladelet with a gibbosity
35  Arched backed bladelet
36  Bladelet with a curved backed end
37  Partially retouched bladelet
38  Continuously retouched bladelet
39  Bladelet with an ouchtata type retouch
40  Single notched bladelet
41  Bladelet broken under a notch
42  Bladelet broken in notch
43  Bladelet with a concave truncation
44  Bladelet with a straight truncation
45  Bladelet with distal retouch
46  Bladelet with an oblique truncation
47  Bladelet broken under an oblique truncation
----------------------------------------------------------------

Class 7. Points without retouched base

48  Obliquely truncated point (proximal)
49  Obliquely truncated point (distal)
50  Short point
51  Unilaterally backed point (proximal)
52  Unilaterally backed point (distal)
53  Chaville point
54  Double backed point (proximal)
55  Double backed point (distal)
----------------------------------------------------------------

Class 8. Crescents

56  Sauveterre point
57  Retouched crescent
58  Crescent of a circle
59  Asymmetric crescent
60  Large crescent
----------------------------------------------------------------

Class 9. Narrow bladelets

61  Narrow backed bladelet
62  Fragment of narrow backed bladelet
63  Truncated backed bladelet
64  Backed bladelet
65  Fragment of backed bladelet
66  Truncated backed bladelet
67  Scalene bladelet
----------------------------------------------------------------

Class 10. Triangles

68  Scalene triangle
69  Truncated scalene triangle
70  Montclus triangle
71  Elongated scalene triangle
72  Elongated scalene triangle with short small truncation
73  Scalene triangle with concave small truncation
74  Muge triangle
75  Elongated muge triangle
76  Elongated isocele triangle
77  Isocele triangle
------------------------------------------------------------------
Class 11. Microliths with surface retouche

78  Mistletoe point
79  Triangle with surface retouch
80  Other microlith with surface retouch
81  Point with rounded base
82  Point with oblique base
------------------------------------------------------------------
Class 12. Points with retouched base

83  Short triangular point
84  Short ogival point
85  Long triangular point
86  Tardenois point with convex base
87  Tardenois point
88  Short triangular point with concave base
89  Short ogival point with concave base
90  Long triangular point with concave base
91  Tardenois point with concave base
------------------------------------------------------------------
Class 13. Trapezes

92   Short rhombic trapeze
93   Long rhombic trapeze
94   Short rectangular trapeze
95   Vielle trapeze
96   Short asymmetric trapeze
97   Long asymmetric trapeze
98   Short symmetric trapeze
99   Long symmetric trapeze
100  Symmetric trapeze with oblique truncation
101  Symmetric trapeze with concave truncation
102  Short montclus trapeze
103  Long montclus trapeze
104  Martinet trapeze
105  Point with flat inverse retouch
------------------------------------------------------------------
106 Indeterminate microlith
------------------------------------------------------------------
(continued)

Table 2 (Continued)

Class 14. Montbani blades and bladelets

107   Unilateral multiply notched blade
108   Unilateral multiply notched bladelet
109   Unilateral irregularly retouched blade
110   Unilateral irregularly retouched bladelet
111   Blade with opposing notches
112   Bladelet with opposing notches
113   Blade with opposing retouch
114   Bladelet with opposing retouch
115   Blade with off-set notches
116   Bladelet with off-set notches
117   Blade with off-set retouch
118   Bladelet with off-set retouch
-------------------------------------------------------------------
119   Tools of neolithic affinity
-------------------------------------------------------------------
Class 15. Tanged and shouldered points

120   Tanged point
121   Shouldered point

Table 3. Table of sites with their $^{14}C$ dates and chronological or cultural attributions

| CODES* | SITES | $^{14}C$ DATES(BC) | ATTRIBUTIONS | BIBLIOGRAPHY |
|---|---|---|---|---|
| MIB | Milheeze 1b | | Tjongerian | Rozoy,1978a,p.107-114 |
| LIA | Milheeze 1a | 8930±125 | Tjongerian | Rozoy,1978a,p.107-114 |
| O17 | Oirschot VII | 4740±65 | Tjongerian | Rozoy,1978a,p.107-114 |
| GE1 | Geldrop 1 | 9010±95 | Ahrenburgian | Rozoy,1978a,p.133;Paddayya,1971 |
| G31 | Geldrop III-1 | | Ahrensburgian | Rozoy,1978a,p.126;Paddayya,1971 |
| G32 | Geldrop III-2 | | Ahrensburgian | Rozoy,1978a,p.138;Paddayya,1971 |
| SCI | Schulen I | | old stage (ca.7500) | Vermeersch,et al.,1982,p.112 |
| NEE | Neerharen | 7220±100 | old stage | Vermeersch,et al.,1982,p.52 |
| SC3 | Schulen III | 9190±70 | old stage (ca.7250) | Vermerrsch,et al.,1982,p.112 |
| | | 6755±50 | (younger than SC1) | |
| AAR | Aardhorst | 6600±75 | middle stage | Rozoy,1978a,p.157 |
| 015 | Oirschot V(GW) | 6080±50 | middle stage | Rozoy,1978a,p.165 |
| | | 5560±60 | | |
| | | 4280±50 | | |
| OVE | Overpelt (coll.Maria Louis) | | middle stage | Gob,1981a,p.298 |
| GPA | Gent Port-Arthur | | late stage (ca.6300) (betw.OIS and LOD) | Rozoy,1978a,p.172 |
| LOD | Lommel (coll.Destexhe) | | late stage (ca.5500) | Rozoy,1978a,p.181 |
| WE1 | Weelde 1 | | final stage | Vermeersch,et al.,1982,p.144ff |
| WE4 | Weelde 4 | | (betw.LOD and OPG) | Vermeersch,et al.,1982,p.203ff |
| WE5 | Weelde 5 | 5040±135 | | |
| | | 3760±80 | | |
| OPG | Opglabbeek | | final stage | Vermeersch,et al.,1974,p.102 Rozoy,1978a,p.183-188 |
| BTH | Brecht-Thomas Heyveld | | final stage | Vermeersch and Lauwers, 1982 |

*Codes refer to sites as depicted on the factorial graphs, Figure 1, and 3.

is why we have chosen to apply correspondence analysis to our problem.
(Benzécri, 1973; Janssen et al., 1983).

Works of Hill, Benzécri, Naouri, and Ihm concerning the continuous model display its possibility to express evolutive phenomena
in the form of polynomial relations between orthogonal axes whose
coordinates are functions of a unique parameter (see bibliography).

Beside its ability to show evolutive phenomena, correspondence
analysis, which is especially meant to treat contingency tables, is
also able to detect if there is some other source of variability
disturbing its serial structure.

## 4. First exploratory analysis : the choice of the variables

A first problem is how to eliminate non-chronological sources
of variability. A comparison of previous experiments shows that it
is possible to isolate some factors when the sample is built. For
instance samples composed of sites close in time but distant in
space show variability dominated by a geographical component. On
the opposite a limited region with important chronological thickness,
like in our case, usually offers chronological results (for instance,
Perin, 1980 ; Djindjian, 1977), especially but not only if they are
described by a list of tools achieving very similar functions and
sensitive mainly to stylistic variation in time. Indeed we erase
in this way stylistic diversity due to geography and functional differences due to possible seasonal or specialised activities achieved
on specific sites by prehistoric men (Binford and Binford, 1966;
Gendel et al., 1985; Broglio et al., 1984).

Another way to eliminate non chronological causes of variability
is to work with percentages and thus to give the same weight to each
assemblage. Indeed the total number of artifacts found in a site
offers no necessary link with its chronological position but may

have some relation with nature of activities performed on the site
and the duration.

We have used a first correspondence analysis to select the most
useful types in terms of the chronological perspective.  The results
of the first two axes (see figure 1) show a rough curve.

The classes of types display different modes of behaviour :
- end-scrapers (types 1-10) are isolated in the upper left part
of the graph.
- burins and borers (types 19-22), which are more frequent in
older times, are scattered on upper and lower parts and many of them
also show coordinates separated from an irregular but continuous
curve, defined mainly by positions of sites and microlithic armatures.
- armatures (types 48-105, 120-121), which presumably formed
only one functional class, are scattered from the left to the right
along an irregular curve.
- Montbani blades and bladelets (types 107-118) are concentrated
on the right half of the scattergram.

Trying to translate the graph in archaeological terms based on
previous studies and on the initial data table, we propose to recognize in it two different evolutions.  The first concerns the functional categories (end-scrapers, buring, armatures) taken each as a
whole.  Their variations mainly affect the older part of the period,
when their respective proportions display rapid changes and then
find a relative equilibrium.  From that point situated at about
7500 BC, the main changes concern proportions inside the group of
armatures and the sudden intrusion of Montbani blades and bladelets
at the end of the late stage of the mesolithic.

Microlithic armatures are generally considered as arrow elements
and thus as belonging to the same functional category (Rozoy, 1978a;
Cahen, 1985).  Their variability of shape through time probably
reflects mainly the fashion and habits of the people manufacturing
them.  This kind of non-functional diversity may be qualified as

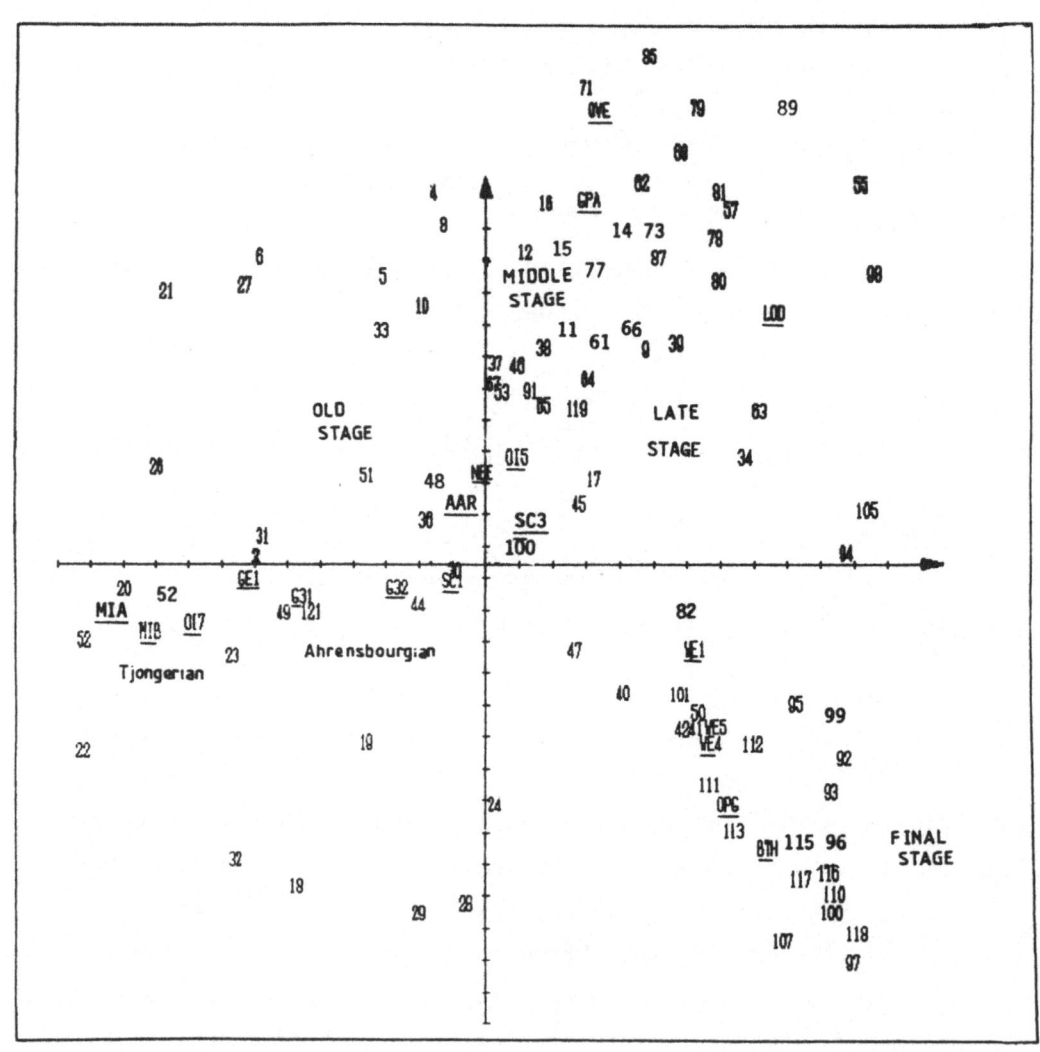

Fig.1

stylistic in that it refers mainly to cultural identity of prehistoric men (Lenoir, 1975; Bromberger, 1979; Stiles, 1979; Plog, 1983;
Wiessner, 1983). Indeed the study of larger areas also shows a geo-
graphic variation among shapes and percentage of armatures whose
combinations seem to be specific of both place and time (Rozoy, 1978a;
Gob, 1976, 1981a, 1985; Kozlowski, 1975; Decormeille and Hinout,
1982).

There is a latent evolutive gradient in the data, but it does
not concern the same variable with the same intensity :
    - It mostly concerns armatures, which form its real base.
    - Burings and borers also have a chronological meaning, but are
unreliable because they may also produce a functional variability.
    - Montbani blades and bladelets appear suddenly. Their global
amount may have both chronological and functional meaning. Moreover
the proportion of types inside this functional group does not seem
to express a chronological variation, contrary to armatures.

So the armatures (types 48-105, 120-121) seem to be the most
reliable variables, because they belong to a common functional group,
whose internal variability has a stylistically homogeneous nature.

Indeed, a second analysis using only the armatures confirmed
the diagnostic significance of these tool types by producing a much
more regular and clearly parabolic curve on the two first axes of
the scattergram. The order of the armatures was then interpretable
in chronological terms :
    - points with unilateral retouches (types 51 and 52) were most
common during the Tjongerian;
    - shouldered, tanged and truncated points (types 120-121, 48-49)
are the product of the Ahrenburgian culture;
    - triangles (types 68-77) were produced mainly during the old
and middle stages of the Mesolithic;
    - over-all retoucned points (types 78-82) stem mainly from the

middle and the late stages of the Mesolithic;

    - crescents (types 56-60) were manufactured mainly at the end
of the middle and during the early late stage of the Mesolithic in
Limburg;

    - trapezes (types 92-104) stem from the late and final stages
of the Mesolithic;

    - microliths with flat inverse retouches (type 105) appear
during the final mesolithic.

    The sites without radiocarbon dates are found at positions on
the curve which correspond to the periods proposed by usual typological methods (Figure 2 and Table 3).

5. Construction of the gradient

    Professor Benzécri (1973) studying applications of correspon-
dence analysis to continuous models, and especially to presence-
absence diagonalisable data tables, showed there are polynomial
relations between the axes.  For instance the coordinates on the
second axis will be a second degree function of the coordinates on
the first axis;  the coordinates on the third axis will be third
degree function ... and so on.

    He has shown that in the case of an ideal diagonalised data
table the relation between the axes may be expressed in the following form :

$$F_n(x) = \lambda_n^{1/2}(2n + 1)^{1/2}P_n(2x - 1)$$

where  $F_n$  is the coordinate of  x  on the axis defined by the n-th
        eigenvector generated by correspondence analysis
      $\lambda_n$  is the eigenvalue associated to the eigenvector  n, for
        n  increasing from 0  ($\lambda_0 = 1$)  when  $\lambda_n$  are taking
        decreasing values
      x   is the value on a gradient varying between 0 and 1
      $P_n$  is the orthogonal polynomial of Legendre number  n.

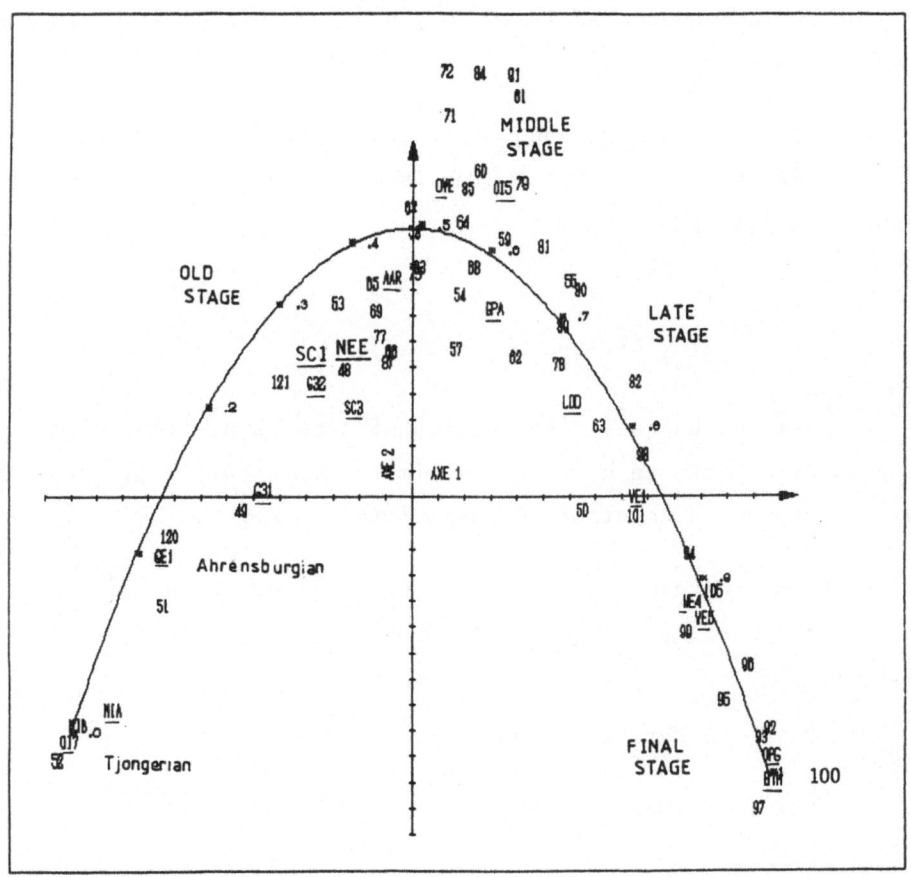

Fig.2

For  n = 1  and  n = 2  we have respectively

$$F_1(x) = \lambda_1^{1/2} 3^{1/2} (2x - 1)$$

$$F_2(x) = \lambda_2^{1/2} 5^{1/2} (3(2x - 1)^2 - 1)/2$$

where in our case estimated values of $\lambda_1$ and $\lambda_2$ are obtained by applying correspondence analysis to the line-profile table of armatures

Computing eigenvalues and eigenvectors might induce arbitrary change in signs of some axes, which is our case for $F_2$. The preceding equations thus become :

$$F_1(x) = (3\lambda_1)^{1/2} (2x - 1)$$

$$F_2(x) = - (5\lambda_2)^{1/2} (3(2x - 1)^2 - 1)/2$$

Because the unequal distribution of types and sites weights along the gradient, we have to reestimate graphically $\lambda_1$ and $\lambda_2$ and to introduce a constant  C  associated to  $F_2$.

Reestimated values of $\lambda_1$ and $\lambda_2$ will be noted $\lambda_1^*$ and $\lambda_2^*$. Table 4 of the values $F_1$, $F_2$ in function of  x  (see also Figure 3) shows that :

Min $F_2$ = - 2 Max $F_2$, which allows to compute  C

C = Max $F_2$ - (Max $F_2$ - Min $F_2$)/3

Minima and maxima of $F_1$ and $F_2$ are available on the Table 6 giving for each site its value on the first and second axis.

We can restimate $\lambda_1^*$, $\lambda_2^*$ from the following equations :

For  $F_2 = C$ :  $F_1 = \pm (\lambda_1^*)^{1/2} \Rightarrow \lambda_1^* = (F_1)^2$

For  Min $F_2$  and  Min $F_1$  or  Max $F_1$ :  $F_2 - C = - (5\lambda_2^*)^{1/2}$

$$\Rightarrow \lambda_2^* = (C - F_2)^2/5$$

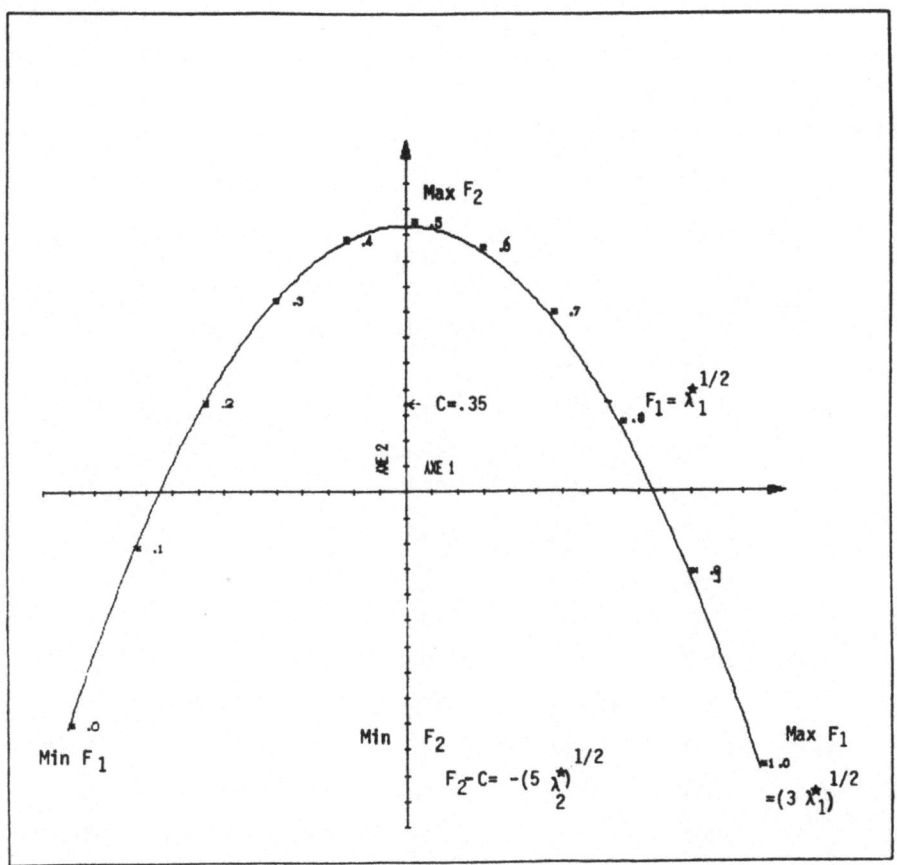

Fig.3

Table 4. Theoretical values of coordinates $F_1, F_2$ in function of x

| x | 2x - 1 | $F_1$ | $F_2$ | remarks |
|---|---|---|---|---|
| 0 | -1 | $-(3\lambda_1)^{1/2}$ | $-(5\lambda_2)^{1/2}+C$ | Min $F_1$ Min $F_2$ |
| 1/2 | 0 | 0 | $(5\lambda_2)^{1/2}/2+C$ | Max $F_2$ |
| 1 | 1 | $(3\lambda_1)^{1/2}$ | $-(5\lambda_2)^{1/2}+C$ | Max $F_1$ Min $F_2$ |
| $1/2 \pm 1/2\sqrt{3}$ | $\pm 1/(3)^{1/2}$ | $\pm \lambda_1^{1/2}$ | $0 + C$ | |

Table 5

| Site | $x_i$ | $F_1$ | $F_1^*$ | $F_2$ | $F_2^*$ | D |
|---|---|---|---|---|---|---|
| AAR | 422 | -.187 | -.113 | .994 | .821 | .187 |
| BTH | 1.000 | 1.361 | 1.340 | -1.081 | -1.117 | .041 |
| G31 | .164 | -.881 | -.596 | .148 | .006 | .318 |
| G32 | .256 | -.633 | -.400 | .577 | .416 | .282 |
| GE1 | .102 | -1.047 | -.974 | -.218 | -.249 | .079 |
| GPA | .643 | .405 | .276 | .847 | .708 | .188 |
| LOD | .762 | .724 | .573 | .435 | .343 | .177 |
| MIB | .007 | -1.302 | -1.310 | -.901 | -.899 | .007 |
| MIA | .018 | -1.271 | -1.175 | -.811 | -.843 | .101 |
| NEE | .292 | -.537 | -.328 | .704 | .534 | .269 |
| O15 | .572 | .214 | .306 | .982 | 1.169 | .208 |
| O17 | .000 | -1.320 | -1.343 | -.954 | -.961 | .024 |
| OPG | .992 | 1.341 | 1.334 | -1.019 | -1.021 | .007 |
| OVE | .516 | .064 | .086 | 1.029 | 1.177 | .149 |
| SC1 | .287 | -.551 | -.350 | .687 | .528 | .256 |
| SC3 | .258 | -.629 | -.259 | .582 | .324 | .451 |
| WE2 | .842 | .937 | .825 | .031 | -.022 | .125 |
| WE4 | .909 | 1.118 | 1.032 | -.393 | -.426 | .092 |
| WE5 | .920 | 1.147 | 1.077 | -.466 | -.491 | .073 |

We find so the following equations :

$$F_1(x) = \lambda_1^{*1/2} 3^{1/2}(2x - 0.985)$$

$$F_2(x) = - [\lambda_2^{*1/2} 5^{1/2}(3(2x - 0.985)^2 - 1)/2] + C$$

where $\lambda_1^* = .6$, $\lambda_2^* = .375$, $C = .35$, and the replacement of 1 by .985 allows to express a small unsymmetricalness of the curve (see figure 2 and 3).

Having the curve, we can compute the minimal distance of each site to the curve and then its orthogonal projections on the gradient. Table 5 gives for each point its gradient value, its $F_1$ and $F_2$ values, the coordinates $F_1^*$ and $F_2^*$ of the projection and the distance between the points and their projections.

We have at our disposal 12 radiocarbon dates, and sometimes two or three convergent or contradictory dates for the same site, which is not unusual situation. We have a look at the factorial graph and if we accept the perspective of a chronological evolution, we notice :

   - two convergent dates at Aardhorst near the top of the curve;
   - at left we see Neerharen at 7220 BC;
   - two dates of Oirschot 5 about 6000-5500 BC;
   - two dates of Weelde 5 the first about 5000 BC and the second much later.

So far these observations are in agreement with the gradient provided by factorial analysis.

But we find a limited inversion between Geldrop 1 and Milheeze 1a and a quite aberrant date for Oirschot 7, which is typologically close to the earlier settlements.

The best coherent dates seem to be those of NEE, AAR, 015 and WE5.

```
NEE    7220
AAR    6675    (mean value between 6600 and 6755)
015    6080
WE5    5040
```

(see figures 2 and 4).

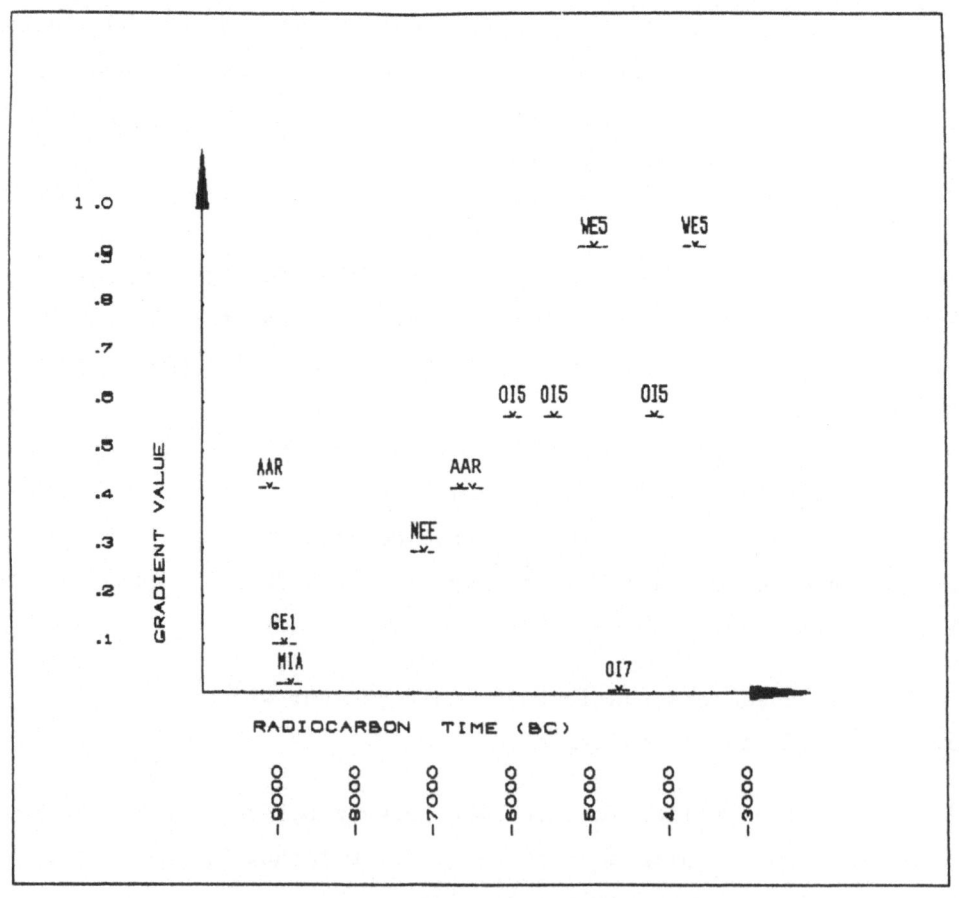

Fig.4

After the selection of the more coherent dates, it is possible to build a regression equation of the form :

$$t_i = T + \text{Cor} \frac{St}{Sx} (x_i - X)$$

where  $t_i$   is the estimated date of the site  i

      T    is the average of the known and accepted dates

      St   their standard deviation

      X    is the average of gradient values of sites with known and accepted dates

      Sx   the standard deviation associated to  X

      $x_i$   the gradient value of the site  i

      Cor  the correlation coefficient  $t_i - x_i$

which gives the following results :

| site | $x_i$ | $t_i'$ | $t_i''$ |
|------|-------|--------|---------|
| AAR | .4225 | 6812 | 6696 |
| BTH | 1.0000 | 4562 | 4714 |
| G31 | .1640 | 7819 | 7584 |
| G32 | .2565 | 7458 | 7266 |
| GE1 | .1020 | 8060 | 7796 |
| GPA | .6435 | 5951 | 5938 |
| LOD | .7625 | 5487 | 5530 |
| MIB | .0070 | 8430 | 8122 |
| MIA | .0185 | 8386 | 8083 |
| NEE | .2920 | 7320 | 7144 |
| O15 | .5725 | 6627 | 6182 |
| O17 | .0005 | 8456 | 8145 |
| OPG | .9925 | 4591 | 4740 |
| OVE | .5165 | 6446 | 6374 |
| SC1 | .2870 | 7340 | 7162 |
| SC3 | .2580 | 7453 | 7261 |
| WE1 | .8420 | 5178 | 5257 |
| WE4 | .9095 | 4915 | 5025 |
| WE5 | .9200 | 4874 | 4989 |

t'   computed dates taking account of NEE, AAR, O15 and WE5

t"   computed dates taking account of NEE, AAR, O15, WE5, GE1 and MIA

We can also compute estimated dates through a second degree equation
of the following form :

$$t_i = A(x_i)^2 + Bx_i + C$$

for  t'  we find  A = - 1671.5    B = 5506.2    C = - 8689.96
for  t"  we find  A = - 3091.5    B = 7503.1    C = - 9321.9
and the following estimated dates :

| site | $x_i$ | $t'_i$ | $t''_i$ |
|------|-------|--------|---------|
| AAR  | .4225  | 6662 | 6704 |
| BTH  | 1.0000 | 4855 | 4910 |
| G31  | .1640  | 7832 | 8175 |
| G32  | .2565  | 7388 | 7601 |
| GE1  | .1020  | 8146 | 8589 |
| GPA  | .6435  | 5839 | 5774 |
| LOD  | .7625  | 5463 | 5398 |
| MIB  | .0070  | 8652 | 9270 |
| MIA  | .0185  | 8589 | 9184 |
| NEE  | .2920  | 7225 | 7395 |
| O15  | .5725  | 6086 | 6040 |
| O17  | .0005  | 8687 | 9318 |
| OPG  | .9925  | 4872 | 4920 |
| OVE  | .5165  | 6292 | 6271 |
| SC1  | .2870  | 7247 | 7423 |
| SC3  | .2580  | 7381 | 7592 |
| WE1  | .8420  | 5239 | 5196 |
| WE4  | .9095  | 5065 | 5055 |
| WE5  | .9200  | 5039 | 5036 |

   The datings proposed here must be considered as provisional
and intended to illustrate the method.  Research in course shows
that other combinations of types may propose dates varying by one or
two centuries from those proposed here.

   Results under hypothesis of decreasing evolution rate seem to
be in larger agreement with chronological attributions based on
traditional methods (compare with Table 3).

6. Discussion and conclusion

   The older sites (MIA, MIB, GE1) are only restitued under hypothesis of decreasing speed of evolution.  In this period the area
studied seems to be at the border of two regional traditions, i.e.
the Tjongerian and the Ahrensburgian, of which geographical and
chronological limits are not clearly known.  If we add that during
this transitional period armature types remain scarce, especially
because it was the time when bow and arrow had just replaced older
hunting techniques (based upon spears and propulsors), we can propose, as an explanation, a higher speed of typological evolution
which progressively decreased and may have remained constant from
about 7500 BC. onwards.  But the presence in the area of two regional traditions encourages us to be careful about this proposition,
where relations between time and geography are not clearly understood.

   The middle and late stage sites do not require such a hypothesis
but remain compatible with it.

   The youngest sites (BTH, OPG) are situated about 4900 BC., which
corresponds to the earliest non-calibrated radiocarbon datings available for the first agriculturalist neolithic occupation in Belgium
(Cahen and Docquier, 1985).

   The nature of this decreasing speed has also to be studied.
We know that radiocarbon disintegration has varied through time.
Calibration techniques based upon dendrochronology do not provide
corrected datings prior to the seventh millenium BC.  The difference
between calibrated and non-calibrated dates decreases from 6500 to
4000 BC. in non-calibrated chronology (Ottaway, 1983; Watkins, 1975;
Klein et al., 1982; Clark, 1979).

   As a consequence of this situation, we can not determine if
variations in evolution rate are due to non-calibration or to a

properly cultural phenomenon. We shall also have to explicitate whether the variation through time of the number of typological categories may have an influence on the structure of the gradient.

A last perspective offered by this approach is the strong similarity between the permanent replacement of a part of types by another without major discontinuity and close markovian phenomenon like the transmutation of isotopes or the geologic succession of species along irreversible single or branching chains.

## 7. References

J.P.Benzécri (1973). L'analyse des données. Vol.2 : L'analyse des correspondances. Paris, Dunod.

L.Binford and S.Binford (1966). A preliminary analysis of functional variability in the Mousterian of Levallois facies, in American Anthropologist, 68(2), 238-295.

A.Broglio, B.Bagolini and R.Lunz (1984). Le Mésolithique des Dolomites, in Il popolamento delle Alpi in éta mesolithica : VIII-V millennio a.c., Preistoria Alpina, 19, 15-36.

C.Bromberger (1979). Technologie et analyse sémantique des objets : pour une sémio-technologie, in L'homme, 19(1), 105-140.

M.Buyse (1983). Les différentes approches conduisant à l'analyse des correspondances. Centre d'analyse des données et processus stochastiques, Université Libre de Bruxelles, 24 p.

D.Cahen (1985). Fonction, industrie et culture, in La signification culturelle des industries lithiques, BAR international series, 239, 39-56.

D.Cahen and J.Docquier (1985). Présence du Groupe de Blicquy en Hesbaye liégeoise, in Helinium, 25(1), 94-122.

R.M.Clark (1979-1980). Calibration, cross-validation and carbon-14 in Journal of the Royal Statistical Society, A, 142(1), 47-62 and 143(1), 177-194.

W.B.Craytor and L.Johnson (1968). Refinements in computerised item
    seriation, in Bulletin of the Museum of Natural History,
    University of Oregon, Vol.10.

A.Decormeille and J.Hinout (1982). Mise en évidence des différentes
    cultures mésolithiques dans le Bassin parisien par l'analyse
    des données, in Bulletin de la Société Préhistorique Française.
    79(3), 81-88.

G.Delibrias and P.-R.Giot (1970). Inadéquation, hétérogénéité et
    contamination des échantillons soumis pour les datations radio-
    carbone, in Bulletin de la Société préhistorique française,
    67(5), 135-137.

F.Djindjian (1977). Etude quantitative des séries aurignaciennes de
    la Ferrassie par l'analyse des données, in Bulletin de la
    Société Préhistorique Française, Etudes et Travaux, 74(1),
    357-361.

D.D.Drennan (1976). A refinement of chronological seriation using
    nonmetric multi-dimensional scaling, in American Antiquity,
    41(3), 290-302.

J.-C.Gardin (1979). Une Archéologie Théorique, Paris, Hachette.

Y.Gasco (1977). Les matrices graphiques ordonnables Bertin et l'inté-
    rêt de leur application en archéologie, in Congrès Préhistorique
    de France, Société Préhistorique Française, XXe session, Paris,
    1974, 295-310.

P.A.Gendel, H.Van de Heyning ang G.Gijselings (1985). Helchteren-
    Sonnisse Heide 2 : a mesolithic site in the Limburg kempen
    (Belgium), in Helinium, 25(1), 1-22.

A.Gob (1976). La notion de "style de débitage" peut-elle servir de
    repère chronologique dans le mésolithique ?, in Congrès Préhis-
    torique de France, XXth session, 311-319.

A.Gob (1976a). Le Mésolithique dans le Bassin de l'Ourthe, Liège,
    Société Wallonne de Palethnologie.

A.Gob (1976b). Paléohistoire de la Belgique, du VIIIe au Vème millé-
    naire b.c. : un état de la question, in Actes du XLVe Congrès
    de la Fédération des Cercles d'Archéologie et d'Histoire de
    Belgique, 2, 117-134.

A.Gob (1985). Extension géographique et chronologique de la culture
    Rhein-Meuse-Schelde (RMS), in Helinium, 25(1), 23-36.

K.Goldmann (1979). Die Seriation chronologischer Leitfunde der Bronze-
    zeit Europas, Berlin. Volker Spiess.

C.H.Gosden (1984). Bohemian Iron Age chronologies and the seriation
    of Radovesice, in Germania, 62(2), 289-309.

M.O.Hill (1973). Reciprocal averaging, in Journal of Ecology, 61,
    237-249.

M.O.Hill (1974). Correspondence analysis : a neglected multivariate
    method, in Applied Statistics, 23, 340-354.

M.O.Hill and H.G.Gauch (1980). Detrended correspondence analysis,
    in Vegetation, 42, 47-58.

P.Ihm (1984). Korrespondenzanalyse und Gausssches Ordinationsmodell,
    in Allgemeines statistisches Archiv, 68(1), 41-62.

P.Ihm and H.van Groenewoud (1984). Correspondence analysis and
    Gaussian ordination, in Compstat lectures, 3, 5-60.

R.M.Jacobi, R.R.Laxton and V.R.Switsur (1980; Seriation and dating
    of mesolithic sites in souhtern England, in Revue d'Archéo-
    métrie, 1, 165-173.

J.Janssen, P.Cullus and P.Van Brussel (1983). Analyse factorielle
    des correspondances binaires (ACOBI), Centre d'analyse des
    données et processus stochastiques, Université Libre de
    Bruxelles, 80 p.

D.G.Kendall (1970). A mathematical approach to seriation, in Philo-
    sophical Transactions of the Royal Society of London, Series A,
    269, 125-135.

D.G.Kendall (1971). Seriation from abundance matrices, in Mathematics
    in the Archaeological and Historical Sciences, F.R.Hodson,
    D.G.Kendall and P.Tautu (eds), Edinburgh, Edinburgh University
    Press, 215-252.

J.Klein, J.C.Lerman, P.E.Damon and E.K.Ralph (1982). Calibration of radiocarbon dates : tables based on the consensus data of the Workshop on Calibrating the radiocarbon time scale, in Radiocarbon, 24, 103-150.

S.K.Kozlowski (1975). Cultural differentiation of Europe from 10th to 5th millenium B.C., Warsaw University Press, 259 p.

L.Lebart, A.Morineau and N.Tabard (1977). Techniques de la Description Statistique : Méthodes et Logiciels pour l'Analyse des Grands Tableaux, Paris, Dunod.

L.Lebart, A.Morineau and J.P.Fénelon (1979). Traitement des Données Statistiques : Méthodes et Programmes, Paris, Dunod.

S.A.Leblanc (1975). Micro-seriation : a method for fine chronological differentiation, in American Antiquity, 40(1), 22-38.

J.N.Lanting and W.G.Mook (1977). The Pre- and Protohistory of the Netherlands in Terms of radiocarbon dates, Groningen.

M.Lenoir (1975). Style et technologie lithique, in Bullein de la Société Préhistorique Française, 72(2), 46-49.

W.H.Marquardt (1978). Advances in archaeological seriation, in M.B.Schiffer (ed.), Advances in archaeological method and theory, 1, 257-314.

J.C.Naouri (1970). Analyse factorielle des correspondances continues, in Publications de l'Institut de Statistique de l'Université de Paris, 19, 1-100.

B.S.Ottaway (ed.) (1983). Archaeology, dendrochronology and radiocarbon calibration curve, University of Edinburgh, Department of archaeology, Occasional paper, 9, 100 p.

K.Paddayya (1971). The late palaeolithic of the Netherlands - a review, in Helinium, 11(3), 257-270.

P.Périn (1980). La Datation des Tombes Mérovingiennes, Paris, Dunod.

W.M.F.Pétrie (1899). Sequences in prehistoric remains, in Journal of the Anthropological Institute, 29, 295-301.

S.Plog (1983). Analysis of style in artifacts, in Annual review of anthropology, 12, 125-142.

J.G.Rozoy (1978a). Les Derniers Chasseurs. Charleville. Société
    Archéologique Champenoise.

J.G.Rozoy (1978b). Typologie de l'Epipaléolithique (Mésolithique)
    franco-belge,in Bulletin de la Société archéologique champenoise,
    Special Volume.

J.G.Rozoy (1980). Le changement dans la continuité : Les débuts de
    l'Epipaléolithique dans l'Europe de l'Ouest, in Veroffenlichungen
    des Museums für Ur- und Frühgeschichte Postdam, 14/15, 11-24.

J.G.Rozoy (1984). La sériation chronologique des nécropoles de
    l'époque de la Tène, in A.Cahen-Delhaye et al. (eds.), Les
    Celtes en Belgique et dans le Nord de la France, Revue du Nord,
    special volume, 101-144.

B.F.Schriever (1983). Scaling of order dependent categorical varia-
    bles with correspondence analysis, in International Statistical
    review, 51, 225-238.

J.-L. Slachmuylder(1985). Seriation by correspondence analysis for
    mesolithic assemblages, in PACT 11 : Data management and mathe-
    matical methods in archaeology, 137-148.

D.Stiles (1979). Paleolithic culture and culture change : experiment
    in theory and method, in Current Anthropology, 20(1), 1-21.

P.M.Vermeersch (1974). Epipaleolithicum en Mesolithicum te Helchteren,
    Sonnisse Heide, in Archaeological Belgica, No 169.

P.M.Vermeersch (1975). Quelques observations sur le capsien du bassin
    de Kasserine (Tunisie steppique), in Anthropologie, 79(2),
    215-242.

P.M.Vermeersch (1982). Quinze années de recherches sur le Mésolithique
    en Basse Belgique : état de la question, in Le Mésolithique
    entre Rhin et Meuse, A.Gob and F.Spier (eds.), Luxembourg,
    Société Préhistorique Luxembourgeoise, 343-353.

P.M.Vermeersch, A.V.Munaut and E.Paulissen (1974). Fouilles d'un site
    du Tardenoisien final à Opglabbeek-Ruiterskuil (Limbourg belge),
    in Quartär. 25, 85-104.

P.M.Vermeersch, R.Lauwers and D.Huyge (1982a). Contributions à
l'étude du mésolithique en Basse Belgique, Tervuren, Musée
Royal de l'Afrique Centrale.

P.M.Vermeersch and R.Lauwers (1982b). Late mesolithic occupation at
Brecht-Thomas Heyveld, in Acta archaeologica Lovaniensia, 21,
1-25.

P.M.Vermeersch (1984). Du paléolithique final au mésolithique dans
le Nord de la Belgique, in Cahen et Haesaerts (eds.), Peuples
chasseurs de la Belgique préhistorique dans leur cadre naturel,
181-193.

T.Warkins (ed.) (1975). Radiocarbon calibration and prehistory,
Edinburgh University Press, 147 p.

P.Wiessner (1983). Style and social information  in Kalahari San
projectile points, in Americal Antiquity, 48 (2), 253-276.

SECTION 2.3

ANALYSIS OF CONTINGENCY DATA USING SPECTRAMAP

P.J. Lewi, J. Van Hoof and G. Calomme

Research Laboratories, Janssen Pharmaceutica NV

B-2340 Beerse, Belgium

## Introduction

A table of contingency data on archeological artifacts has been submitted for analysis by Slachmuylder (1985).

The data made available represent a two-way structure comprising 121 artifacts, grouped into fifteen classes, found in nineteen settlements of northern Belgium and southern Holland.

Eight artifacts have been eliminated from the table because their total occurrence in the nineteen settlements is zero. Radiocarbon dating is available for seven of the nineteen settlements. It has been noted by the submitter that some of the radiocarbon datings may be imprecise.

The problem formulated by the submitter consists in the assignment of reasonable datings to the remaining settlements.

Spectral map analysis (SMA) has been used in the solution of the problem. This technique is especially useful when a size component is present in the data and when an interpretation of the data is feasible in terms of characteristic ratios.

Reasonable datings for the settlements have been derived and
tested in a stepwise manner. First, a global encompassing analysis
established a continuous variable which relates best to prehistoric
time. Subsequently, detailed analyses have been carried out in
order to test the robustness of the solution.

## 1. Method

Spectral map analysis (SMA) is related to factorial analysis of
correspondences (Benzécri, 1973). Both methods reduce the rank of
the data matrix by one. They apply individual weighting to rows and
columns and produce a joint representation of the interactions
between row- and column-items. SMA aims at the visualisation of
ratios rather than of distances of chi-square.

The details of the method are described elsewhere in this
volume (Lewi, 1985). Spectramap is the name of a program that performs SMA.

Briefly, SMA includes logarithmic re-expression of the data,
row- and column-wise centering, factorization and simultaneous plotting of row- and column-markers in the plane of the dominant factors
(Lewi, 1982, 1984). Orthogonal projection of row-representations
upon colum-representations reproduces the doubly-centered data.
Weight coefficients are assigned to individual row- and column-items
according to their size or importance.

## 2. Global analysis of artifacts vs settlements

The original data are reproduced in Table 1. The labels of the
artifacts refer to the serial numbers in the original report of the
submitter.

Interactions between artifacts and settlements are made visible
in the Spectramap of Figure 1. In this map, circles represent artifacts and settlements are identified by squares. Areas of circles

Table 1. Original data on lithic artifacts in a small west-european area of prehistoric settlements (Slachmuylder, 1985).

A : AARDHORST
B : BRECHT-THOMAS HEYVELD
C : GELDROP III-1
D : GELDROP III-2
E : GELDROP 1

F : GRAND PORT-ARTHUR
G : LOMMEL-DESTEXHE
H : MILHEEZE 1A
I : MILHEEZE 1B
J : NEERHAREN

K : OIRSCHOT V
L : OIRSCHOT VII
M : OPGLABBEEK
N : OVERPELT
O : SCHULEN I

P : SCHULEN III
Q : WEELDE 1
R : WEELDE 4
S : WEELDE 5

| Group | | A | B | C | D | E | F | G | H | I | J | K | L | M | N | O | P | Q | R | S | TT |
|---|---|---|---|---|---|---|---|---|---|---|---|---|---|---|---|---|---|---|---|---|---|
| 1 | 1 : 1 | 8 | 0 | 7 | 5 | 3 | 2 | 0 | 2 | 0 | 3 | 15 | 4 | 1 | 0 | 4 | 1 | 7 | 5 | 4 | 71 |
| | 2 : 2 | 9 | 1 | 14 | 7 | 8 | 1 | 1 | 21 | 17 | 1 | 7 | 13 | 2 | 1 | 5 | 1 | 3 | 2 | 1 | 116 |
| | 3 : 3 | 23 | 2 | 30 | 20 | 2 | 0 | 0 | 23 | 41 | 3 | 14 | 13 | 3 | 0 | 5 | 0 | 5 | 6 | 2 | 192 |
| | 4 : 4 | 14 | 6 | 16 | 10 | 4 | 5 | 0 | 12 | 19 | 4 | 34 | 16 | 3 | 15 | 13 | 3 | 12 | 12 | 10 | 208 |
| | 5 : 5 | 5 | 1 | 7 | 2 | 1 | 4 | 0 | 3 | 10 | 1 | 27 | 6 | 0 | 0 | 5 | 1 | 2 | 3 | 2 | 80 |
| | 6 : 6 | 0 | 0 | 0 | 0 | 0 | 1 | 0 | 2 | 3 | 0 | 0 | 0 | 0 | 0 | 0 | 0 | 0 | 0 | 0 | 6 |
| | 7 : 7 | 0 | 2 | 4 | 2 | 0 | 0 | 0 | 13 | 8 | 2 | 17 | 3 | 1 | 4 | 6 | 9 | 17 | 8 | 9 | 106 |
| | 8 : 8 | 1 | 5 | 6 | 1 | 1 | 2 | 0 | 9 | 5 | 2 | 17 | 8 | 0 | 6 | 9 | 4 | 4 | 8 | 3 | 91 |
| | 9 : 9 | 1 | 0 | 0 | 1 | 0 | 0 | 0 | 1 | 0 | 0 | 3 | 0 | 1 | 1 | 0 | 0 | 0 | 0 | 1 | 10 |
| | 10 : 10 | 1 | 1 | 0 | 0 | 0 | 0 | 0 | 2 | 2 | 0 | 7 | 2 | 0 | 1 | 2 | 0 | 1 | 1 | 1 | 23 |
| 2 | 11 : 11 | 1 | 1 | 1 | 1 | 0 | 0 | 0 | 1 | 2 | 0 | 3 | 2 | 2 | 2 | 2 | 0 | 3 | 2 | 2 | 24 |
| | 12 : 12 | 3 | 2 | 2 | 3 | 0 | 0 | 0 | 3 | 1 | 1 | 7 | 4 | 0 | 4 | 3 | 2 | 7 | 10 | 6 | 47 |
| | 13 : 13 | 1 | 1 | 1 | 0 | 0 | 0 | 0 | 0 | 0 | 0 | 2 | 2 | 0 | 0 | 0 | 0 | 0 | 0 | 0 | 8 |
| | 14 : 14 | 3 | 3 | 0 | 3 | 0 | 0 | 0 | 0 | 0 | 1 | 3 | 4 | 3 | 3 | 5 | 2 | 10 | 9 | 9 | 58 |
| | 15 : 15 | 5 | 11 | 2 | 2 | 4 | 5 | 0 | 3 | 1 | 2 | 2 | 0 | 1 | 0 | 5 | 3 | 15 | 12 | 8 | 86 |
| | 16 : 16 | 27 | 15 | 12 | 24 | 3 | 0 | 2 | 12 | 7 | 6 | 42 | 13 | 7 | 2 | 39 | 27 | 61 | 52 | 36 | 385 |
| | 17 : 17 | 4 | 1 | 1 | 0 | 0 | 1 | 0 | 0 | 1 | 0 | 13 | 2 | 0 | 0 | 1 | 1 | 2 | 0 | 2 | 31 |
| | 18 : 18 | 0 | 0 | 0 | 0 | 0 | 0 | 0 | 1 | 0 | 0 | 2 | 0 | 0 | 0 | 0 | 0 | 0 | 0 | 0 | 3 |
| 3 | 19 : 19 | 9 | 0 | 4 | 2 | 0 | 0 | 0 | 9 | 4 | 0 | 8 | 6 | 1 | 1 | 4 | 0 | 6 | 2 | 1 | 61 |
| 4 | 20 : 20 | 0 | 0 | 2 | 0 | 0 | 2 | 0 | 3 | 4 | 0 | 6 | 7 | 0 | 0 | 0 | 4 | 0 | 0 | 0 | 23 |
| | 21 : 21 | 19 | 0 | 7 | 12 | 1 | 0 | 0 | 46 | 41 | 1 | 14 | 23 | 0 | 0 | 13 | 2 | 1 | 1 | 1 | 183 |
| | 22 : 22 | 10 | 0 | 15 | 4 | 6 | 2 | 1 | 86 | 63 | 0 | 4 | 43 | 0 | 0 | 9 | 1 | 1 | 1 | 1 | 245 |
| | 23 : 23 | 3 | 0 | 0 | 0 | 1 | 0 | 0 | 4 | 12 | 0 | 0 | 2 | 1 | 0 | 3 | 0 | 0 | 0 | 1 | 29 |
| | 24 : 24 | 2 | 0 | 0 | 3 | 0 | 1 | 0 | 0 | 1 | 1 | 0 | 0 | 0 | 2 | 0 | 0 | 0 | 0 | 0 | 14 |
| 5 | 25 : 25 | 0 | 0 | 6 | 1 | 2 | 0 | 0 | 17 | 6 | 0 | 5 | 18 | 1 | 0 | 3 | 2 | 0 | 1 | 0 | 62 |
| | 26 : 26 | 5 | 0 | 0 | 3 | 1 | 0 | 0 | 4 | 6 | 0 | 1 | 3 | 0 | 0 | 1 | 0 | 1 | 0 | 0 | 27 |
| | 27 : 27 | 0 | 2 | 6 | 8 | 2 | 0 | 0 | 7 | 7 | 0 | 2 | 4 | 2 | 2 | 2 | 2 | 4 | 2 | 3 | 57 |
| | 28 : 28 | 6 | 3 | 11 | 6 | 0 | 0 | 0 | 6 | 4 | 1 | 0 | 4 | 0 | 1 | 0 | 0 | 11 | 8 | 4 | 54 |
| | 29 : 29 | 6 | 3 | 5 | 7 | 0 | 0 | 0 | 1 | 1 | 1 | 2 | 5 | 4 | 0 | 4 | 3 | 5 | 3 | 1 | 49 |
| | 30 : 30 | 34 | 16 | 15 | 19 | 0 | 2 | 1 | 17 | 7 | 5 | 9 | 20 | 7 | 3 | 5 | 2 | 15 | 11 | 9 | 199 |
| | 31 : 31 | 11 | 0 | 1 | 3 | 0 | 0 | 0 | 1 | 9 | 0 | 1 | 6 | 1 | 0 | 6 | 0 | 0 | 0 | 0 | 40 |

(Continued)

P. J. LEWI ET AL.

Table 1 (Continued)

| | | 1 | 2 | 3 | 4 | 5 | 6 | 7 | 8 | 9 | 10 | 11 | 12 | 13 | 14 | 15 | 16 | 17 | 18 | 19 | Total |
|---|---|---|---|---|---|---|---|---|---|---|---|---|---|---|---|---|---|---|---|---|---|
| 6 | 32 :: 32 | 3 | 2 | 0 | 1 | 2 | 0 | 30 | 16 | 5 | 0 | 5 | 0 | 2 | 1 | 5 | 2 | 5 | 4 | 6 | 84 |
| | 33 :: 33 | 11 | 2 | 2 | 0 | 0 | 2 | 14 | 8 | 4 | 2 | 13 | 0 | 1 | 3 | 0 | 3 | 0 | 3 | 0 | 72 |
| | 34 :: 34 | 0 | 2 | 0 | 0 | 0 | 0 | 0 | 0 | 0 | 0 | 0 | 0 | 0 | 0 | 3 | 0 | 0 | 0 | 0 | 9 |
| | 35 :: 35 | 0 | 0 | 0 | 0 | 0 | 1 | 1 | 2 | 3 | 0 | 3 | 0 | 0 | 5 | 4 | 1 | 0 | 0 | 0 | 13 |
| | 36 :: 36 | 41 | 0 | 13 | 0 | 18 | 22 | 4 | 6 | 3 | 4 | 25 | 0 | 13 | 6 | 1 | 9 | 4 | 9 | 0 | 185 |
| | 37 :: 37 | 2 | 4 | 0 | 3 | 2 | 0 | 2 | 2 | 3 | 0 | 13 | 5 | 4 | 2 | 5 | 3 | 9 | 5 | 0 | 48 |
| | 38 :: 38 | 7 | 1 | 2 | 0 | 2 | 4 | 0 | 0 | 3 | 3 | 5 | 0 | 5 | 1 | 0 | 4 | 1 | 3 | 0 | 31 |
| | 39 :: 39 | 3 | 0 | 4 | 1 | 4 | 0 | 0 | 0 | 2 | 0 | 3 | 0 | 6 | 0 | 0 | 10 | 2 | 1 | 0 | 60 |
| | 40 :: 40 | 6 | 1 | 2 | 0 | 0 | 1 | 0 | 0 | 3 | 2 | 7 | 1 | 2 | 1 | 0 | 2 | 0 | 15 | 11 | 33 |
| | 41 :: 41 | 27 | 0 | 0 | 1 | 1 | 0 | 0 | 0 | 1 | 0 | 2 | 0 | 19 | 3 | 3 | 36 | 5 | 2 | 4 | 161 |
| | 42 :: 42 | 5 | 15 | 1 | 0 | 6 | 6 | 0 | 0 | 0 | 3 | 1 | 0 | 1 | 0 | 0 | 3 | 0 | 0 | 12 | 27 |
| | 43 :: 43 | 19 | 0 | 5 | 2 | 20 | 0 | 0 | 0 | 0 | 1 | 0 | 0 | 0 | 2 | 0 | 9 | 1 | 0 | 1 | 85 |
| | 44 :: 44 | 7 | 4 | 14 | 5 | 8 | 1 | 0 | 0 | 0 | 0 | 2 | 0 | 9 | 1 | 0 | 5 | 3 | 5 | 2 | 65 |
| | 45 :: 45 | 9 | 1 | 3 | 2 | 14 | 2 | 0 | 0 | 0 | 1 | 1 | 2 | 6 | 3 | 0 | 24 | 0 | 5 | 0 | 112 |
| | 46 :: 46 | 6 | 9 | 11 | 5 | 34 | 3 | 1 | 0 | 0 | 3 | 1 | 1 | 4 | 0 | 0 | 15 | 2 | 2 | 5 | 136 |
| | 47 :: 47 | 58 | 5 | 21 | 5 | 30 | 3 | 1 | 0 | 0 | 1 | 0 | 3 | 21 | 0 | 0 | 10 | 19 | 9 | 11 | 201 |
| 7 | 48 :: 48 | 3 | 0 | 10 | 0 | 16 | 0 | 2 | 0 | 0 | 1 | 2 | 1 | 1 | 2 | 0 | 2 | 2 | 3 | 3 | 43 |
| | 49 :: 49 | 2 | 0 | 6 | 5 | 3 | 0 | 0 | 0 | 0 | 0 | 12 | 8 | 8 | 1 | 0 | 19 | 12 | 1 | 0 | 54 |
| | 50 :: 50 | 1 | 0 | 9 | 5 | 1 | 0 | 0 | 0 | 0 | 0 | 0 | 0 | 0 | 1 | 0 | 1 | 0 | 17 | 7 | 103 |
| | 51 :: 51 | 4 | 0 | 0 | 0 | 0 | 0 | 0 | 0 | 4 | 0 | 4 | 4 | 0 | 0 | 0 | 0 | 0 | 3 | 0 | 72 |
| | 52 :: 52 | 2 | 0 | 0 | 0 | 0 | 0 | 0 | 0 | 0 | 1 | 0 | 0 | 2 | 0 | 0 | 0 | 0 | 0 | 1 | 11 |
| | 53 :: 53 | 1 | 0 | 0 | 0 | 0 | 0 | 0 | 0 | 0 | 0 | 0 | 0 | 1 | 0 | 0 | 0 | 0 | 0 | 0 | 5 |
| | 54 :: 54 | 0 | 0 | 0 | 0 | 0 | 0 | 0 | 0 | 0 | 0 | 1 | 0 | 0 | 0 | 0 | 0 | 0 | 0 | 0 | 2 |
| | 55 :: 55 | 0 | 0 | 0 | 0 | 0 | 0 | 0 | 0 | 0 | 0 | 0 | 0 | 0 | 0 | 0 | 1 | 0 | 0 | 0 | 4 |
| 8 | 56 :: 57 | 0 | 0 | 0 | 0 | 0 | 0 | 0 | 0 | 0 | 0 | 0 | 2 | 1 | 1 | 0 | 0 | 0 | 0 | 0 | 9 |
| | 57 :: 58 | 0 | 0 | 0 | 0 | 0 | 0 | 0 | 0 | 0 | 0 | 1 | 1 | 0 | 0 | 0 | 0 | 0 | 0 | 0 | 1 |
| | 58 :: 59 | 0 | 0 | 0 | 0 | 1 | 0 | 0 | 0 | 0 | 0 | 0 | 0 | 0 | 0 | 0 | 1 | 0 | 0 | 0 | 2 |
| | 59 :: 60 | 0 | 0 | 1 | 0 | 0 | 0 | 0 | 0 | 0 | 0 | 0 | 20 | 2 | 0 | 0 | 0 | 0 | 0 | 0 | 31 |
| 9 | 60 :: 61 | 1 | 1 | 0 | 2 | 6 | 0 | 0 | 0 | 0 | 3 | 2 | 1 | 1 | 5 | 0 | 14 | 2 | 0 | 6 | 50 |
| | 61 :: 62 | 0 | 0 | 1 | 1 | 2 | 1 | 0 | 0 | 0 | 0 | 1 | 0 | 4 | 0 | 0 | 10 | 1 | 6 | 7 | 28 |
| | 62 :: 63 | 0 | 0 | 0 | 0 | 0 | 0 | 0 | 0 | 0 | 0 | 0 | 0 | 0 | 1 | 0 | 1 | 0 | 4 | 0 | 17 |
| | 63 :: 64 | 4 | 6 | 2 | 3 | 0 | 0 | 0 | 0 | 0 | 0 | 1 | 0 | 0 | 0 | 0 | 9 | 0 | 6 | 1 | 74 |
| | 64 :: 65 | 1 | 2 | 0 | 1 | 0 | 0 | 0 | 0 | 0 | 0 | 4 | 4 | 3 | 3 | 0 | 1 | 0 | 4 | 0 | 74 |
| | 65 :: 66 | 1 | 0 | 2 | 0 | 0 | 0 | 0 | 0 | 0 | 1 | 0 | 0 | 0 | 0 | 0 | 8 | 3 | 6 | 1 | 31 |
| | 66 :: 67 | 0 | 0 | 0 | 0 | 0 | 1 | 0 | 0 | 0 | 0 | 0 | 0 | 0 | 1 | 0 | 4 | 0 | 5 | 0 | 79 |
| | 67 :: 68 | 0 | 2 | 3 | 0 | 0 | 0 | 0 | 0 | 0 | 0 | 0 | 0 | 0 | 2 | 0 | 0 | 1 | 2 | 1 | 15 |
| | 68 :: 69 | 0 | 0 | 0 | 0 | 0 | 0 | 0 | 0 | 0 | 0 | 0 | 0 | 0 | 0 | 0 | 0 | 0 | 0 | 0 | 13 |
| | 69 :: 71 | 0 | 0 | 0 | 0 | 0 | 0 | 0 | 0 | 0 | 0 | 0 | 0 | 0 | 0 | 0 | 0 | 0 | 4 | 0 | 6 |
| | 70 :: 72 | 0 | 0 | 0 | 0 | 0 | 2 | 0 | 0 | 0 | 0 | 0 | 0 | 0 | 0 | 0 | 0 | 0 | 4 | 1 | 25 |
| | 71 :: 73 | 4 | 0 | 3 | 0 | 0 | 1 | 0 | 0 | 0 | 0 | 0 | 1 | 0 | 0 | 0 | 0 | 7 | 2 | 0 | 1 |
| | 72 :: 74 | 1 | 1 | 0 | 1 | 0 | 0 | 0 | 0 | 0 | 1 | 4 | 4 | 0 | 3 | 0 | 9 | 16 | 0 | 1 | 13 |
| | 73 :: 77 | 1 | 0 | 3 | 3 | 0 | 0 | 0 | 0 | 0 | 3 | 3 | 3 | 0 | 0 | 1 | 3 | 8 | 2 | 0 | 35 |
| 11 | 74 :: 78 | 0 | 0 | 0 | 1 | 0 | 0 | 0 | 0 | 0 | 1 | 1 | 0 | 0 | 3 | 1 | 3 | 11 | 0 | 1 | 30 |
| | 75 :: 79 | 0 | 0 | 0 | 0 | 0 | 0 | 0 | 0 | 0 | 0 | 0 | 0 | 0 | 0 | 0 | 2 | 0 | 1 | 0 | 16 |
| | 76 :: 80 | 2 | 0 | 0 | 1 | 0 | 0 | 0 | 0 | 0 | 0 | 0 | 0 | 0 | 0 | 0 | 3 | 13 | 1 | 1 | 22 |
| | 77 :: 81 | 0 | 0 | 0 | 0 | 0 | 0 | 0 | 0 | 0 | 0 | 0 | 0 | 1 | 0 | 1 | 7 | 0 | 1 | 1 | 24 |

| | | A | B | C | D | E | F | G | H | I | J | K | L | M | N | O | P | Q | R | S | TT |
|---|---|---|---|---|---|---|---|---|---|---|---|---|---|---|---|---|---|---|---|---|---|
| **12** | 79 :: 83 | 1 | 0 | 0 | 0 | 0 | 2 | 0 | 0 | 1 | 0 | 2 | 0 | 0 | 8 | 2 | 0 | 1 | 2 | 0 | 19 |
| | 80 :: 84 | 0 | 0 | 0 | 0 | 0 | 0 | 0 | 0 | 0 | 0 | 2 | 0 | 0 | 1 | 0 | 0 | 0 | 0 | 0 | 3 |
| | 81 :: 85 | 8 | 0 | 0 | 0 | 0 | 2 | 1 | 0 | 0 | 0 | 1 | 0 | 0 | 4 | 0 | 0 | 1 | 0 | 0 | 17 |
| | 82 :: 86 | 1 | 0 | 0 | 0 | 0 | 0 | 0 | 0 | 0 | 0 | 0 | 0 | 0 | 0 | 0 | 0 | 0 | 0 | 0 | 1 |
| | 83 :: 87 | 6 | 0 | 0 | 0 | 0 | 0 | 2 | 0 | 0 | 0 | 4 | 1 | 0 | 2 | 0 | 1 | 0 | 0 | 1 | 17 |
| | 84 :: 89 | 0 | 0 | 0 | 0 | 0 | 1 | 1 | 0 | 0 | 0 | 0 | 0 | 0 | 0 | 0 | 0 | 1 | 0 | 0 | 3 |
| | 85 :: 90 | 1 | 0 | 0 | 0 | 0 | 0 | 0 | 0 | 0 | 0 | 0 | 0 | 0 | 0 | 0 | 0 | 0 | 0 | 0 | 1 |
| | 86 :: 91 | 0 | 0 | 0 | 0 | 0 | 0 | 0 | 0 | 0 | 0 | 1 | 0 | 0 | 0 | 0 | 0 | 0 | 0 | 0 | 1 |
| **13** | 87 :: 92 | 0 | 12 | 0 | 0 | 0 | 1 | 2 | 0 | 0 | 0 | 1 | 0 | 5 | 0 | 0 | 0 | 10 | 8 | 0 | 45 |
| | 88 :: 93 | 5 | 13 | 2 | 1 | 0 | 0 | 5 | 0 | 0 | 0 | 0 | 0 | 18 | 0 | 0 | 0 | 13 | 16 | 5 | 86 |
| | 89 :: 94 | 0 | 0 | 0 | 0 | 0 | 0 | 0 | 0 | 0 | 0 | 0 | 0 | 0 | 0 | 0 | 0 | 44 | 35 | 12 | 107 |
| | 90 :: 95 | 1 | 21 | 3 | 1 | 0 | 1 | 2 | 0 | 0 | 0 | 1 | 0 | 4 | 0 | 1 | 0 | 11 | 4 | 14 | 60 |
| | 91 :: 96 | 0 | 3 | 0 | 0 | 0 | 0 | 0 | 0 | 0 | 0 | 0 | 0 | 0 | 0 | 0 | 0 | 8 | 9 | 8 | 29 |
| | 92 :: 97 | 0 | 23 | 0 | 0 | 0 | 0 | 0 | 0 | 0 | 0 | 1 | 0 | 0 | 0 | 0 | 0 | 2 | 2 | 8 | 38 |
| | 93 :: 98 | 0 | 1 | 0 | 0 | 0 | 3 | 8 | 0 | 1 | 0 | 0 | 0 | 0 | 0 | 0 | 0 | 2 | 4 | 3 | 21 |
| | 94 :: 99 | 0 | 4 | 0 | 0 | 0 | 0 | 1 | 0 | 0 | 0 | 1 | 0 | 0 | 0 | 0 | 1 | 4 | 5 | 5 | 20 |
| | 95 :: 100 | 0 | 0 | 0 | 0 | 0 | 0 | 0 | 0 | 0 | 0 | 2 | 0 | 2 | 1 | 0 | 0 | 2 | 1 | 1 | 6 |
| | 96 :: 101 | 1 | 0 | 0 | 0 | 0 | 0 | 0 | 0 | 0 | 0 | 0 | 0 | 0 | 0 | 0 | 0 | 0 | 0 | 1 | 2 |
| | 97 :: 105 | 0 | 1 | 0 | 0 | 0 | 0 | 2 | 0 | 0 | 0 | 0 | 0 | 0 | 0 | 0 | 0 | 3 | 2 | 4 | 12 |
| | 98 :: 106 | 2 | 0 | 0 | 0 | 0 | 0 | 2 | 0 | 0 | 0 | 0 | 0 | 0 | 0 | 1 | 14 | 3 | 0 | 2 | 41 |
| **14** | 99 :: 107 | 0 | 5 | 0 | 0 | 0 | 0 | 0 | 1 | 1 | 0 | 0 | 1 | 3 | 0 | 11 | 0 | 1 | 6 | 0 | 11 |
| | 100 :: 108 | 0 | 0 | 1 | 0 | 0 | 0 | 0 | 0 | 0 | 0 | 4 | 0 | 2 | 0 | 1 | 0 | 1 | 29 | 2 | 13 |
| | 101 :: 109 | 0 | 18 | 1 | 0 | 0 | 0 | 0 | 0 | 0 | 1 | 0 | 0 | 7 | 0 | 0 | 5 | 23 | 25 | 14 | 102 |
| | 102 :: 110 | 0 | 19 | 0 | 0 | 0 | 0 | 0 | 0 | 0 | 0 | 2 | 0 | 0 | 0 | 0 | 0 | 25 | 10 | 14 | 95 |
| | 103 :: 111 | 2 | 2 | 0 | 0 | 0 | 0 | 0 | 2 | 0 | 0 | 3 | 0 | 0 | 0 | 0 | 0 | 8 | 7 | 7 | 27 |
| | 104 :: 112 | 0 | 1 | 0 | 0 | 0 | 0 | 0 | 0 | 0 | 0 | 3 | 0 | 4 | 0 | 0 | 0 | 7 | 5 | 3 | 22 |
| | 105 :: 113 | 0 | 6 | 0 | 0 | 0 | 0 | 0 | 0 | 0 | 1 | 0 | 0 | 1 | 0 | 2 | 2 | 7 | 9 | 6 | 34 |
| | 106 :: 114 | 0 | 6 | 0 | 0 | 0 | 0 | 0 | 0 | 0 | 0 | 1 | 0 | 0 | 0 | 0 | 0 | 6 | 3 | 4 | 34 |
| | 107 :: 115 | 0 | 3 | 0 | 0 | 0 | 0 | 0 | 0 | 0 | 1 | 2 | 0 | 1 | 0 | 0 | 0 | 7 | 3 | 3 | 17 |
| | 108 :: 116 | 0 | 0 | 0 | 0 | 0 | 0 | 0 | 0 | 0 | 0 | 5 | 0 | 2 | 0 | 1 | 1 | 4 | 3 | 4 | 12 |
| | 109 :: 117 | 0 | 15 | 0 | 0 | 0 | 0 | 0 | 0 | 0 | 1 | 0 | 0 | 0 | 0 | 0 | 0 | 5 | 6 | 1 | 33 |
| | 110 :: 118 | 0 | 6 | 0 | 0 | 0 | 3 | 0 | 0 | 0 | 0 | 0 | 0 | 0 | 0 | 0 | 12 | 3 | 1 | 0 | 16 |
| | 111 :: 119 | 0 | 0 | 12 | 0 | 0 | 0 | 0 | 0 | 0 | 0 | 0 | 0 | 0 | 0 | 0 | 1 | 2 | 0 | 0 | 20 |
| **15** | 112 :: 120 | 0 | 0 | 12 | 1 | 9 | 0 | 0 | 0 | 0 | 0 | 0 | 0 | 0 | 0 | 0 | 0 | 0 | 0 | 0 | 24 |
| | 113 :: 121 | 0 | 0 | 6 | 19 | 0 | 0 | 0 | 0 | 0 | 0 | 0 | 0 | 0 | 0 | 0 | 0 | 0 | 0 | 0 | 26 |
| | TT : TABLE TOTAL | 572 | 306 | 316 | 355 | 82 | 101 | 95 | 407 | 367 | 108 | 582 | 287 | 122 | 152 | 308 | 163 | 642 | 514 | 348 | 5827 |

are proportional to the total number of occurrences of the corresponding artifacts in all settlements. Areas of squares are proportional to the total number of artifacts found at the corresponding
sites. Note that contours of symbols vary in thickness from a very
thin to a very fat outline. Variation in thickness of the contour
codes for a third factor which is oriented perpendicularly to the
plane of the map. Finally, some contours are broken, indicating
significant contributions from a fourth or higher-order factor.
Artifacts and settlements that are represented by broken symbols
cannot be fully represented by a three-factor model of the spectral
map. These outliers may possess characteristics which are extraneous
to the main context of the data.

Settlements that are not well-represented are Lomme, Overpelt, Gand, Schulen-3 and Opglabeek. It is not surprising that
the outliers are mostly settlements where only a small number of
artifacts has been found. Spectramap first attempts to fit the most
important sites, at the cost of the less important ones. Twenty-five
artifacts, or about 20% of the total, cannot be adequately represented by the three-factor model of the spectral map. These outliers
represent less abundant artifacts and Spectramap assigns less weight
to them.

The first three factors contribute 50% to the variance due to
interactions between artifacts and settlements. Together with the
variances due to the differences in abundance among artifacts and
settlements, the Spectramap reproduces 78% of the total variance
of the table.

In Figure 1 we find that artifacts that occur specifically in
one or more of the settlements have their representative circles
shifted in the direction of the corresponding squares. Likewise the
same circles are repelled on the map by the squares that represent
sites where these artifacts are rare or non-existing. This way we

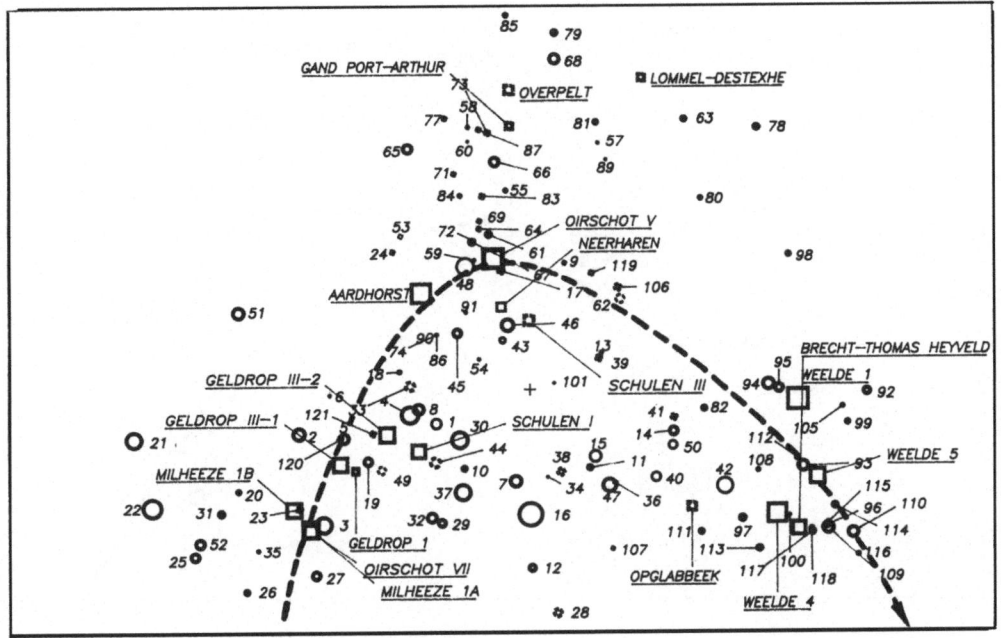

Fig.1

find that artifact 22 (burin), on the lower left is most prevalent
in the Milheeze sites, Geldrop-1 and Oirschot-7.  Object 42 (bladelet), on the lower right, is abundant in Weelde and Brecht.  Also
object 48 (point), somewhat above the center, appears to be specific
for Aardhorst and Neerharen, among others.  Objects that are represented near the barycenter of the map (indicated by a small cross)
are common to the majority of the settlements.

The continuous broken line on figure 1 represents the time-
related variable that bests fits to the available radiocarbon datings.
The fitting has been constructed by hand.

In order to study the arrangement of settlements on the map
with respect to the radiocarbon datings, we reproduced them separately.  Figure 2 is identical to figure 1, except that the artifacts
have been omitted.

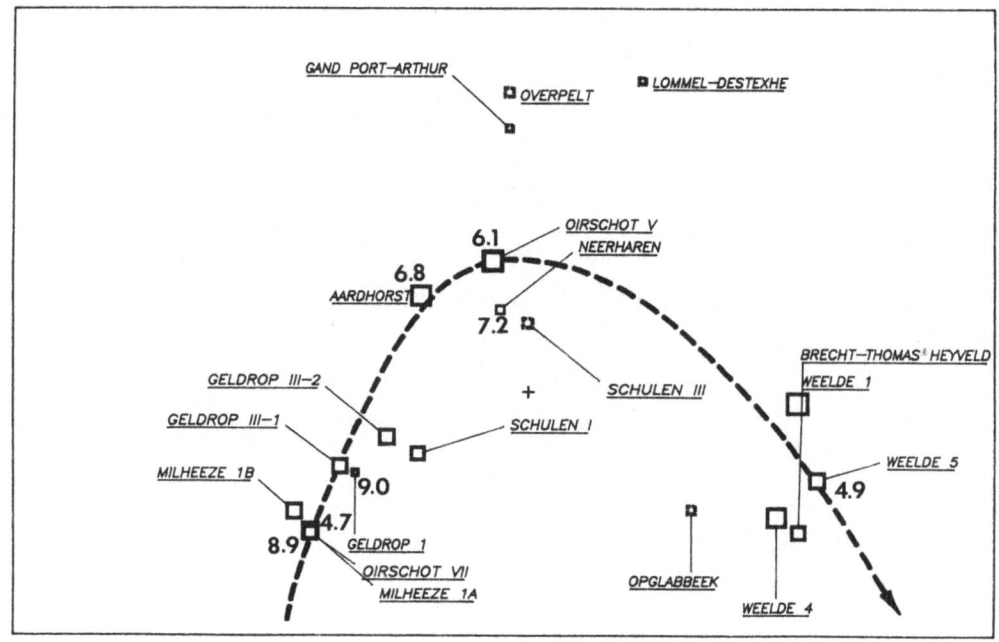

Fig.2

## 3. Time-related variable

If we focus on the settlements that are well-represented on the Spectramap of figure 2, a rough pattern appears.  Indeed, the radio-carbon data suggest a curved line that joins Milheeze-1A (8.9 Ky), to Aardhorst (6.8 Ky), Oirschot-5 (6.1 Ky) and to Weelde-5 (4.9 Ky). For the sake of simplicity we substituted median values when more than one radiocarbon dating is available at a particular settlement. (1 Ky means 1000 years).

The time-variable also agrees with the location of the two minor sites Schulen-3 (7.2 Ky) and Geldrop-1 (9.0 Ky).  With respect to Oirschot-7 (4.7 Ky), however, we find a notable discrepancy.  This and further analyses suggest that the radiocarbon dating for Oirschot-7 is probably in error by a factor of about 2.

Perpendicular projection on the continuous line provides an approximate seriation of the settlements and artifacts with respect

Table 2. Seriation of prehistoric settlements with respect to a computed time-variable. Separation of settlements into early, middle and recent epochs. Radiocarbon datings are indicated between brackets. (1 Ky means 1000 years).

| | | |
|---|---|---|
| 9.0 Ky | Oirschot-7 | (4.7 Ky?) |
| | Milheeze-1A | (8.9 Ky) |
| | Milheeze-1B | |
| | Geldrop-1 | (9.0 Ky) |
| | Geldrop 3-1 | |
| | Geldrop 3-2 | |
| | Schulen-1 | |
| 6.8 Ky | Aardhorst | (6.8 Ky) |
| | Neerharen | |
| | Oirschot-5 | (6.1 Ky) |
| | Schulen-3 | (7.2 Ky) |
| | Gand, Overpelt, Lommel | |
| 4.9 Ky | Opglabbeek | |
| | Weelde-1 | |
| | Weelde-5 | (4.9 Ky) |
| | Weelde-4 | |
| | Brecht | |

to prehistoric time (Table 2).

Roughly we distinguish an early epoch (9 Ky), a middle epoch (7 Ky) and a more recent epoch (5 Ky).

The curvature of the time-variable is due to a number of artifacts that are specific for the sites associated with the middle epoch from 6 to 7 Ky. We will show that these objects belong to classes 8 to 12 (crescents, narrow bladelets, triangles and points with retouched base), many of which are not well-represented on the map of figure 1.

This phenomenon will be more clearly discussed in the following section.

The interval between early and middle epochs, as measured along the time-line of the Spectramap, is not significantly different in length from the interval between middle and recent epochs. The corresponding time differentials amount to 2.2 and 1.9 Ky respectively. Considering the large variability of the datings, we assume that these differentials are also approximately equal. In spectral map analysis equal intervals along the time-line usually indicate steady but non-accelerating diversification with advancing time.

In summary, using spectral mapping, we have demonstrated the existence of a time-variable with the following properties :
1.  approximate fit through the major dated settlements,
2.  one major misclassification (Oirschot-7),
3.  clear separation of early (9 Ky), middle (7 Ky) and recent (5 Ky) epochs,
4.  indication of specialization in minor sites (Gand, Overpelt, Lommel) during the middle epoch (7 Ky).

## 4. Analysis of classes of artifacts vs settlements

Figure 3 shows a spectral map of grouped artifacts and of settlements. The grouping of artifacts has been performed on the basis of geometrical and functional characteristics (Slachmuylder, 1985).

The total degree of reproduction of the data in this map (92%) is much higher than in the global analysis (78% of the total variance in the table). This effect results from collapsing 113 objects into 15 classes. As a result of the improved degree of reproduction, all objects and sites are well-represented within a three-factor model.

Comparing figures 2 and 3, one readily observes that the time-variable is conserved. Indeed, the arrangement of the settlements appears to be almost identical.

Classes 8 to 12 (crescents, narrow bladelets, triangles and

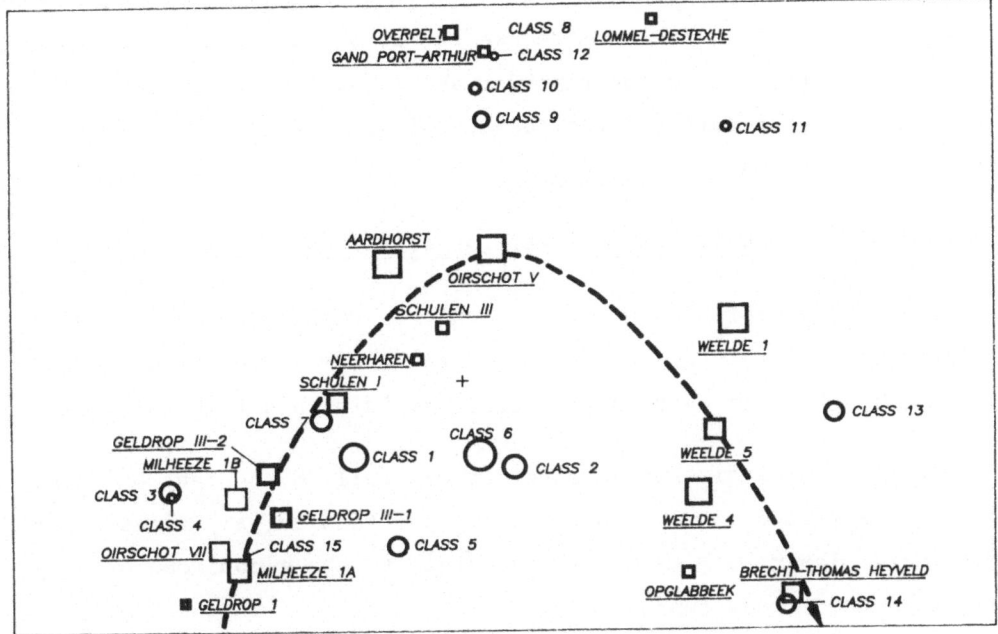

Fig.3

points with retouched base) are almost exclusively associated with the small sites Overpelt, Gand and Lommel. Classes 3 and 4 (borers, burins and some common tools) are specifically tied to the earlier epoch (9 Ky). Classes 13 and 14 (trapezes, notched blades and bladelets) are most typical for the recent period (Ky). All other classes are common to two epochs (e.g. classes 1 and 7) or to all three epochs (e.g. classes 6 and 2).

It is surprising that the numbering of the classes correlates closely with their dating along the curved time-line. Classes 1-7 appear in the early epoch, classes 8-12 in the middle epoch and classes 13-14 are associated to the most recent epoch on the time curve. Our analysis reveals a prior classification in the data. Only class 15 forms an exception.

We conclude that our time-variable withstands grouping of objects into broad classes. Furthermore, we can associate patterns of these classes with respect to the time-variable. We also have identified concentrations at particular sites and disseminations of artifacts over several sites.

## 5. Analysis of subsets of artifacts vs settlements

We have performed a number of analyses with selected artifacts. The aim was to test the robustness of the time-variable and to find a minimal subset of objects that could explain this time-variable.

A first analysis was carried out over all 38 abundant objects, i.e. those with a total occurrence of more than 50, all sites taken together. The result is almost identical to the global one of figure 1 and is not reproduced.

Then, we further split the abundant subset into those 18 artifacts that are most concentrated at one or more settlements. The remaining 20 abundant objects are those that are disseminated over several settlements. As can be seen from figure 4, these 18 abundant and concentrated artifacts still give a fair reproduction of the time-variable. On the other hand, the 20 disseminated objects do not produce a meaningful pattern and their result is not shown.

We conclude that the time-variable is robust and withstands systematic selection of objects. Furthermore we can identify specialized objects for the purpose of seriation and dating of settlements.

## 6. Analysis of artifacts vs selected settlements

A final test of robustness of the time-variable consists in separating the settlements into functional and positioned ones. The functional sites are those for which radiocarbon dating is available

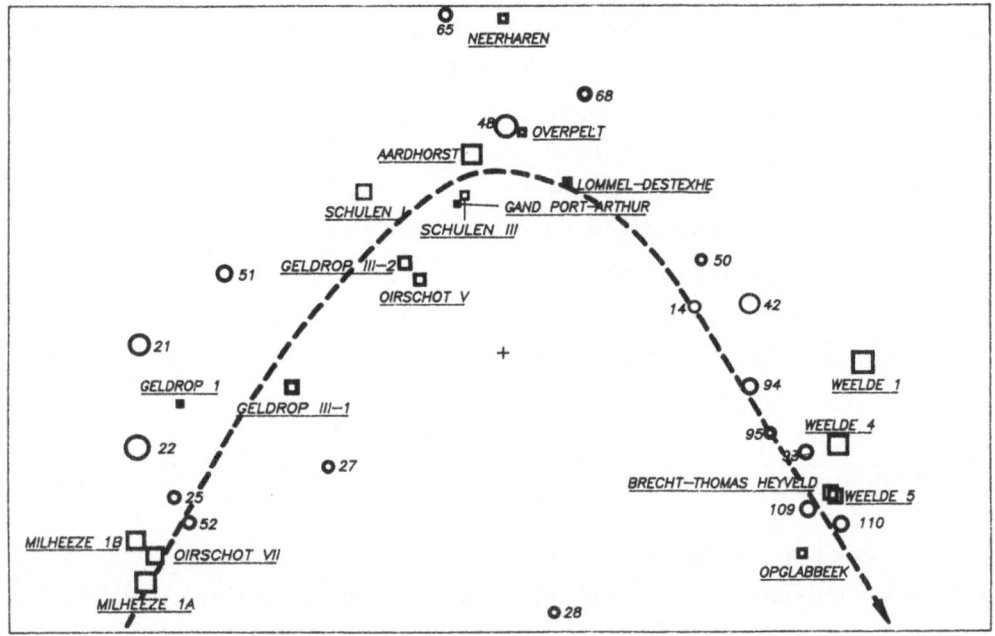

Fig.4

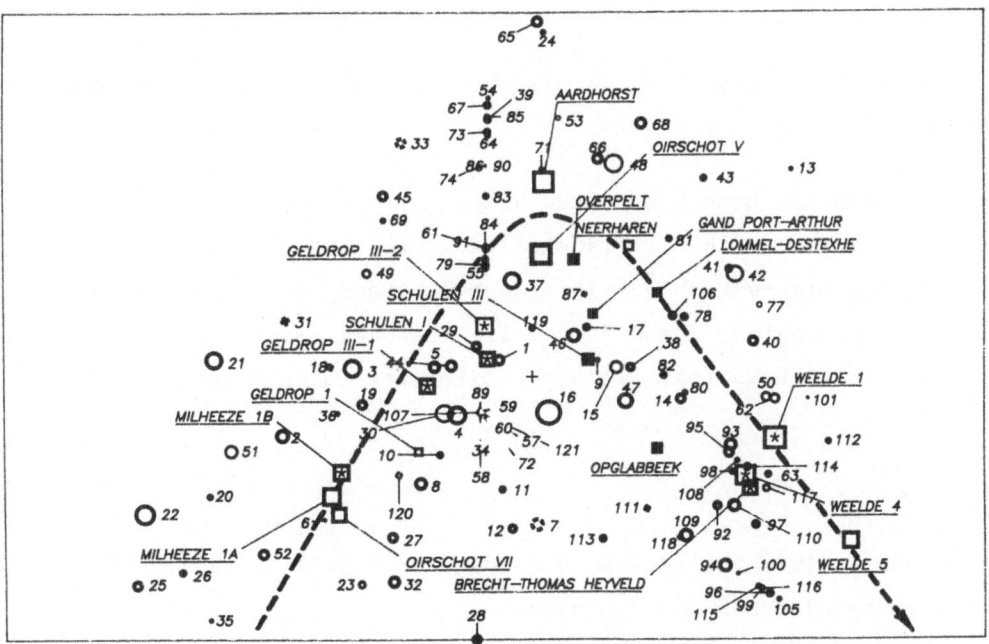

Fig.5

Only these have been included into the analysis while the others have been positioned afterwards on the result. Hence, our new analysis bears on 113 objects and 7 functional settlements.

The result is shown in figure 5, which reproduces the time-line correctly. Settlements, which have been positioned afterwards, are marked with an asterisk on the map. If we compare the seriation along the time-variable of the map, we find a close approximation to the previous one, with a few minor displacements. (Schulen-1 and Schulen-3 are closer).

## 7. Discussion and conclusion

Spectral map analysis of artifacts and settlements produces a continuous time-variable. This variable allows to seriate most of the sites with known radiocarbon dating. Although the classification is rough, it is compatible with the known variability of the dating. The analysis revealed one probable error in the dating.

The result appeared to be very robust with respect to aggregation and division of the data.

Although the number of dated sites is limited (seven out of nineteen) we proved that it is sufficient for the construction of a reproducible time-variable. Three epochs stand out clearly along the time-line at about 9, 7 and 5 Ky. The approach also allows to identify those artifacts that are best suited for dating prehistoric settlements.

## References

J.P.Benzécri (1973). L'analyse des données. Vol.II. L'analyse des correspondances, Dunod, Paris.

P.J.Lewi (1982). Multivariate analysis in industrial practice. Research Studies Press (J.Wiley), Chicester, Engl.

P.J.Lewi (1984). Multivariate data representation in medicinal che-
    mistry. In : Chemometrics. Mathematics and statistics in che-
    mistry. B.R.Kowalski, Ed., Reidel Publ., Dordrecht, The Neth.,
    355-376.

P.J.Lewi (1985). Multivariate and longitudinal data on growing
    children, analyzed by Spectramap. Proceedings 4th Symposium on
    Data Analysis, J.Janssen e.a., Eds., Plenum, London, pp in this
    volume.

J.-L.Slachmuylder (1985). Prehistoric and lithic artifacts from a
    small west-european area.  Problem submitted to the 4th Sympo-
    sium on Data Analysis, Brussels.

SECTION 2.4

CONCLUSION

J. G. Rozoy

Charleville, France

From my point of view as a prehistorian, the mathematical
methods of data analysis, applied to chronological seriation, have
demonstrated here brightly their validity, their utility and their
coherence : The quasi-similarity of the results of two different
systems is striking.  It is also the effect of the choice of the
method proper to that specific problem.  The demonstration of the
fastness of the time-variable by the study of separated lots of
types is also noticeable, and quite useful.  Not only these procedures reproduce the time order on which we had no doubts, and even
in details, but also they give us, either confirmations of decisions
on points which are sometimes strongly discussed, either even the
beginning of new quite promising progresses.

Confirmations : the possibility (sometimes denied) to give
dates through typology; regularity in the evolution of the types of
armatures within a same prehistorical era;  the difficult character
in transition periods;  the more chronological value of armatures
and more common value of common tools;  the aberrant radiocarbon
datation of Oirschot VII;  the less accuracy in short series (around
100 tools).  Confirmation also, which I personally appreciate much,

169

in the operational value of the proposed typology and more especially of its grouping into classes.  It is here validation by external
criteria, which is particularly precious as it is added to that of
radiocarbon and should put an end to any kind of polemic.  Of
course, validation is reciprocal.

New progresses : J.L.Slachmuylder's hope, based on the fact
that the speed of evolution seems to be constant within some limits,
to propose relatively accurate dates for sites where there is no
possibility of radiocarbon datation, and also to observe evolutive
accelerations or decelerations in transition periods.  The sociological value of theses processes will have to be discussed, of
course.

For short, the first one of the three problems I set in my
presentation is on the right way to be solved entirely, with lots of
quite valuable subordinate consequences.  The other two remain now,
and, close at hand, the comparisons and differentiations between
regions.  The chronological estimation was a preliminary request to
solve these regional problems.

It is not evidence, at least for non statisticians, that this
new question can be solved by the same mathematical methods as the
previous one.  One can certainly trust those who were able to find
the right procedure the first time, to choose properly the means
which should be applied to the other two themes.  And, then, they
will have to fight on a "terra incognita", as regionalisation has
not been much recognized for the study of two million years;  moreover, the distinctions I made myself for Epipaleolithic are strongly
contradicted by a noticeable part of European searchers.  Mathematical analyses might then act in a decisive manner, allowing at last
to understand far better the (physical and psychic) evolution of
human species.

CHAPTER 3. A COMPARISON OF RESULTS OF EUROPEAN ELECTIONS

PRESENTATION OF THE PROBLEM

J. Janssen, P. Cullus and P. Van Brussel

CADEPS - Brussels

## 1. Aim of the study

The aim of the problem is to make a comparative transnational study of results of two European elections of 1979 and 1984, in order to measure the importance's evolution of the great political groups. Data, which are presented in tabular form, give, for each of the 10 countries of the E.E.C. (Greece, Netherland, England, Ireland, Denmark, Belgium, Italy, Germany, France, Luxemburg), the results (in percentage) of the European elections of 1979 (1st row) and 1984 (2nd row). They also give the results of a reference intermediate national parliamentary election for each country (3rd row) (°).

## 2. Presentation of the data

Political parties have been grouped in 13 political families according to the classification of J. Van Laer (°°). The results of each family are given in the following order :

- Extreme left-wing
- Communists
- Socialists
- Various left-wing

173

- Ecologists
- Regionalists
- "Christian-democrats"
- "Radical-liberal"
- "Conservative-liberal"
- Protestants
- "Néo-poujadistes"
- Extreme right-wing
- Various right-wing

The study could begin with a separate analysis of each election before performing the comparative study.

(°)  Date of reference intermediate elections :
     Greece (10-18-81), Netherland (09-08-82), England (06-09-83),
     Ireland (11-24-82), Denmark (01-10-84), Belgium (11-08-81),
     Italy (09-26-83), Germany (03-06-83), France (04-26-81),
     Luxemburg (06-10-79).

(°°) "200 Millions de voix" : Une géographie des familles politiques
     européennes. J.Van Laer, Laboratoire de Géographie Humaine
     de l'Université Libre de Bruxelles et Société Royale de
     Géographie.

## 3. The data matrix

| Country | Sub | V1 | V2 | V3 | V4 | V5 | V6 | V7 | V8 | V9 | V10 | V11 | V12 | V13 |
|---|---|---|---|---|---|---|---|---|---|---|---|---|---|---|
| GRE | 1 | 0 | 2.0 | 0 | 0 | 31.3 | 0 | 0 | 0 | 0 | 0 | 44.3 | 18.1 | 0 |
|  | 2 | 0 | 2.3 | 0 | 0 | 38.1 | 0 | 0 | 0 | 0 | 0 | 42.4 | 14.9 | 0 |
|  | 3 | 0 | 0 | 0 | 0 | 35.8 | 0 | 0 | 0 | 0 | 0 | 48.7 | 12.6 | 0 |
| HOL | 1 | 0 | 0 | 0 | 3.3 | 16.1 | 0 | 35.6 | 0 | 0 | 9.0 | 30.4 | 1.7 | 3.3 |
|  | 2 | 0 | 0 | 0 | 5.2 | 18.9 | 0 | 30.0 | 0 | 0 | 2.3 | 33.7 | 1.8 | 3.8 |
|  | 3 | 0 | 0 | 0 | 2.7 | 23.1 | 0 | 29.3 | 0 | 1.3 | 4.3 | 30.4 | 1.8 | 3.9 |
| RYU | 1 | 0 | 0 | 0 | 0 | 49.3 | 12.6 | 0 | 1.9 | 0 | 0 | 32.7 | 0 | 0 |
|  | 2 | 0 | 0 | 0 | 0 | 40.8 | 19.2 | 0 | 3.2 | 0 | 0 | 36.5 | 0 | 0 |
|  | 3 | 0 | 0 | 0 | 0 | 42.4 | 25.3 | 0 | 1.1 | 0 | 0 | 27.6 | 0 | 0 |
| IRN | 1 | 67.8 | 0 | 0 | 0 | 0 | 0 | 0 | 0 | 0 | 0 | 14.5 | 0 | 0 |
|  | 2 | 71.8 | 0 | 0 | 0 | 0 | 0 | 0 | 0 | 0 | 0 | 8.4 | 0 | 0 |
|  | 3 | 84.4 | 0 | 0 | 0 | 0 | 0 | 0 | 3.3 | 0 | 0 | 9.4 | 0 | 0 |
| DNK | 1 | 6.2 | 0 | 5.8 | 1.8 | 28.9 | 3.3 | 0 | 0 | 0 | 24.3 | 25.6 | 3.5 | 0 |
|  | 2 | 6.6 | 0 | 3.5 | 2.8 | 33.2 | 3.1 | 0 | 0 | 0 | 22.3 | 28.7 | 1.3 | 0 |
|  | 3 | 4.6 | 1.4 | 3.6 | 0 | 35.5 | 3.5 | 0 | 0 | 0 | 0 | 43.1 | 2.6 | 0 |
| BEL | 1 | 0 | 5.4 | 0 | 0 | 16.3 | 0 | 37.7 | 13.6 | 3.4 | 0 | 23.4 | 2.7 | .9 |
|  | 2 | 0 | 6.5 | 0 | 0 | 18.1 | 0 | 27.4 | 12.1 | 8.2 | 0 | 30.4 | 1.1 | .7 |
|  | 3 | 0 | 6.9 | 2.7 | 0 | 21.5 | 0 | 26.5 | 14.5 | 4.8 | 0 | 25.1 | 2.3 | 1.1 |
| ITA | 1 | 0 | 0 | 0 | 0 | 3.6 | 2.6 | 36.5 | .6 | 0 | 3.7 | 15.3 | 29.6 | 1.8 |
|  | 2 | 0 | 0 | 0 | 0 | 3.1 | 3.0 | 33.0 | 1.6 | 0 | 3.4 | 14.7 | 33.3 | 1.4 |
|  | 3 | 0 | 0 | 0 | 0 | 3.0 | 5.2 | 33.6 | .6 | 0 | 2.3 | 15.9 | 30.5 | 1.5 |
| ALL | 1 | 0 | 0 | 0 | 0 | 49.2 | 6.0 | 0 | 0 | 3.2 | 0 | 40.8 | .4 | 0 |
|  | 2 | 0 | 0 | 0 | 0 | 46.0 | 4.8 | 0 | 0 | 8.2 | 0 | 37.4 | 0 | 0 |
|  | 3 | 0 | 0 | 0 | 0 | 48.8 | 7.0 | 0 | 0 | 5.6 | 0 | 38.2 | .2 | 0 |
| FRA | 1 | 3.2 | 1.3 | 0 | 0 | 43.9 | 0 | 0 | 0 | 4.4 | 0 | 23.5 | 20.5 | 3.1 |
|  | 2 | 2.6 | 11.0 | 0 | 0 | 43.0 | 0 | 0 | 0 | 3.4 | 3.3 | 20.8 | 11.2 | 2.1 |
|  | 3 | 4.0 | 0 | 0 | 0 | 46.3 | 0 | 0 | 0 | 3.9 | 2.2 | 26.8 | 15.3 | 2.3 |
| LUX | 1 | 0 | 0 | 0 | 0 | 0 | 28.1 | 36.1 | 0 | 0 | 0 | 28.7 | 5.0 | .5 |
|  | 2 | 0 | 0 | 0 | 0 | 0 | 21.2 | 35.3 | 0 | 6.1 | 0 | 30.3 | 0 | .4 |
|  | 3 | 0 | 0 | 0 | 0 | 0 | 21.3 | 34.5 | 0 | .9 | 0 | 30.2 | 0 | 0 |

4. Number of participants

|       | 1979      | 1984      | INTER.    |
|-------|-----------|-----------|-----------|
| GRE   | 5753478   | 5956060   | 5530885   |
| HOL   | 5667303   | 5297621   | 8226924   |
| RYU   | 13446091  | 13998188  | 29906280  |
| IRN   | 1339072   | 1120416   | 1665353   |
| DNK   | 1754850   | 2001875   | 3360327   |
| BEL   | 5442867   | 6430779   | 5919250   |
| ITAL  | 35042601  | 35098046  | 36901166  |
| ALL   | 27847109  | 24851371  | 38940687  |
| FRA   | 20242347  | 20180934  | 29038036  |
| LUX   | 170759    | 173888    | 188909    |

SECTION 3.2

THE COMPARATIVE STUDY OF RESULTS OF ELECTIONS

P. Cullus

CADEPS - Brussels

1. <u>Introduction</u>

The data can be taken as a succession of three data matrices, the results of the three elections, crossing the countries (individuals) and the political trends (variables).

Instead of the presentation of three "Normed Principal Component Analysis" * (each by year) we use the technique of the juxtaposition and of the mean table with supplementary variables of individuals to have a better global view of the political evolution during the three elections.

These are a good tool to extract information from this data and gather it by means of graphic output.

2. <u>Presentation of the tables to analyse</u>

Data table  X  is crossing a set of individuals (the 10 countries)

---

* We use the techniques described by Lebart, Morineau and Tabard (1977).

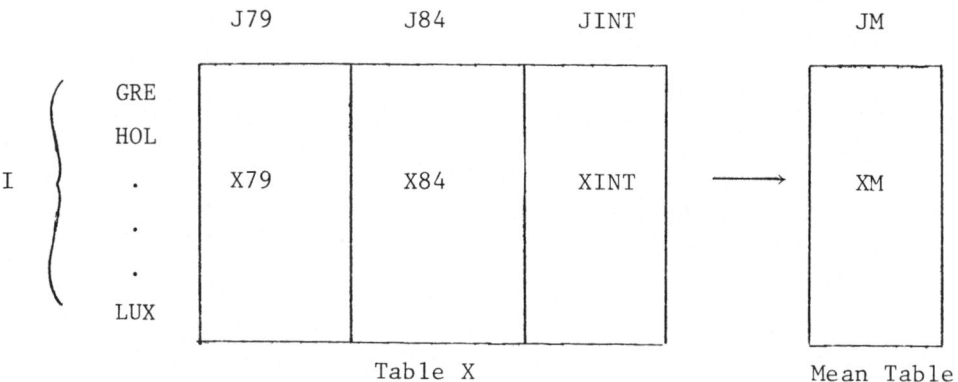

called I, and a set of variables (the 39 political groups), called J. By subdivision of the columns, J is composed by 3 groups : J79, J84 and JINT (representing, respectively, the great political groups results for the elections of 1979, 1984 and for the intermediate national election), so that 3 subtables, X79, X84 and XINT can be defined.

By the same operation on the rows, table X can be transformed into table X' which is crossing a set of individuals ("30" coun-tries), called I', with the set of the 13 political groups. Table I' can also be divided in 3 groups, I'79, I'84, I'INT, representing the countries scores in 79, 84 and for the intermediate national election.

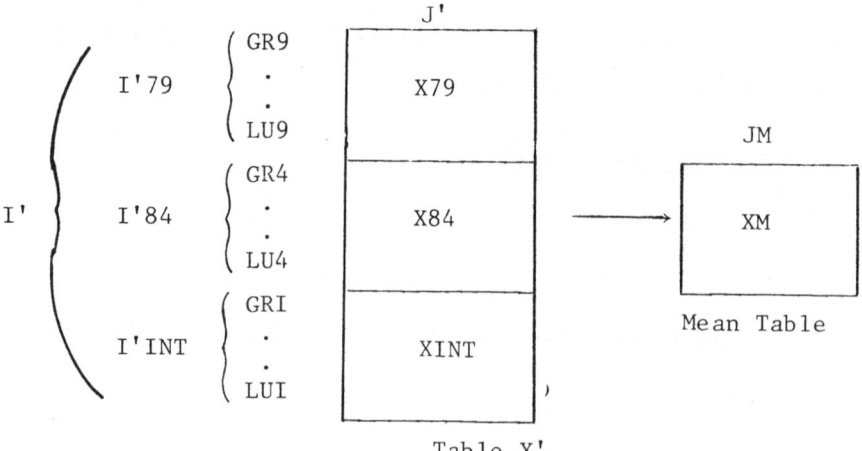

## 3. The factorial analysis

### 3.1. The performed analysis

As basic method for solving the problem, we preferred the "Normed Principal Components Analysis" (NPCA) to the Correspondence Analysis (see Cullus and Janssen, 1985) in order to avoid weighting the tables.  As a matter of fact, the participation to European elections is very different from one country to another one.

#### 3.1.1. The partial analysis

We first performed 3 partial analysis on the different groups, namely 3 NPCA of tables X79, X84 and XINT (see graphics 1, 2 and 3).

#### 3.1.2. The global analysis

- In order to visualize the evolution of the great political groups, two methods haven been used :

+ NPCA of table X (see graphic 4)

+ NPCA of table XM (considering that the 3 tables have the same weight) with J79, J84 and JINT as supplementary variables groups (see graphic 5).

- In order to visualize the evolution of the countries, two methods have also been used :

+ NPCA of table X' (see graphic 6)

+ NPCA of mean table XM (with the same weight for the different tables) with I'79, I'84 and I'INT as supplementary individuals groups (see graphic 7).

The theoretical aspect of those methods have been treated by Cullus (forthcoming).

### 3.2. Summarized presentation of the results

Here, the percentage of variance explained by the factorial axes are given :

Table 3

|              | AXIS 1 | AXIS 2 | AXIS 3 | AXIS 4 | AXIS 5 | AXIS 6 |
|--------------|--------|--------|--------|--------|--------|--------|
| NPCA of X79  | 22.38  | 20.09  | 16.01  | 13.76  | 12.36  | 8.34   |
| NPCA of X84  | 21.53  | 20.54  | 16.59  | 14.85  | 9.28   | 9.00   |
| NPCA of XINT | 28.23  | 19.25  | 14.83  | 13.53  | 10.70  | 7.53   |
| NPCA of XM   | 22.88  | 19.40  | 16.21  | 14.20  | 11.19  | 9.62   |
| NPCA of X    | 22.69  | 17.34  | 15.72  | 14.07  | 11.65  | 10.48  |
| NPCA of X'   | 22.13  | 20.00  | 15.75  | 14.68  | 11.25  | 10.04  |

## 3.3. Interpretation of the axes

a. NPCA of table X79 (see graphic 1)

- Axis 1 : especially expresses the particular and antagonistics
situations of 2 countries : Denmark and Italy.  The "poujadistes"
and various left-wing trends are essentially found in Denmark.  At
the other side, extreme left-wing, communists, "Christian-Democrats"
and extreme right-wing are characteristics of Italy.

- Axis 2 : especially displays the so-called "bipolarisation"
of the electorate in Germany and United-Kingdom, where the "left-
wing" trend is represented by the socialists, the "right-wing" by
the conservative trend.  On the contrary, the splitting up of the
electorate in Denmark, Netherlands and Italy clearly appears.

- Axis 3 : distinguishes Ireland from the other countries.  The
most important political parties there have to be classed in the
various right-wing.

- Axis 4 : explains the particular situation of Benelux countries
(where Christian right-wing, extreme left-wing and Ecologists can be
found in the same time, with an important regionalistic trend in

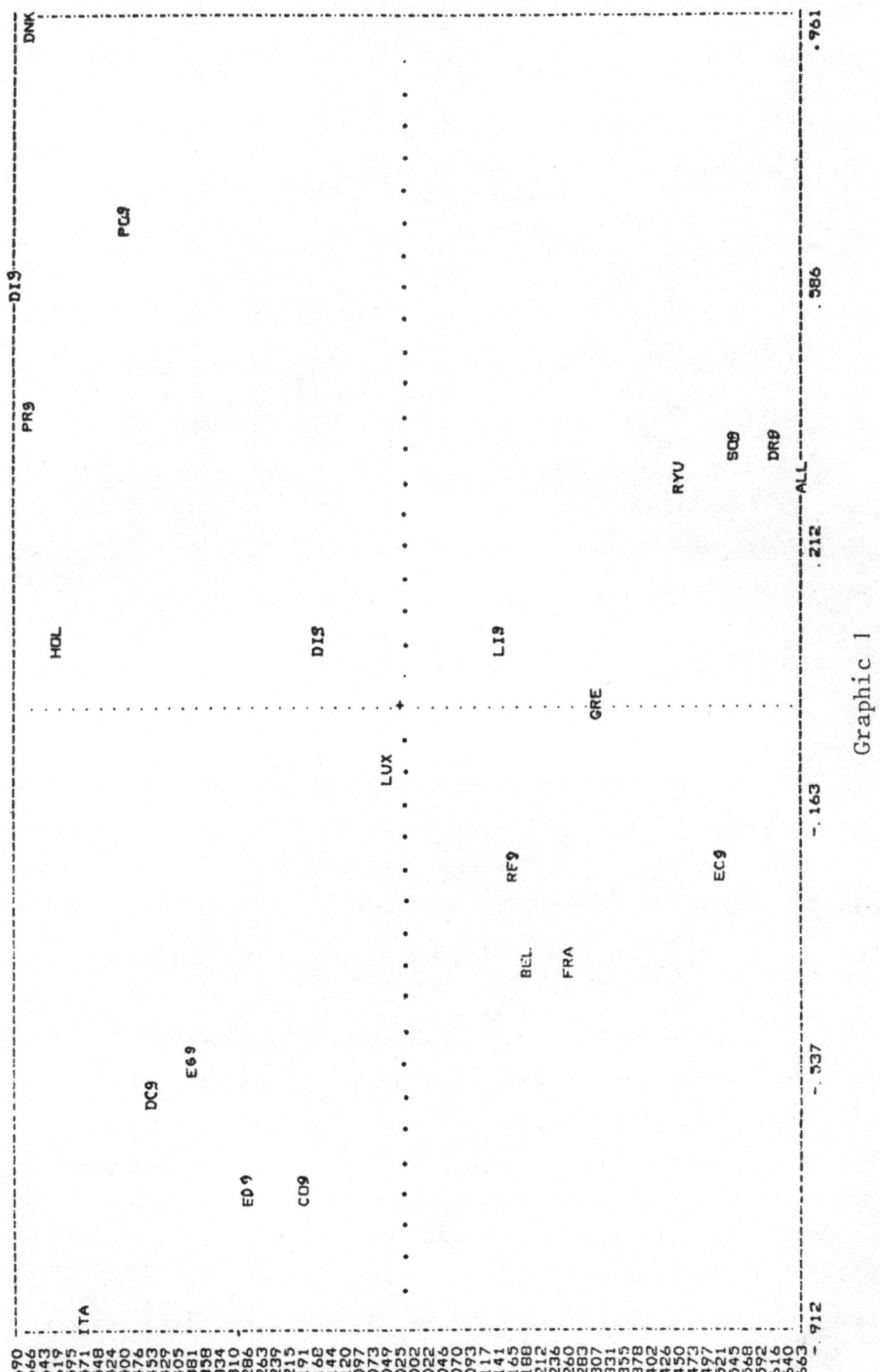

Graphic 1

Belgium).  At the opposite side, those various families are not to
be found in Greece, where the 2 main trends are extreme right-wing
and communists.

b. NPCA of table X84 (see graphic 2)

- Axis 1 : displays again the particular importance of political
trends in several countries of the EEC.  It opposes more clearly
Denmark ("poujadistes" and various left-wing) against the other,
especially Belgium and Italy because of presence of christian, regionalistic and ecologists trends in these countries.

- Axis 2 : can be interpreted in the same way as in 79.  The
presence of ecologists in Germany and Belgium, and the growth of
the extreme right-wing's importance in France have also to be mentioned, so that it brings France and Italy closer together.

- Axis 3 : has the same interpretation as in 1979 : Ireland
stands out clearly by its various right-wing trend.

- Axis 4 : essentially divides the European right-wing.  The
conservative and extreme right is to be found in France and Greece,
the Christian-democrats and Radical-liberals in the countries of
Benelux.

c. NPCA of table XINT (see graphic 3)

Here, the results have to be analysed very carefully, considering the fact that the national elections did not take place in the
same time and that they went off sometimes in a very particular
context (e.g. the results of the various left-wing in Denmark).

- Axis 1 : distinguishes essentially the countries where the
so-called "bipolarisation" is very strong (Denmark, Germany, United-
Kingdom : left-wing represented by the Socialists, right-wing by
the conservative trend) and those where the splitting up of the
electorate appears very clearly (Italy, Netherland, France).  In

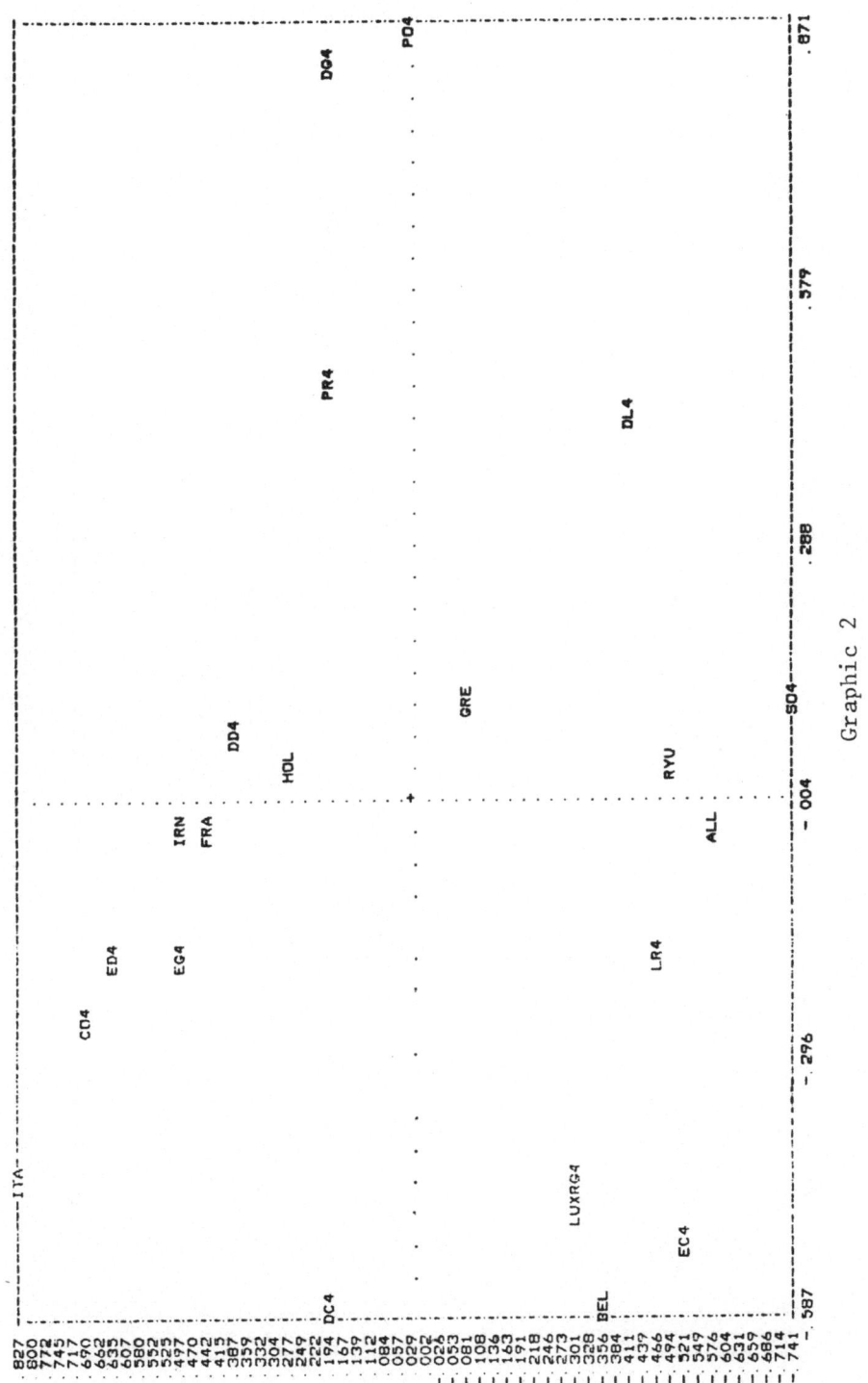

Graphic 2

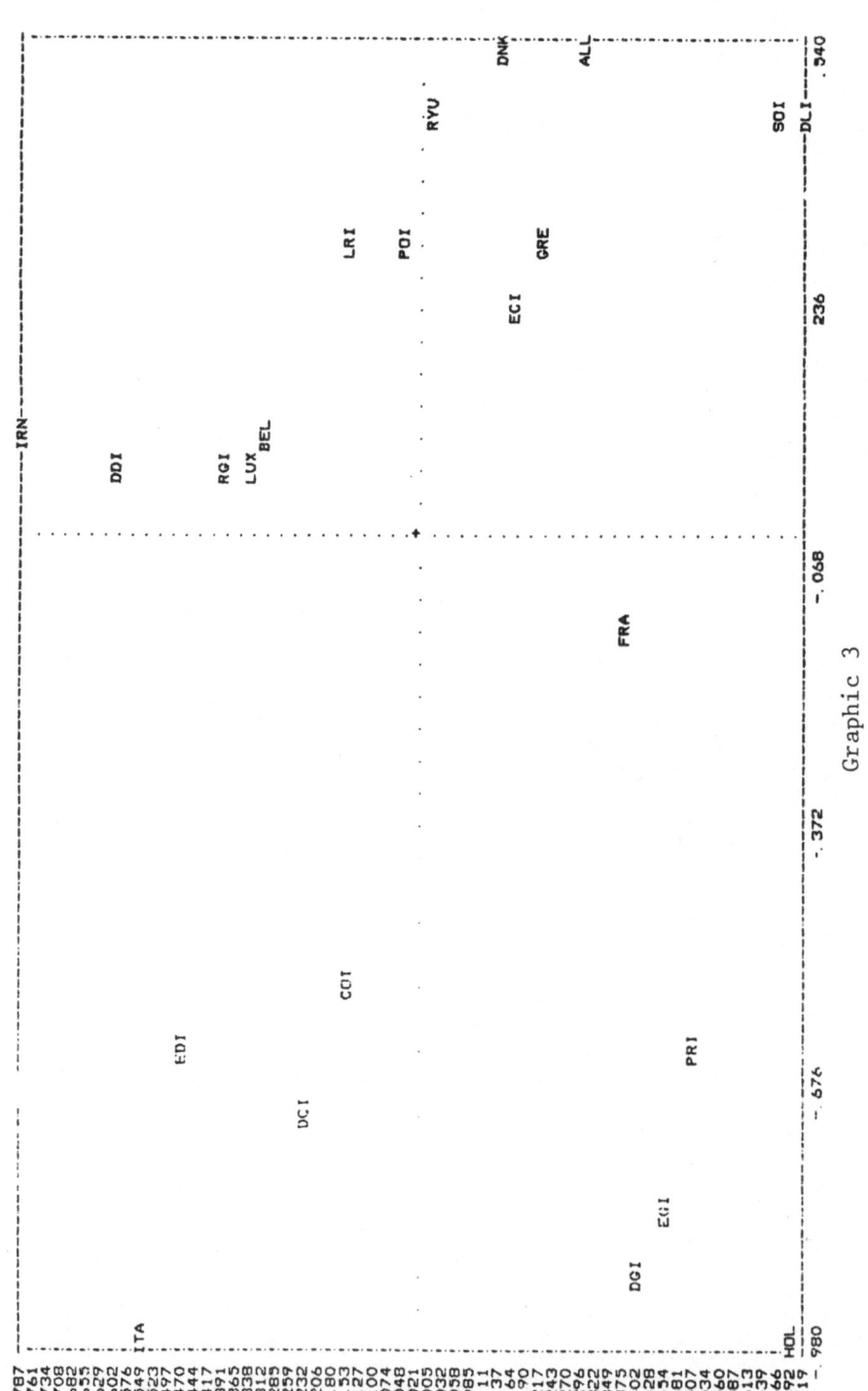

Graphic 3

those 3 countries, the left trend has to include the presence of
communists and extreme left-wing, the right trend that of extreme
right-wing (such as in France and Italy).

- Axis 2 : distinguishes France and Netherlands from Italy.
In the first two countries, conservative and socialist's trend are
more important than extreme right-wing and regionalists which are
more particular to Italy.

- Axis 3 : indicates the very eccentric position of Belgium
where important results of Ecologists and Regionalists are to be
found.

- Axis 4 : opposes the particular situations of Italy (Communists and extreme right-wing) and Ireland (various right-wing).

d. NPCA of tables X and XM

- Axis 1 : displays the very strong "bipolarisation" of Germany,
United-Kingdom and (more weakly) Denmark, where the left trend is
represented by a very strong socialist current and the right trend
(minority party), by the conservative liberals.  The extreme right
or left-wing are almost non-existent.  On the contrary, the splitting
up of the political trends in Italy and, more weakly, in the Netherlands where several various left and right-wing parties can be found,
and where the most important part of the left trend is taken by the
Communists, that of the right trend by the Christian-democrats.

- Axis 2 : distinguishes particular situations of small political parties, such as in Denmark and the Netherlands (where various
left-wing and protestants trends can be found) or on the other side,
in Belgium (Ecologists and Regionalists), Germany, Luxemburg and
United-Kingdom (with a strong presence of "Radical-liberals").

- Axis 3 : displays the particular position of Ireland where
the high score of 2 important parties of the various right-wing

group is the only case of the importance of this trend in the ten
countries.


    - Axis 4 : separates Benelux countries (characterized by the
presence of Liberals, Regionalists, Protestants, Ecologists, extreme
left-wing and Christian democrats) from France, Greece and, more
weakly, Italy (characterized by the presence of Communists, extreme
and conservative right-wing).


Evolution of the political trends (see graphics 4 and 5).


    Graphic 4 displays the transformations resulting from the 3
elections.  Several stable situations can be observed.  In the same
way, a rather big difference is to be observed between the european
elections and the intermediate national ones.  Let us note the evolution of the extreme right-wing, and on the opposite side that of
the Ecologists.
The same remarks can be made concerning graphic 5, where the situation of the political trends are displayed in relation with their
mean.


Evolution of the countries (see graphics 6 and 7).


    The trajectories of the countries can here be observed.  Let us
mention the following facts :
    - a few stable situations : Italy, Ireland
    - a difference between european and intermediate elections :
Denmark, Greece, France, Netherland and Belgium
    - a growth of the importance of the Ecologists during the time,
and the apparition of extreme right-wing in France.
Graphic 7 allows us to the position of the countries in relation
with their mean.

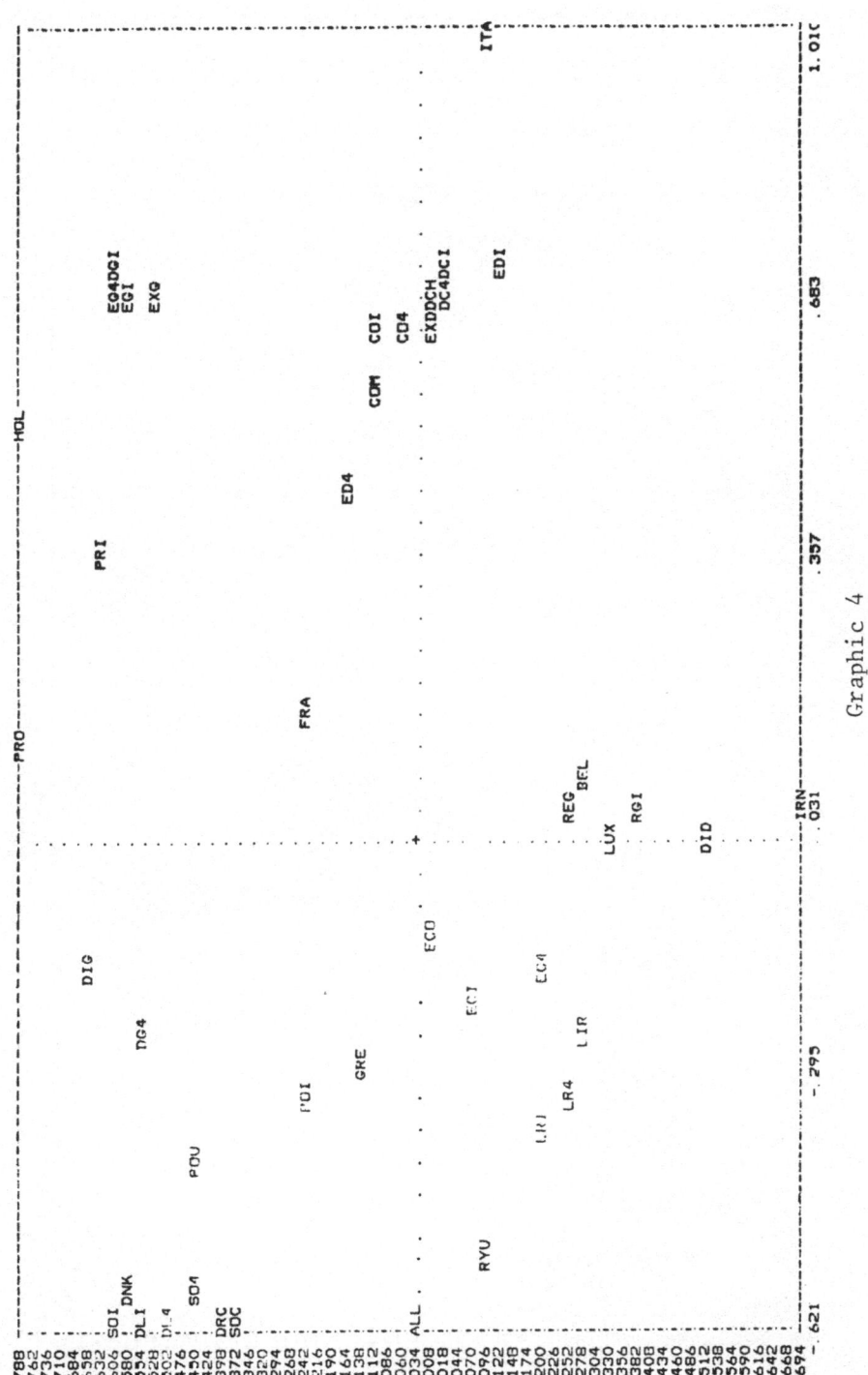

Graphic 4

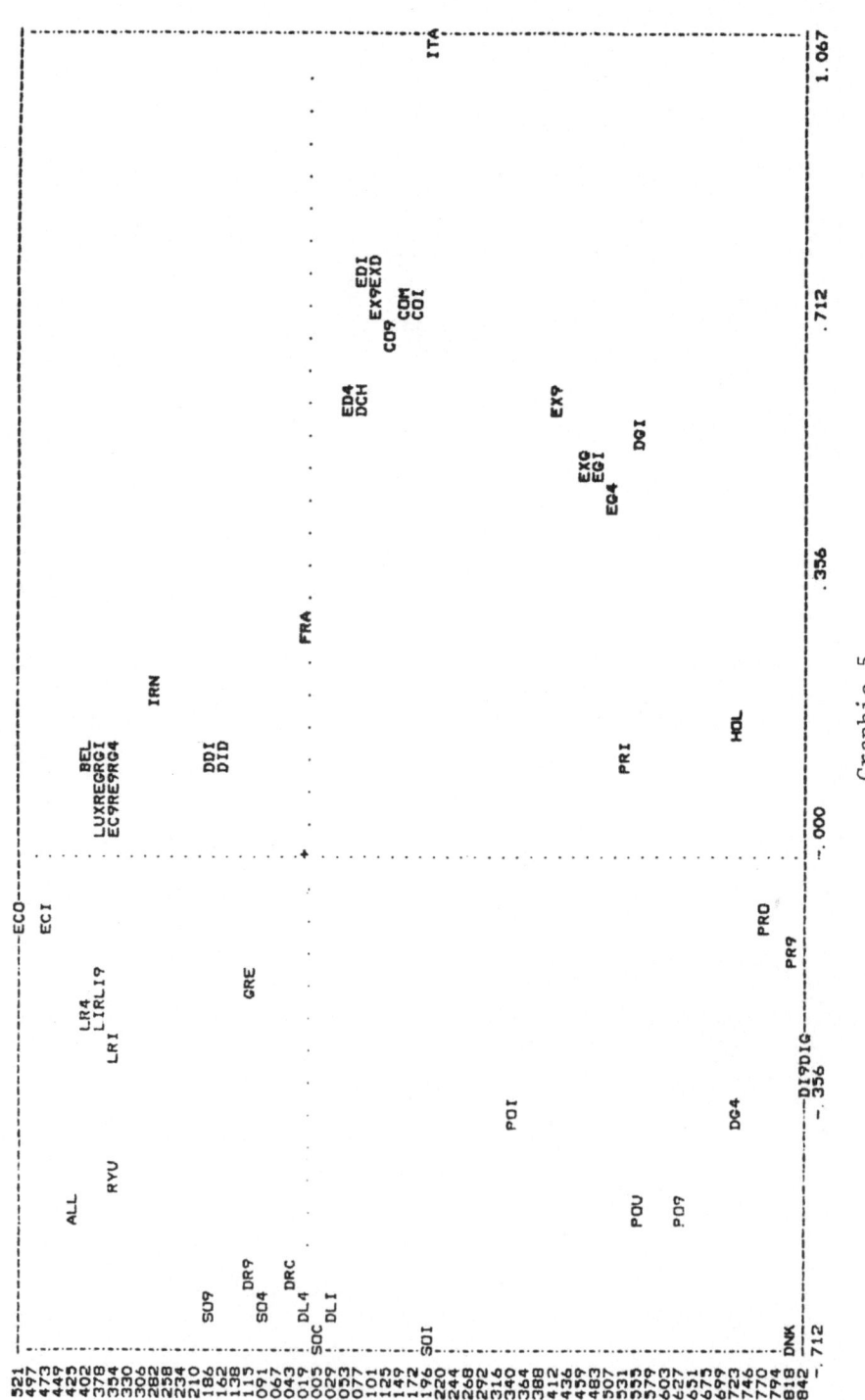

Graphic 5

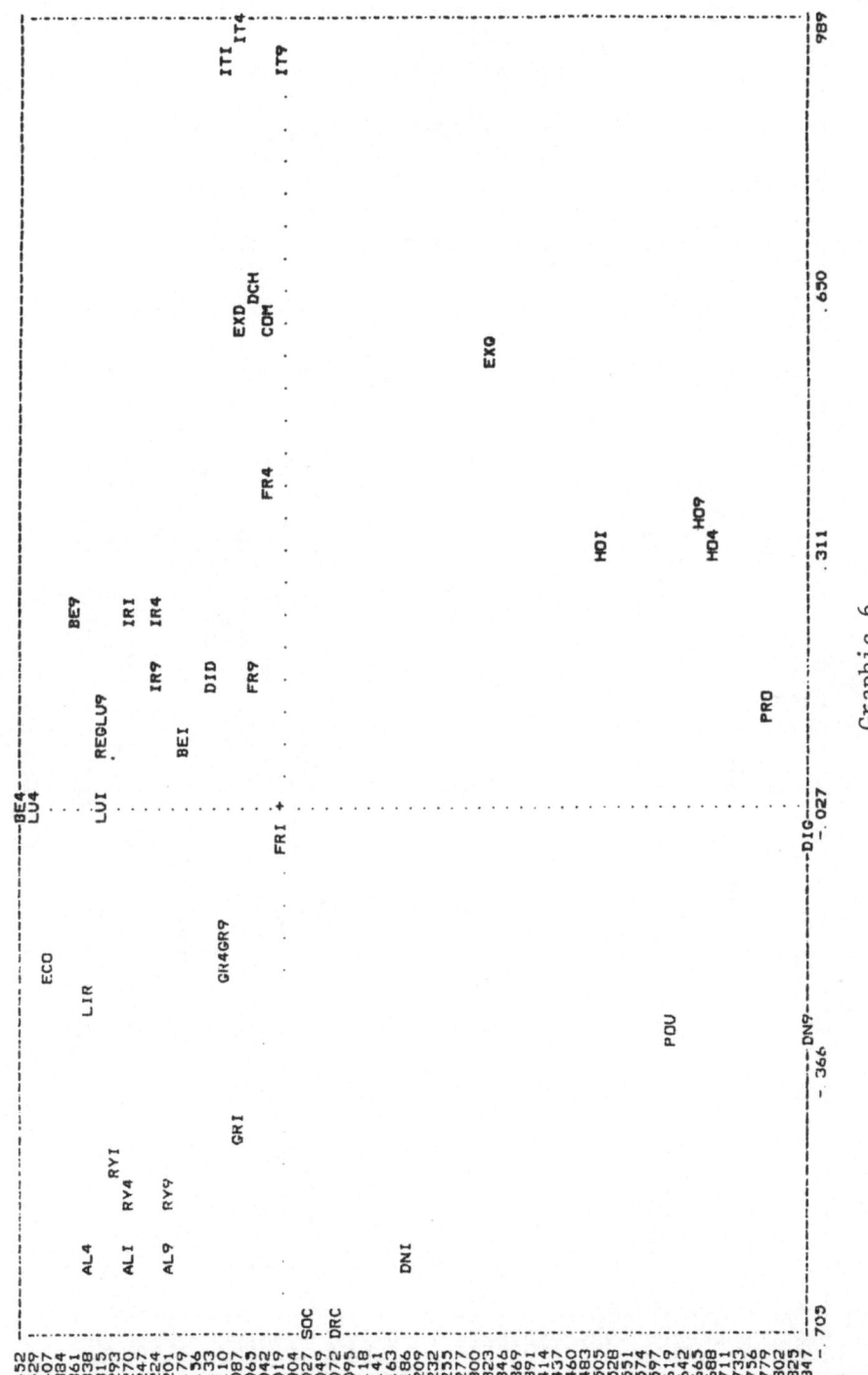

Graphic 6

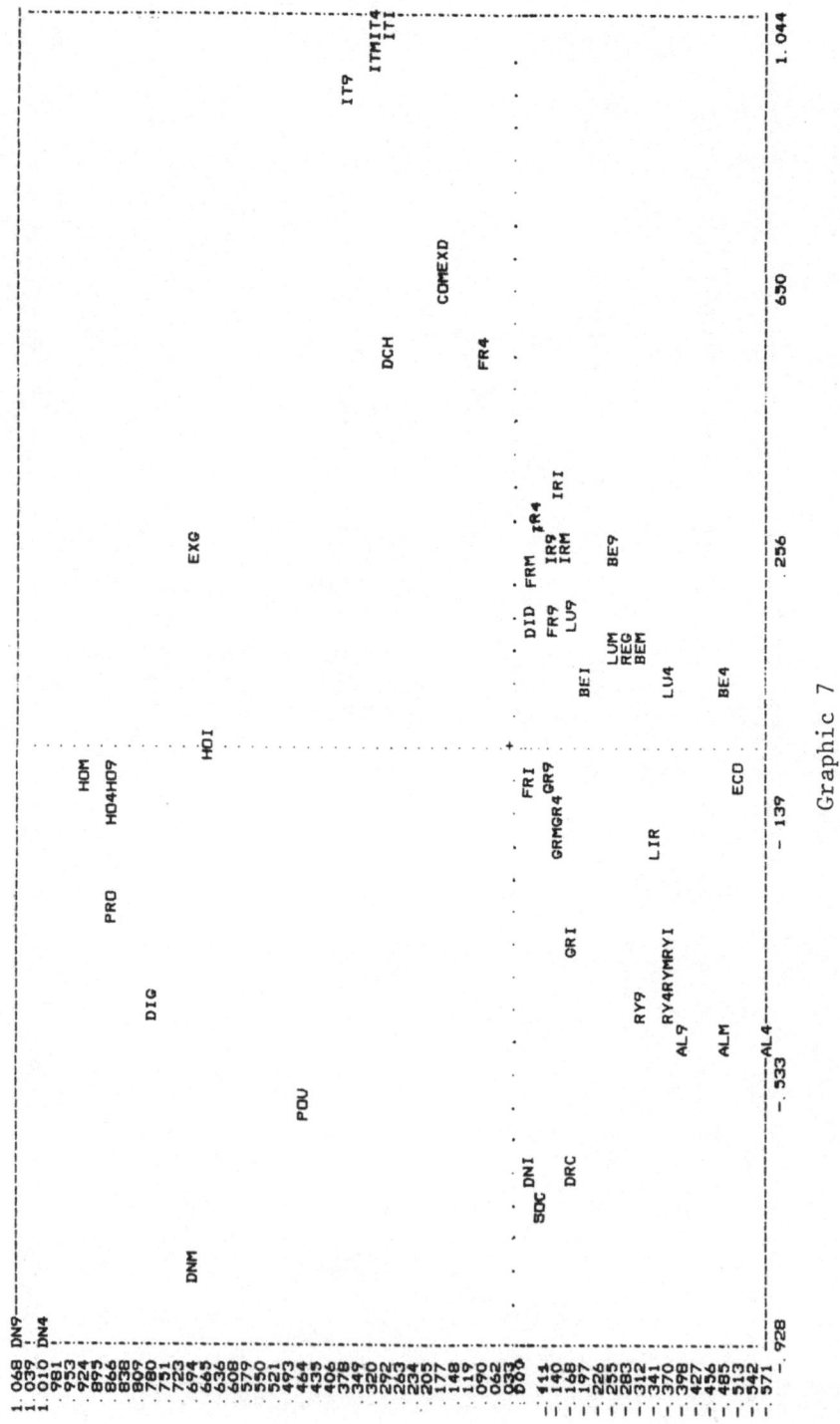

Graphic 7

4. Conclusion

The great trends of the study show that european elections are very close, while intermediate elections have a rather different department due to the presence of rather different electoral lists. At a political point of view, the big trend is the so-called "bipolarisation" of some countries, while splitting up of the electorate appears clearly in others.

It would be interesting to take into account, for further analysis, some explanatory variables such as socio-economic parameters, in order to display the differences observed in our study. Nevertheless, those variables would not be able to explain everything because some historical political context's elements are to be taken into account, so that the study would become more critical.

References

P.Cullus and J.Janssen (1985). A comparative study of results of
    successive Belgian parliamentary elections, in Data Analysis in
    Real Life Environment : Ins and Outs of solving Problems, North-
    Holland, Amsterdam.

P.Cullus. Comparaison de tableaux de données, CADEPS Technical Report
    (forthcoming).

L.Lebart, A.Morineau and N.Tabard (1977). Techniques de la descrip-
    tion statistique : méthodes et logiciels pour l'analyse des
    grands tableaux, Dunod, Paris.

L.Lebart, A.Morineau and J.-P.Fenélon (1979). Traitement des données
    statistiques : méthodes et programmes. Dunod, Paris.

SECTION 3.3

SIMILARITY STRUCTURE ANALYSIS OF EUROPEAN ELECTIONS

L. Guttman and S. Levy

The Hebrew University of Jerusalem and The Israel

Institute of Applied Social Research

The problem is to compare the political party preferences of
the ten European countries of the E.E.C. as expressed in three
election periods.  Two of the points of time are the European elections of 1979 and 1984.  The third period is intermediate to these
for each country, most of the interim elections being held between
1981 and 1983.  (The data are given by Janssen et al.).

The main political parties for each country were grouped by
thirteen political ideologies according to a system of J.Van Laer,
to enable cross-national comparisons.  Hence we shall speak of
thirteen "parties", as if each ideology was the name of a party.

The ordering of the thirteen parties (ideologies) from political "left" to political "right", according to J.Van Laer, is as
follows :

| | |
|---|---|
| 1 - Extreme left-wing | 8 - Radical-Liberal |
| 2 - Communists | 9 - Conservative-Liberal |
| 3 - Socialists | 10 - Protestants |
| 4 - Various left-wing | 11 - Neo-poujadistes |
| 5 - Ecologists | 12 - Extreme right-wing |
| 6 - Regionalists | 13 - Various right-wing |
| 7 - Christian-democrats | |

Table 1

| | | 31 Extreme left-wing | 32 Communist | 33 Socialist | 34 Various left-wing | 35 Ecologist | 36 Regionalists | 37 Christian democrats | 38 Radical-liberal | 39 Conservative-liberal | 40 Protestants | 41 Neo-poujadistes | 42 Extreme right-wing | 43 Various right-wing |
|---|---|---|---|---|---|---|---|---|---|---|---|---|---|---|
| Greece | 1 | 0 | 18.1 | 44.3 | 0 | 0 | 0 | 0 | 0 | 31.3 | 0 | 0 | 2.0 | 0 |
| | 2 | 0 | 14.9 | 42.4 | 0 | 0 | 0 | 0 | 0 | 38.1 | 0 | 0 | 2.3 | 0 |
| | 3 | 0 | 12.6 | 48.7 | 0 | 0 | 0 | 0 | 0 | 35.8 | 0 | 0 | 0 | 0 |
| Netherlands | 4 | 3.3 | 1.7 | 30.4 | 9.0 | 0 | 0 | 35.6 | 0 | 16.1 | 3.3 | 0 | 0 | 0 |
| | 5 | 3.8 | 1.8 | 33.7 | 2.3 | 1.3 | 0 | 30.0 | 0 | 18.9 | 5.2 | 0 | 0 | 0 |
| | 6 | 3.9 | 1.8 | 30.4 | 4.3 | 0 | 0 | 29.3 | 0 | 23.1 | 2.7 | 0 | 0 | 0 |
| England | 7 | 0 | 0 | 32.7 | 0 | 0 | 1.9 | 0 | 12.6 | 49.3 | 0 | 0 | 0 | 0 |
| | 8 | 0 | 0 | 36.5 | 0 | 0 | 3.2 | 0 | 19.2 | 40.8 | 0 | 0 | 0 | 0 |
| | 9 | 0 | 0 | 27.6 | 0 | 0 | 1.1 | 0 | 25.3 | 42.4 | 0 | 0 | 0 | 0 |
| Ireland | 10 | 0 | 0 | 14.5 | 0 | 0 | 0 | 0 | 0 | 0 | 0 | 0 | 0 | 67.8 |
| | 11 | 0 | 0 | 8.4 | 0 | 0 | 0 | 0 | 0 | 0 | 0 | 0 | 0 | 71.8 |
| | 12 | 0 | 0 | 9.4 | 0 | 0 | 3.3 | 0 | 0 | 0 | 0 | 0 | 0 | 84.4 |
| Denmark | 13 | 0 | 3.5 | 25.6 | 24.3 | 0 | 0 | 0 | 3.3 | 28.9 | 1.8 | 5.8 | 0 | 6.2 |
| | 14 | 0 | 1.3 | 28.7 | 22.3 | 0 | 0 | 0 | 3.1 | 33.2 | 2.8 | 3.5 | 0 | 6.6 |
| | 15 | 0 | 2.6 | 43.1 | 0 | 0 | 0 | 0 | 5.5 | 35.5 | 0 | 3.6 | 0 | 4.6 |
| Belgium | 16 | 9 | 2.7 | 23.4 | 0 | 3.4 | 13.6 | 37.7 | 0 | 16.3 | 0 | 0 | 0 | 0 |
| | 17 | 7 | 1.1 | 30.4 | 0 | 8.2 | 12.1 | 27.4 | 0 | 18.1 | 0 | 0 | 0 | 0 |
| | 18 | 1.1 | 2.3 | 25.1 | 0 | 4.8 | 14.5 | 26.5 | 0 | 21.5 | 0 | 2.7 | 1.4 | 0 |
| Italy | 19 | 1.8 | 29.6 | 15.3 | 3.7 | 0 | 6 | 36.5 | 2.6 | 3.6 | 0 | 0 | 5.4 | 0 |
| | 20 | 1.4 | 33.3 | 14.7 | 3.4 | 0 | 1.6 | 33.0 | 3.0 | 3.1 | 0 | 0 | 6.5 | 0 |
| | 21 | 1.5 | 30.5 | 15.9 | 2.3 | 0 | 6 | 33.6 | 5.2 | 3.0 | 0 | 0 | 6.9 | 0 |
| Germany | 22 | 0 | 4 | 40.8 | 0 | 3.2 | 0 | 0 | 6.0 | 49.2 | 0 | 0 | 0 | 0 |
| | 23 | 0 | 0 | 37.4 | 0 | 8.2 | 0 | 0 | 4.8 | 46.0 | 0 | 0 | 0 | 0 |
| | 24 | 0 | 2 | 38.2 | 0 | 5.6 | 0 | 0 | 7.0 | 48.8 | 0 | 0 | 0 | 0 |
| France | 25 | 3.1 | 20.5 | 23.5 | 0 | 4.4 | 0 | 0 | 0 | 43.9 | 0 | 0 | 1.3 | 3.2 |
| | 26 | 2.1 | 11.2 | 20.8 | 3.3 | 3.4 | 0 | 0 | 0 | 43.0 | 0 | 0 | 11.0 | 2.6 |
| | 27 | 2.3 | 15.3 | 26.8 | 2.2 | 3.9 | 0 | 0 | 0 | 46.3 | 0 | 0 | 0 | 4.0 |
| Luxemburg | 28 | 5 | 5.0 | 28.7 | 0 | 0 | 0 | 36.1 | 28.1 | 0 | 0 | 0 | 0 | 0 |
| | 29 | 4 | 0 | 30.3 | 0 | 6.1 | 0 | 35.3 | 21.2 | 0 | 0 | 0 | 0 | 0 |
| | 30 | 0 | 0 | 30.2 | 0 | 9 | 0 | 34.5 | 21.3 | 0 | 0 | 0 | 0 | 0 |

The ten countries are : Greece, Netherlands, England, Ireland, Denmark, Belgium, Italy, Germany, France, Luxemburg.

The data matrix is in Table 1 above (copied from Janssen et al.) It gives the percentage of the total vote for each party for each of the three points of time. The data matrix consists of proportions arrayed into thirty rows (three for each of the ten countries) by thirteen columns (political parties). The total for each row is in principle 100, representing the total vote over the thirteen parties

for the country at the given election.  The data design can be expressed by the following mapping sentence :


MAPPING SENTENCE FOR THE OBSERVATIONS

The proportion of the voters of country [c] for party (ideology)

$$[p] \quad \text{in the election at period} \left\{ \begin{array}{l} 1979 \\ \text{intermediate} \\ 1984 \end{array} \right\} \rightarrow \left\{ \begin{array}{l} 0 \\ \text{---} \\ 100 \end{array} \right\} \text{proportion.}$$


The problem is to portray the matrix graphically in a manner
which will bring out the differential trends over time of the political orientations of the ten countries.  To this end, the percentages of the data matrix will be treated as <u>similarity coefficients</u>:
the higher the proportion voting for a party, the more similar or
"closer" are the party and the country.  Absence of a party from a
country is represented by a zero proportion, or the greatest dissimilarity.  Given this interpretation, an appropriate data analytic
technique is Similarity Structure Analysis (SSA : formerly called
Smallest Space Analysis. See Guttman 1968; Lingoes, 1968, 1973;
Runkel and McGrath, 1972).  In such a geometrical approach, each of
the 43 items - namely, 30 country elections and 13 parties - is represented as a point in a Euclidean space.  The higher the proportion
for a party in an election, the closer the election point is to the
party point in the space.  The space of smallest dimensionality is
calculated which will permit such an inverse relationship between
geometrical distance and similarity coefficients.

## 1. Analysis of a rectangular data matrix

As already mentioned, the data matrix here is not only asymmetric, but rectangular : 30 rows by 13 columns.  But the SSA output
supplies a square and symmetric matrix of distances, with 43 rows
and columns.  The graphic output is of 43 points located in an m-
dimensional space (m  minimal);  and there is a (symmetric) distance

between each pair of points, or 43 × 43 distances.  We are thus
dealing with a special case of SSA denoted by SSAR- Similarity
Structure (Smallest Space) Analysis for Rectangular matrices.

In SSAR, similarity coefficients are given only <u>between</u> rows
and columns of the data matrix, and not <u>within</u> rows nor <u>within</u>
columns.  Regardless, distances are <u>produced</u> within rows and within
columns.  This raises both theoretical and algorithmic questions.
We shall discuss these briefly in the context of the present voting
problem.

The given election data illustrate these properties.  They
relate the thirty election items (10 countries at 3 points of time)
with the thirteen parties.  There is no direct relationship given of
one country with another country, nor between elections in the same
country, nor of one party with another party.  The input data are
only for the relationships between the two different kinds of objects
– elections vis a vis parties – as shown schematically in figure 1
below :

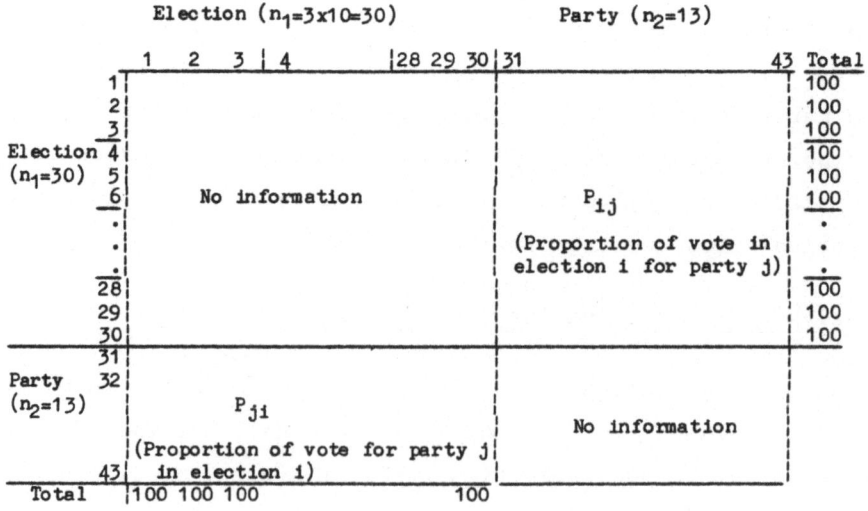

Fig.1

The fact that each country had three elections is indicated in
figure 1 by the breaks within the first 30 captions.  The first three
elections (1,2,3) were for the country listed first, elections 4-6
for the next country, etc., and elections 28-30 for the tenth country.  The SSA per se takes no cognizance of this; but knowing the
design of the elections enables us to consider each country as a
trajectory over its three points of time.  This will be seen later
in the output.

The parties are labelled from 31 (extreme left-wing) to 43
(various right-wing), as presented in the data matrix above.

The main diagonal submatrices - within the 30 elections and
within the 13 parties - are of "missing" information, and given
weight zero in the WSSA algorithm, making this equivalent to SSAR-I.

## 2. The three dimensional (cylindrical) portrayal

According to SSAR results, the correspondence between political
parties and the E.E.C. countries is revealed within the framework of
a cylinder-like three-dimensional structure.  This correspondence
is seen more clearly by going on to four dimensions, the extra dimensionality allowing for "noise", as presented in figures 2a and 2b
below.

Figure 2a is a projection on two dimensions out of four, revea-
ling the axis of the cylinder by discriminating between two strata
along the axis : elections and parties.  This distinction is basical-
ly technical, separating the rows of the rectangular input matrix
(elections) from columns (parties).  Looking at the projection on the
remaining two dimensions implies "holding constant" the first projection - and hence the axis - and turns out to allow for a clearer
content correspondence with regions as in figure 2b.

It should be noted, however, that even allowing for "noise" in

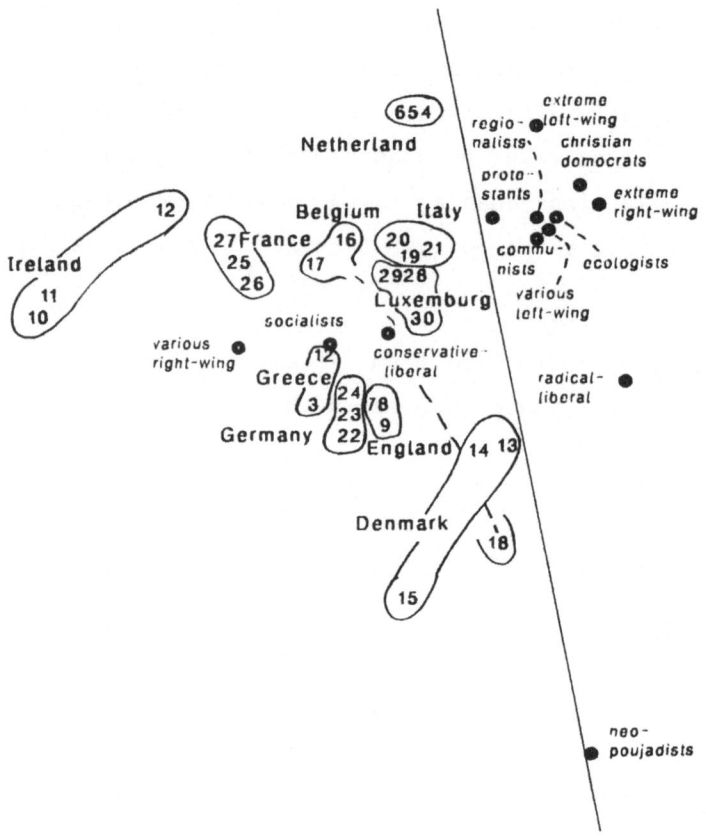

Fig.2a

figure 2a, three of the parties are located in the region of elections, namely the socialist, conservative-liberal and right-wing
parties. This indicates that these parties have special rôles in
their respective countries. There are not enough data here for
going on to higher dimensionality and trying to explicate such special features. Here we may merely note that the socialist ideology
is especially strong in Greece and Germany, and correspondingly this
party is located very close to Greece and Germany in the region of
elections. Similarly, the fact that various right-wing parties pre-
vail in Ireland is reflected by their location in figure 2a.

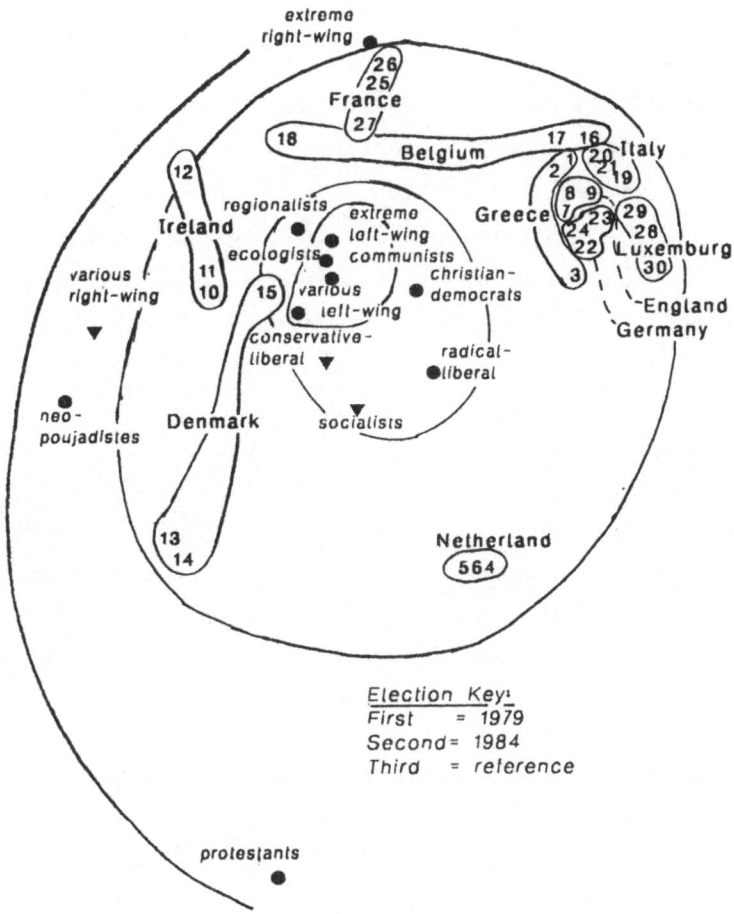

Fig.2b

The situation with respect to conservative-liberal is more complex. This party is especially strong not only in England, but also in Germany and France. Figure 2a draws it especially close to England; the remaining projection suffices to portray the strength of relation to Germany and France.

All told, then, discrimination between elections and parties in the first two-dimensional projection is not purely technical, but is confounded by special dominance of certain ideologies in some of

the E.E.C. countries not sufficiently revealed in the second projection.

The distinction between rows and columns – elections and parties –
appears again in the second projection (figure 2b), but this time in
a modulating – not axial – fashion.  The right-wing parties appear in
an outer ring, the remaining parties in a central circle;  and the
countries (as triplets of elections) in the intermediate ring.  There
is further modulation from "left" to "right" in the central region.
In the very inner circle, close to the origin, are located the extreme
"left" parties.  The next band contains various kinds of "central"
political parties including socialist which are not extreme-left.
Among these are Christian-democrats, liberals, and regionalists.

The triplets of points of the ten countries are located in different directions of the intermediate band, as marked by heavy lines.
Thus, it is possible to see the trajectory of the votes of each
country over the three points of time.

As already mentioned, the three dimensional structure revealed
in the two two-dimensional projections of figures 2a and 2b is cylinder-like.  The actual structure can be sketched schematically as in
figure 3.  At the top of the structure is the outer band of right-
wing parties, which actually occupy only one side of the band.  Also
at the top of the structure in the innermost circles, are the left-
wing and central political parties.  Three of the parties also stretch
down into the region of the countries and their elections.  The election region is a hollow cylinder, with the ten countries going off
into different directions from the axis, being "pulled" outwardly or
inwardly according to their inclinations to the right wing and left
wing parties.  The details are shown in the projection for figure 2b.
In figure 3, a further differentiation among the countries is shown
vertically along the axis, according to the special emphasis on the
three parties which required this third dimensions.  The details are

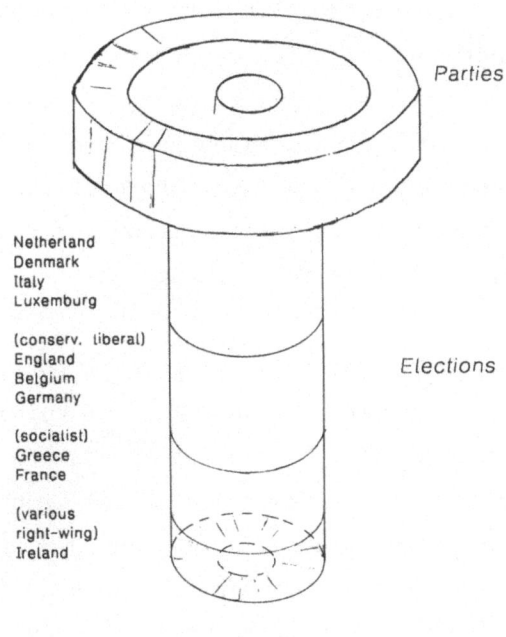

Parties

Netherland
Denmark
Italy
Luxemburg

(conserv. liberal)
England
Belgium
Germany

(socialist)
Greece
France

(various
right-wing)
Ireland

Elections

Fig.3

given in the projection of figure 2a.

## 3. The differential meanings of rows and columns

It should be remarked here that usually SSAR poses a methodological substantive problem.  The rectangular data matrices encountered
in practice usually have a different meaning for rows and for columns.
For example, supposing the columns of table 1 were of export rates
of thirteen different commodities for the countries and year listed
there.  SSAR would portray both the commodities and the countries as
points in the same space.  But commodities and countries are two different kinds of entities.  Treating them as points in the same space
disregards their substantive differences, and may not be the optimal
way for studying their interrelationships.  For example, Multidimensional Structuple Analysis (MSA) treats the rows of a rectangular
matrix as points in a space, but columns as partitions of the space -
the categories of the columns corresponding to regions of the space.

Other ways of looking at a rectangular matrix are possible which
distinguish sharply between the differential meanings of the rows
and the columns. Some popular computer techniques fail to make this
distinction, including correspondence-analysis and variations thereon,
wherein rows and columns are treated as vectors of the same space.
There may be no proper scientific justification for doing so, and
the data analysis may not be the most appropriate.

    In the present example, there is a rationale for portraying
rows and columns as if they were the same kind of thing. Both parties
and countries are social groups, so the interpretation of any point
in the joint space is that it represents a social group. The distances in the output - as in figures 2a, 2b, and 3 - are always
between social groups, no matter whether they have a row label or a
column label. Accordingly, the seeming break of modulation in figure
2b is more apparent than real. The apparent break is between the
right wing parties in the periphery and the left parties in the
middle. However the countries are mixtures of these extremes, and
hence should fall in between them - as they do in figure 2b.

    A further, more technical comment, is on the algorithm for SSAR
itself. Often, the rows appear as points grouped around the center
of the space and columns further in the periphery (or vice versa).
This may be an indication of using too low dimensionality, no matter
what the coefficient of alienation, stress, or any other index of
fit may say. The technical feature of "missing information" illustrated in figure 1 plays a dominant rôle in estimating (or misestima-
ting) goodness of fit. The "unknown" elements may be more numerous
than those in the given rectangular matrix, especially when there is
a different number of rows than columns. As in our example, at least
two dimensions may be required to eliminate the technical difference
between rows and columns. The number of further dimensions required
depends on the substantive nature of the problem. Thus, the study of

rectangular matrices SSAR-wise may generally be expected to require four or more dimensions.

## 4. The political trends

Returning to the election problem proper, we can say that some countries are characterized by a difference between their European and national political trends. This stands out especially for Denmark, Ireland, Belgium and to some extent France. While the trend for Denmark and France is from European political right to internal political left, the political trends for Ireland, and to some extent for Belgium, are in the opposite direction, namely : from political left to political right. It should be noted, that for Belgium - unlike Ireland - the internal political trend towards the right is rather from the political "center" than from political "left". The remaining countries are not characterized by such differences between their European and national political trends.

As already mentioned, the countries are distributed towards the different directions of the political parties, showing the relations between countries and parties. Countries at the right hand of figure 2b are oriented mainly towards political center parties, especially Christian-democrats, liberals, and to some extent socialist parties. These countries are England, Germany, Luxemburg, Belgium (the European elections), Italy and Greece. Hence, six out of the ten E.E.C. countries under research have similar political "center" trends, with almost no difference between internal and European political trends - except for Belgium, which internally turns more towards political right. France, Ireland and Denmark relate towards political right, though - as described above - both Denmark and France are nationally more oriented towards political left. The remaining country - the Netherlands - is quite remote from the others, and is located between "Center" parties, including socialist on the one hand and religious oriented parties (Protestants) on the other, with a stability over time. Indeed, Netherlands and to a lesser extent also

Denmark, are the only countries which pull towards the religious political orientations (Protestants).

These two main political trends, differentiating among the "Center-socialist" and political "right" countries, were also demonstrated previously in the analysis of the technical separation between countries and parties (figure 2a) for the few countries in which these orientations played an especially strong rôle.

To sum up, the SSAR technique revealed the political European and national trends in the E.E.C. countries over time, by technically separating the countries from parties on the one hand, and by relating countries to different types of parties on the other.

References

L.Guttman (1968). A general nonmetric technique for finding the smallest coordinate space for a configuration of points. Psychometrika, 33, 469-506.

J.C.Lingoes (1968). The multivariate analysis of qualitative data. Multivariate Behavioral Research 3, 61-94. Also in J.C.Lingoes, E.E.Roskam, I.Borg (Editors), Geometric Representations of Relational Data. Mathesis Press, Ann-Arbor, 1979, 575-608.

J.C.Lingoes (1973). The Guttman-Lingoes Nonmetric Program Series. Ann Arbor, Michigan : Mathesis Press.

P.J.Runkel and J.E.McGrath (1972). Research on Human Behavior : A Systematic Guide to Method. New York : Holt, Rinehart and Winston.

J.Van Laer (1984). 200 Millions de Voix. Une Géographie de Familles Politiques Européennes. Laboratoire de Géographie Humaine de l'Université Libre de Bruxelles et Société Royale Belge de Géographie.

SECTION 3.4

CONCLUSION

P. Van Brussel

CADEPS, Brussels, Belgium

Both approaches of the problem issued on the two following main facts :

- the similarity of the results of the two European elections and their differences for some countries (F, DK, B), with the intermediate national election.

- the grouping of electors around centrist oriented political parties, especially in countries (GB, D, Lux) where only small differences exist between European and national elections (cfr. Guttman).  This fact is the so-called "bipolarisation" of the voting people.  Its consequence for the concerned countries is that right and left oriented political side, a large diversity of electors appears in the Netherlands (cfr. Cullus and Guttman) and in Italy and Denmark (cfr. Cullus).

Several supplementary informations are given by interpreting the PCA axes, especially concerning the presence of political parties of small importance.. On the other side, Buttman's solution seems to be more condensed and easier to read.

205

CHAPTER 4. CLASSIFICATION OF HETEROGENEOUS DATA
RELATED TO MICROCOMPUTERS

PRESENTATION OF THE PROBLEM

S. Chah

Centre Scientifique IBM-France

36, av. R.Poincaré, F-75116 Paris, France

## The problem

Forty Microcomputers described by 10 qualitative variables and 4 quantitative variables are given.  The problem is how to cluster this set of heterogeneous data.  The informations (description variables) has been collected in a special issue of "L'Ordinateur Individuel Revue 84/85".

## The qualitative variables

If no information is provided then :

The "0" modality indicates that the feature does not exist;

The "1" modality indicates that the feature is optional;

The "2" modality indicates that the feature exists.

1.  Color Monitor : {0,1,2}.
2.  Disk Operating System CP/M : {0,1,2}.
3.  Disk Operating System MS-DOS : {0,1,2}.
4.  Disk Operating System "Other" : {0,1,2}.
5.  Processor : {1(8 Bits), 2(16 Bits), 3(32 Bits)}.
6.  Parallel Interface : {0,1,2}.
7.  Serial Interface : {0,1,2}.
8.  IEEE 488 Interface : {0,1,2}.

9.  Hard Disk : {O(if it does not exist),1(5 Mo),2(10 Mo)}.

10. Number of diskette drives : {1,2}.

The quantitative variables

1.  The prices (F.F.).

2.  Random Access Memory : configuration (Kb).

3.  Random Access Memory : maximum (Kb).

4.  Mass Storage : Diskette Unit (Kb).

Table of data

| | | | | | | | | | | | | | | |
|---|---|---|---|---|---|---|---|---|---|---|---|---|---|---|
| PAP | 0 | 0 | 2 | 0 | 2 | 1 | 1 | 1 | 0 | 1 | 20000 | 192 | 512 | 720 |
| QX 10 | 1 | 2 | 0 | 0 | 1 | 2 | 1 | 1 | 0 | 2 | 23500 | 192 | 250 | 320 |
| MACINTOSH | 0 | 0 | 0 | 2 | 3 | 0 | 2 | 0 | 0 | 1 | 26000 | 128 | 512 | 400 |
| TI PC | 2 | 2 | 2 | 0 | 2 | 2 | 1 | 0 | 0 | 1 | 26300 | 128 | 768 | 320 |
| PAP (2) | 0 | 0 | 2 | 0 | 2 | 1 | 1 | 1 | 0 | 2 | 27200 | 192 | 512 | 720 |
| APRICOT | 0 | 0 | 2 | 0 | 2 | 2 | 2 | 1 | 0 | 2 | 28400 | 256 | 768 | 315 |
| Z 150 | 0 | 0 | 2 | 0 | 2 | 2 | 2 | 0 | 0 | 2 | 28500 | 320 | 640 | 360 |
| GOUPIL 3 | 0 | 2 | 0 | 0 | 1 | 2 | 2 | 0 | 0 | 2 | 29700 | 64 | 1024 | 360 |
| APPLE 3 | 0 | 0 | 0 | 2 | 1 | 1 | 2 | 1 | 1 | 1 | 35000 | 256 | 256 | 140 |
| TANDY 2000 | 1 | 0 | 2 | 0 | 2 | 2 | 2 | 0 | 0 | 2 | 30200 | 128 | 768 | 720 |
| IBM PC | 1 | 1 | 2 | 0 | 2 | 2 | 1 | 0 | 0 | 2 | 36100 | 128 | 640 | 320 |
| TI PC (2) | 2 | 2 | 2 | 0 | 2 | 2 | 1 | 0 | 0 | 2 | 39000 | 256 | 768 | 320 |
| APPLE 2E | 1 | 1 | 0 | 2 | 1 | 1 | 1 | 1 | 1 | 1 | 39400 | 128 | 832 | 140 |
| TELE PC | 1 | 0 | 2 | 0 | 2 | 2 | 2 | 1 | 2 | 1 | 59200 | 256 | 640 | 360 |
| PAP (3) | 0 | 0 | 2 | 0 | 2 | 1 | 1 | 1 | 2 | 1 | 47400 | 192 | 512 | 720 |
| IBM PC XT | 1 | 1 | 2 | 0 | 2 | 2 | 1 | 0 | 2 | 1 | 51000 | 128 | 640 | 320 |
| Z150(2) | 0 | 2 | 2 | 0 | 2 | 2 | 2 | 0 | 2 | 2 | 51500 | 320 | 640 | 360 |
| TANDY 2000 (2) | 1 | 0 | 2 | 0 | 2 | 2 | 2 | 0 | 2 | 2 | 52200 | 128 | 768 | 720 |
| VICTOR S1 | 1 | 2 | 2 | 0 | 2 | 2 | 2 | 1 | 2 | 2 | 66000 | 256 | 896 | 1228 |
| T 200 | 0 | 2 | 0 | 0 | 1 | 2 | 2 | 0 | 0 | 2 | 22500 | 64 | 64 | 256 |
| AS 100 | 2 | 1 | 2 | 0 | 2 | 2 | 2 | 2 | 0 | 2 | 32000 | 128 | 512 | 640 |
| MZ 35 | 1 | 0 | 2 | 2 | 1 | 2 | 2 | 0 | 0 | 2 | 34000 | 136 | 372 | 400 |
| BASIS 108 | 1 | 0 | 0 | 2 | 1 | 2 | 2 | 1 | 0 | 2 | 28500 | 384 | 384 | 160 |

| LISA 2 | 0 | 0 0 0 2 3 2 1 1 0 1 | 35500 | 512 | 512 | 400 |
|---|---|---|---|---|---|---|
| EUROPE PC | 1 | 1 0 2 0 2 2 2 0 2 1 | 47400 | 128 | 1024 | 327 |
| PSI 80 | 0 | 0 2 0 0 1 2 2 0 0 2 | 47800 | 80 | 256 | 308 |
| CORONA PC 2 | 1 | 1 1 2 0 2 2 2 1 2 2 | 45000 | 256 | 512 | 320 |
| OPLITE | 1 | 1 0 2 0 2 2 2 0 0 2 | 33500 | 256 | 640 | 360 |
| HORIZON | 0 | 0 2 0 0 1 2 2 0 0 2 | 35000 | 64 | 576 | 360 |
| FOXY | 1 | 1 1 2 0 2 2 2 1 2 1 | 51000 | 256 | 1024 | 360 |
| SKS 2500 | 0 | 0 2 0 0 1 1 2 1 0 2 | 32000 | 64 | 256 | 800 |
| ZEPHYR | 0 | 0 2 0 0 1 2 2 0 0 2 | 41400 | 64 | 64 | 640 |
| MBC 4050 | 0 | 0 2 0 0 2 2 2 0 0 2 | 35600 | 256 | 1024 | 640 |
| SANCO 8000 | 0 | 0 2 0 0 1 2 2 0 0 2 | 26100 | 70 | 192 | 400 |
| IPC MODEL 15 | 0 | 0 2 0 0 1 0 2 0 0 2 | 43000 | 64 | 512 | 782 |
| DESKTOP 10 | 0 | 0 1 2 0 2 0 2 1 0 2 | 44800 | 128 | 768 | 360 |
| LISA 2-S | 0 | 0 0 0 2 3 2 1 1 1 1 | 47400 | 512 | 1024 | 400 |
| NEC PC 8000 | 0 | 0 0 0 2 1 2 1 1 0 2 | 31800 | 32 | 64 | 320 |
| M 20 | 0 | 0 0 0 2 2 2 1 0 2 | 21600 | 128 | 512 | 286 |
| TRS 80 MODEL 12 | 0 | 0 0 2 1 2 2 0 0 1 | 32000 | 80 | 768 | 422 |

SECTION 4.2

TRIPLED COMPARISON IN AUTOMATIC CLASSIFICATION

S. Chah

Centre Scientifique IBM-France

36, Av. R.Poincaré, F-75116 Paris, France

1. Introduction

In order to classify a given set, assume that the only available information about the resemblance between the elements of the
set to be classified is of the following type : for each triplet
(i,j,k) one knows which of the elements i and j is "more
similar" to k. This structure is called a Triordonnance defined
on the set to be classified. This report deals with new techniques
to solve the above mentioned problem.

For instance, in speech recognition, the triordonnance induced
on a given phonem set by the confusion matrix can be defined in the
following terms : for each triplet (i,j,k) of phonems one knows
which of the pronounced phonems i and j was most confused with
the phonem k. Using this type of information it is possible to
make clustering on the given phonems set.

The case where more than one triordonnance are given is raised
and called the Triordonnance Aggregation Problem. We show that this
model can also be used to solve the problem of heterogeneous variables coding, the problem of association between heterogeneous variables, and finally the problem of classification of a set of data

213

described with qualitative, quantitative and ordinal variables.

## 2. The triordonnance notion

### 2.1. Definition

Let  E  be a set of  n  objects to be classified.  F  represents the set of  E  pairs and  G  denotes the following set :

$$G = \{(\{i,j\},\{j,k\}) \mid i,j,k \in E\}$$

A Triordonnance defined on  E  is a partial preorder relation defined on  F  comparing only couples of type  G.  The set of triordonnances defined on  E  is denoted  $\Omega(E)$.

### 2.2. A triordonnance coding

In order to represent a given  P  triordonnance, we will use
the following coding :

$$T(i,j,k) = \begin{cases} 1 & \text{if } \{i,j\} >_p \{j,k\} \\ 0 & \text{if } \{i,j\} =_p \{j,k\} \\ -1 & \text{if } \{i,j\} <_p \{j,k\} \end{cases}$$

$\Phi(E)$  denotes the set of represents of  $\Omega(E)$  elements according to
the previous coding.

### 2.3. Classification of triordonnances

The aim of this paragraph is to introduce the notions of triordonnances induced on  E  by a similarity index, a binary relation,
or a description variable.

### 2.3.1. The case of a similarity index

S  being a similarity index defined on  $E \times E$, the following
partial preorder :

$$(\{i,j\},\{j,k\}) \in G \qquad \{i,j\} \quad P_s \quad \{j,k\} \Leftrightarrow S(i,j) \geqslant S(j,k)$$

is called the triordonnance induced on  E  by the  S  similarity
index, denoted  $P_s$  The  $T_s$  corresponding coding is given by :

$$T_S(i,j,k) = \begin{cases} 1 & \text{if} \quad S(i,j) > S(j,k) \\ 0 & \text{if} \quad S(i,j) = S(j,k) \\ -1 & \text{if} \quad S(i,j) < S(j,k) \end{cases}$$

## 2.3.2. The case of a binary relation

R being a binary relation defined on E, the following partial preorder :

$$(\{i,j\},\{j,k\}) \in G \qquad \{i,j\} \ P_R \ \{j,k\} \Leftrightarrow iRj \quad \text{and} \quad jRk$$

(ties if (iRj and jRk) or (ij and jk)
is called the triordonnance induced on E by the R relation and denoted $P_R$. The $T_R$ corresponding coding is given by :

$$T_R(i,j,k) = \begin{cases} 1 & \text{if} \quad iRj \quad \text{and} \quad jRk \\ 0 & \text{if} \quad (iRj \text{ and } jRk) \quad \text{or} \quad (i\overline{R}j \text{ and } j\overline{R}k) \\ -1 & \text{if} \quad i\overline{R}j \quad \text{and} \quad jRk \end{cases}$$

$i\overline{R}j$ denotes the fact that i is not in relation with j.

## Remark

$$T_R(i,j,k) = Y(i,j) - Y(j,k) = \text{Sign}(Y(i,j) - Y(j,k))$$

$$Y(i,j) = \begin{cases} 1 & \text{if} \quad iRj \\ \\ 0 & \text{if not.} \end{cases}$$

The set of triordances, induced on E by the set of equivalence relations defined on E, is a subset of $\Omega(E)$, denoted $\Omega_e(E)$ (or $\Omega_e$). The corresponding $\Phi(E)$ subset is denoted $\Phi_e$.

## 2.3.3. The case of a qualitative variable

V being a qualitative description variable. R denotes the equivalence relation induced on E by the V variable. The triordonnance defined on E associated with the R binary relation is also called the triordonnance induced on E by the V qualitative variable.

## 2.3.4. The case of an ordinal variable

V being an ordinal variable, r denotes the ranking application defined on E by the V ordinal variable, the following partial preorder :

$$(\{i,j\},\{j,k\}) \in G \qquad \{i,j\} \ P_V \ \{j,k\} \Leftrightarrow |r(i)-r(j)| \leqslant |r(j)-r(k)|$$

is called the triordonnance induced on E by the V ordinal variable. The $T_V$ corresponding coding is given by :

$$T_V(i,j,k) = \begin{cases} 1 & \text{if} \quad |r(i) - r(j)| < |r(j) - r(k)| \\ 0 & \text{if} \quad |r(i) - r(j)| = |r(j) - r(k)| \\ -1 & \text{if} \quad |r(i) - r(j)| > |r(j) - r(k)| \end{cases}$$

## 2.3.5. The case of a quantitative variable

V being a quantitative variable, V(i) denotes the value of the V variable for the i object, the following partial preorder :

$$(\{i,j\},\{j,k\}) \in G \qquad \{i,j\} \ P_V \ \{j,k\} \Leftrightarrow |V(i)-V(j)| \leqslant |V(j)-V(k)|$$

is called the triordonnance induced on E by the V quantitative variable. The $T_V$ corresponding coding is given by :

$$T_V(i,j,k) = \begin{cases} 1 & \text{if} \quad |V(i) - V(j)| < |V(j) - V(k)| \\ 0 & \text{if} \quad |V(i) - V(j)| = |V(j) - V(k)| \\ -1 & \text{if} \quad |V(i) - V(j)| > |V(j) - V(k)| \end{cases}$$

## Remark

n   denotes the E cardinal.

m   denotes the number of description variables.

$$N = \sum_{i=1}^{n} \sum_{j=1}^{n} 1 = n^2 \quad \text{if we consider the set of the couples} \quad (i,j).$$

$$N = \sum\sum_{i \leqslant j} 1 = \frac{n(n+1)}{2} \quad \text{if we consider the set of pairs} \\ \{\{i,j\} \mid i,j \in E\}.$$

$$N = \underset{i<j}{\Sigma\Sigma} \; 1 = \frac{n(n-1)}{2} \quad \text{if we consider the set of pairs}$$

$$\{\{i,j\} \mid i \neq j\}.$$

$$M = \overset{n}{\underset{i=1}{\Sigma}} \; \overset{n}{\underset{j=1}{\Sigma}} \; \overset{n}{\underset{k=1}{\Sigma}} \; 1 = n^3 \; \text{if we consider the set of triplets}$$

$$(i,j,k)$$

## Important remark

If the pairs $(i,j)$ and $(j,i)$ do not play the same rôle then the triordonnance must be defined as a partial preorder defined on the E couples set, comparing only the elements of the following G set :

$$G = \{((i,j),(j,k)) \mid i,j,k \in E\}$$

## 3. Adjustment measures between triordonnances

In order to define some adjustment measures between the given P and Q triordonnances two approaches were proposed by the author. The first is based upon covariance and correlation coefficients and the second is based upon contingency criteria. Only the first approach will be presented here, the second one is developed in Chah (1985a). The $T_p$ and $T_q$ coding of P and Q are given by :

$$T_p(i,j,k) = \begin{cases} 1 & \text{if } \{i,j\} \underset{p}{>} \{j,k\} \\ 0 & \text{if } \{i,j\} \underset{p}{=} \{j,k\} \\ -1 & \text{if } \{i,j\} \underset{p}{<} \{j,k\} \end{cases}$$

$$T_q(i,j,k) = \begin{cases} 1 & \text{if } \{i,j\} \underset{q}{>} \{j,k\} \\ 0 & \text{if } \{i,j\} \underset{q}{=} \{j,k\} \\ -1 & \text{if } \{i,j\} \underset{q}{<} \{j,k\} \end{cases}$$

## 3.1. Adjustment measure associated with the covariance

To the covariance coefficient corresponds an adjustment measure defined on $[\Omega(E)]^2$, denoted $\Psi_{Cov}$, and given by

$$\Psi_{Cov}(P,Q) = Cov(T_p,T_q) = \frac{\Sigma\Sigma\Sigma\ T_p(i,j,k)T_q(i,j,k)}{M}$$

In fact   $E(T_p) = E(T_q) = 0$

## 3.2. Adjustment measure associated with the correlation

To the correlation coefficient corresponds an adjustment measure between triordonnances, denoted $\Psi_{Cor}$, given by

$$\Psi_{Cor}(P,Q) = Cor(T_p,T_q) = \frac{Cov(T_p,T_q)}{\sqrt{Var(T_p)}\ \sqrt{Var(T_q)}}$$

$$= \frac{\Sigma\Sigma\Sigma\ T_p(i,j,k)T_q(i,j,k)}{\sqrt{\Sigma\Sigma\Sigma\ T_p^2(i,j,k)}\ \sqrt{\Sigma\Sigma\Sigma\ T_q^2(i,j,k)}}$$

## 4. The triordonnances aggregation problem

Being given  m  triordonnances denoted $P_r$  (r = 1,m)  and an $\Omega(E)$  subset denoted $\Omega_t$, the problem of finding the  $P^*$  element of  $\Omega_t$  which best fits, in the sense of an adjustment measure between triordonnances of type  $\Psi_c$, the given triordonnances  ($P_r$, r = 1,m)  is called the Triordonnances Aggregation Problem and can be written as :

$$\underset{P\in\Omega_t}{Max}\ \sum_{r=1}^{m}\ \Psi_c(P,P_r) \Leftrightarrow \underset{T\in\Phi_t}{Max}\ \sum_{r=1}^{m}\ C(T,T_r)$$

C  denotes either the covariance, or the correlation coefficient in the case of the first approach, or a contingency criteria in the case of the second approach (which is not developed here).

The object of the following chapters is to show that the Triordonnances Aggregation model can be used to solve the three following problems :

1. The classification of a set of data described with homogeneous variables (the same type of variables) and provided with a  S similarity index.

2. The classification of a set of data described with heterogeneous variables (qualitative, quantitative and ordinal variables).
The proposed approach do not use the classical coding (transformation of quantitative variables on qualitative one) and quantification (the inverse) techniques.

3. The classification of a  E  set for which the only available
information is a triordonnance (exp : the classification of phonems
form the confusion matrix).

## 5. Classification of homogeneous data

### 5.1. The proposed classification model

S  being a similarity index defined on  $E \times E$,  $P_s$  denotes the
Triordonnance induced on  E  by the  S  similarity index.  In order
to classify the  E  set (described with  m  homogeneous variables)
we propose to calculate the  R*  partition (or the  P*  associated
triordonnance) which best fits, in the sense of an adjustment measure between triordonnances of type  $\Psi_c$, the given  $P_s$  triordonnance.  The proposed model is a particular application of the general Triordonnances Aggregation problem and can be written as :

$$\underset{P \in \Omega_e}{\text{Max}} \ \Psi_c(P, P_s) \quad \Leftrightarrow \quad \underset{T \in \Phi_e}{\text{Max}} \ C(T, T_s)$$

### 5.2. Classification criterion associated with the covariance

#### Proposition

The problem of looking for the  R*  partition (i.e. the corresponding  P*  triordonnance) which fits best the  $P_s$  triordonnance, in the sense of the  $\Psi_{Cov}$  adjustment measure, is given by :

$$\underset{P \in \Omega_e}{\text{Max}} \ \Psi_{Cov}(P, P_s) \quad \Leftrightarrow \quad \underset{T \in \Phi_e}{\text{Max}} \ Cov(T, T_s)$$

This optimization problem is equivalent to the following linear
program :

$$\text{Max} \quad \underset{i \neq j}{\Sigma\Sigma} \; [\underset{k}{\Sigma} \; T_s(i,j,k) - \underset{k}{\Sigma} \; T_s(k,i,j)] Y(i,j)$$

$$Y(i,j) = Y(j,i)$$

$$Y(i,j) + Y(j,k) - Y(i,k) \leqslant 1 \quad i \neq j \neq k$$

$$Y(i,j) \in \{0,1\}$$

Y (the unknown of the problem) denotes the paired comparison representation of the R partition :

$$Y(i,j) = \begin{cases} 1 & \text{if} \quad iRj \\ 0 & \text{if not.} \end{cases}$$

The constraints designates that the Y variable represents a partition (i.e. reflexive, symmetric and transitive relation). This program can be solved using either linear programming techniques or the same heuristic used in the similarity aggregation methods (Marcotorchino & Michaud, 1982).

In fact :

$$T(i,j,k) = Y(i,j) - Y(j,k)$$

$$ME(T) = \underset{i \; j \; k}{\Sigma \; \Sigma \; \Sigma} \; T(i,j,k) = \underset{i \; j \; k}{\Sigma \; \Sigma \; \Sigma} \; [Y(i,j) - Y(j,k)] = 0.$$

$$2ME(T_s) = \underset{i \; j \; k}{\Sigma \; \Sigma \; \Sigma} \; T_s(i,j,k) + \underset{i \; j \; k}{\Sigma \; \Sigma \; \Sigma} \; T_s(k,i,j)$$

$$= \underset{i \; j \; k}{\Sigma \; \Sigma \; \Sigma} \; [T_s(i,j,k) + T_s(k,j,i)] = 0.$$

$$M \; Cov(T,T_s) = \underset{i \; j \; k}{\Sigma \; \Sigma \; \Sigma} \; T(i,j,k) T_s(i,j,k)$$

$$= \underset{i \; j \; k}{\Sigma \; \Sigma \; \Sigma} \; Y(i,j) T_s(i,j,k) - \underset{i \; j \; k}{\Sigma \; \Sigma \; \Sigma} \; Y(j,k) T_s(i,j,k)$$

$$= \underset{i \; j \; k}{\Sigma \; \Sigma \; \Sigma} \; Y(i,j) T_s(i,j,k) - \underset{i \; j \; k}{\Sigma \; \Sigma \; \Sigma} \; Y(i,j) T_s(k,i,j)$$

$$= \underset{i \; j \; k}{\Sigma \; \Sigma} [\underset{k}{\Sigma} \; T_s(i,j,k) - \underset{k}{\Sigma} \; T_s(k,i,j)] Y(i,j)$$

## 5.2.1. Interpretation of the obtained criterion

## Corollary

The linear program (5.2) can be written as :

$$
\begin{cases}
\text{Max } \Sigma\Sigma[B(i,j) - \bar{B}(i,j)]Y(i,j) \\
Y \text{ is a Partition}
\end{cases}
$$

$$B(i,j) = 2b(i,j) + f(i,j)$$
$$\bar{B}(i,j) = 2\bar{b}(i,j) + \bar{f}(i,j)$$

$$b(i,j) = \text{Card}\{k \mid k \in E \text{ and } S(i,j) > \max\{S(i,k),S(j,k)\}\}$$
$$\bar{b}(i,j) = \text{Card}\{k \mid k \in E \text{ and } S(i,j) < \min\{S(i,k),S(j,k)\}\}$$
$$f(i,j) = \text{Card}\{k \mid k \in E \text{ and } S(i,j) = \max\{S(i,k),S(j,k)\}\}$$
$$\bar{f}(i,j) = \text{Card}\{k \mid k \in E \text{ and } S(i,j) = \min\{S(i,k),S(j,k)\}\}$$

If we denote by : $B_i$ the opened sphere of center $i$ and diameter the dissimilarity between $i$ and $j$ $(d(i,j) = 1 - S(i,j))$

$B_j$ the opened sphere of center $j$ and diameter the dissimilarity between $i$ and $j$.

$Fr(B_i \cup B_j)$ the set of $E$ element belonging to the $B_i \cup B_j$ boundary.

$Fr(B_i \cap B_j)$ the set of $E$ element belonging to the $B_i \cap B_j$ boundary then :

$$b(i,j) = n - [\text{Card}\{(B_i \cup B_j\} + \text{Card}\{Fr(B_i \cup B_j)\}]$$
$$\bar{b}(i,j) = \text{Card}\{B_i \cap B_j\}$$
$$f(i,j) = \text{Card}\{Fr(B_i \cup B_j)\}$$
$$\bar{f}(i,j) = \text{Card}\{Fr(B_i \cap B_j)\}$$

In other words :

The contribution of an observer (E element) outside of the $B_i$ and $B_j$ spheres is + 2.

The contribution of an observer over $(B_i \cup B_j)$ boundary is + 1.

The contribution of an observer inside the $(B_i \cap B_j)$ is - 2.

The contribution of an observer over the $(B_i \cap B_j)$ boundary is - 1.

The contributions of the others element is null.

i.e.

|        |   |   |       | 0    |    |       |   |   |        |
|--------|---|---|-------|------|----|-------|---|---|--------|
| 222222 | 1 |   |       | N    |    |       | 1 |   | 222222 |
| 22222  | 1 |   |       | -1   | -1 |       | 1 |   | 22222  |
| 2222   | 1 |   |       | -1 -2 -1 |   |       | 1 |   | 2222   |
| 222    | 1 |   |       | -1 -2 | -1 |       | 1 |   | 222    |
| 22     | 1 |   |       | -1   | -2 | -1    | 1 |   | 22     |
| 2      | 1 | 0 | *i    | -2   |    | *j  0 | 1 |   | 2      |
| 22     | 1 |   |       | -1   | -2 | -1    | 1 |   | 22     |
| 222    | 1 |   |       | -1 -2 | -1 |       | 1 |   | 222    |
| 2222   | 1 |   |       | -1 -2 -1 |   |       | 1 |   | 2222   |
| 22222  | 1 |   |       |      |    |       | 1 |   | 22222  |
| 222222 | 1 |   |       | N'   |    |       | 1 |   | 222222 |
|        |   |   |       | 0    |    |       |   |   |        |

Noting that the points  N  and  N'  belong in the same time to
$f(i,j)$  and  $\bar{f}(i,j)$, their contributions are consequently null.


## 5.2.2. The corresponding paired comparison rôle

Without taking into account the linear constraint "Y  is a
partition", the  Y*  variable which maximizes the economic function:

$$\sum\sum_{i \neq j} [B(i,j) - \bar{B}(i,j)]Y(i,j)$$

can be obtained by using the following paired comparison rôle :

$i,j \in E \qquad B(i,j) > \bar{B}(i,j)$  then  $Y^*(i,j) = 1$.

Because of effects of type :

$B(i,j) > \bar{B}(i,j)$  and  $B(j,k) > \bar{B}(j,k)$  but  $B(k,i) < \bar{B}(k,i)$

The obtained  Y*  solution is not a partition.  The linear constraint
"Y  is a Partition" must be taken into account.  If there is no ties
$(f(i,j) = \bar{f}(i,j) = 0)$, then the previous paired comparison rôle can
be enounced in the following terms : If the number of  E  element

outside of the  $B_i$  and  $B_j$  union is greater than the number of  E
element inside of the  $B_i$  and  $B_j$  intersection then  i  and  j  are
assembled in the optimal partition.

## 5.2.3. On some aspects of the new criterion

According to the paired comparison rôle (5.2.2), every element
of the set to be classified is either an observer (judge) or an
observed point (element submitted to a judgement).  In fact in order
to classify each pair  {i,j}  (i.e. to know if  i  and  j  are assembled or separated in the optimal partition) all the elements to be
classified are "consulted".  The opinion (or the contribution) of
the consulted element, k, depends of his position relative to the
{i,j}  pair.  If the  k  observer is "further" from  {i,j}  then he
is favourable (his contribution is + 2) to put  i  and  j  in the
same class, else he is not (his contribution is − 2).  If the sum
of all contributions is positive then  i  and  j  should be assembled in the optimal partition, else they should be separated.

## 5.3. Criterion taking into account the distancing of an observer

The problem :

Assume that two observers  K  and  K'  are outside of the  $B_i$
and  $B_j$  spheres (5.2.1), and that the  K'  position is further than
that of  K  from the observed pair  {i,j}.  According to the criterion (5.2),  K  and  K'  have the same contribution (i.e. + 2).  Our
objective is to build a new criterion for which the contribution of
the K'  observer is more important than that of  K.

## Proposition

If we ponder each term of the criterion :

$$\sum_{i=1}^{n} \sum_{j=1}^{n} \sum_{k=1}^{n} T_s(i,j,k)T(i,j,k)$$

by the quantity  $|S(i,j) - S(j,k)|$ , we obtain the new following
criterion :

$$\sum_{i=1}^{n} \sum_{j=1}^{n} \sum_{k=1}^{n} |S(i,j) - S(j,k)| T_s(i,j,k) T(i,j,k)$$

$$= \sum_{i=1}^{n} \sum_{j=1}^{n} \sum_{k=1}^{n} (S(i,j) - S(j,k)) T(i,j,k)$$

The problem of looking for the $Y^*$ partition (i.e. the corresponding $T^*$ triordonnance) which fits best the given $T_s$ triordonnance, in the sense of the new criterion, is equivalent to the following program :

$$\text{Max } 2n \sum_{i \neq j} \{S(i,j) - [\frac{S(i,\cdot) + S(\cdot,j)}{2n}]\} Y(i,j)$$

$$Y(i,j) = Y(j,i)$$
$$Y(i,j) + Y(j,k) - Y(i,k) \leqslant 1 \quad i \neq j \neq k$$
$$Y(i,j) \in \{0,1\}$$

$$S(i,\cdot) = \sum_{k} S(i,k), S(\cdot,j) = \sum_{k} S(k,j)$$

Remark

According to this criterion, the contributions of observers $K$ and $K'$ (outside of the $B_i$ and $B_j$ spheres) are : $S(i,j) - S(j,k)$ for the $K$ observer, and $S(i,j) - S(j,k')$ for the $K'$ observer. As $S(j,k) > S(j,k')$, the contribution of $K'$ is more important han that of $K$.

5.3.1. Statistical interpretation

Proposition

The new criterion can be interpreted statistically as a covariance :

$$\sum_{i} \sum_{j} \sum_{k} (S(i,j) - S(j,k)) T(i,j,k) = \text{Cov}(\xi, T)$$

$$i,j,k \in E \quad \xi(i,j,k) = S(i,j) - S(j,k)$$

In fact : $E(\xi) = E(T) = 0$.

Proposition

To the correlation coefficient corresponds a classification criterion, related with the (5.3) criterion, given by :

$$Cor(\xi,T) = \frac{\sum\limits_{i=1}^{n}\sum\limits_{j=1}^{n}\sum\limits_{k=1}^{n} T(i,j,k)\xi(i,j,k)}{\sqrt{\Sigma\Sigma\Sigma \; T^2(i,j,k)} \; \sqrt{\Sigma\Sigma\Sigma \; \xi^2(i,j,k)}}$$

## 5.4. Classification criterion associated with the correlation coefficient

Proposition

The problem of looking for the $R^*$ partition (i.e. the $P^*$ corresponding Triordonnance) which fits best the given $P_s$ triordonnance in the sense of the $\Psi_{Cor}$, can be written as :

$$\underset{P\in\Omega_e}{Max} \; \Psi_{Cor}(P,P_s) = \underset{T\in\Phi_e}{Max} \; Cor(T,T_s)$$

$$= \underset{T\in\Phi_e}{Max} \; \frac{\Sigma\Sigma\Sigma \; T(i,j,k)T_s(i,j,k)}{\sqrt{\Sigma\Sigma\Sigma \; T^2(i,j,k)} \; \sqrt{\Sigma\Sigma\Sigma \; T_s^2(i,j,k)}}$$

## 6. Classification of heterogeneous data

The problem :

Being given a E set described with m heterogeneous variables (i.e. qualitative, quantitative and ordinal variables) the question is how to build classification criteria which deals with heterogeneous data without favouring one type of variable to another as done in the classical approach with coding (transformation of quantitative variables in qualitative one) or quantification (the reverse). Our objective is to show that the Triordonnances Aggregation model can be used to solve the above mentioned problem. We propose to represent each $V_r$ description variable (r = 1,m) by the corresponding $P_r$ triordonnance and to calculate the $R^*$ partition (or the $P^*$ associated triordonnance) which best fit the $P_r$ triordonnances (r = 1,m). In the sense of an adjustment measure of type $\Psi_c$ the problem can be written as :

$$\underset{P \in \Omega_e}{\text{Max}} \ \sum_{r=1}^{m} \Psi_c(P_r, P) \quad \Leftrightarrow \quad \underset{T \in \Phi_e}{\text{Max}} \ \sum_{r=1}^{m} C(T_r, T)$$

$T_r \ (r = 1, m)$ denotes the $P_r$ coding.

## 6.1. Classification criterion associated with the covariance

The problem of looking for the triordonnance of type $\Omega_c$ which best fits the triordonnances $T_r$ $(r = 1, m)$, in the sense of the $\Psi_{Cov}$ adjustment measure, is given by :

$$\underset{P \in \Omega_e}{\text{Max}} \ \sum_{r=1}^{m} \Psi_{Cov}(P, P_r) \quad \Leftrightarrow \quad \underset{T \in \Phi_e}{\text{Max}} \ \sum_{r=1}^{m} \text{Cov}(T, T_r)$$

## Proposition

The above optimization problem is equivalent to the following linear program :

$$\underset{i \neq j}{\text{Max}} \ \Sigma\Sigma \ \{ \sum_{r=1}^{m} [\sum_k T_r(i,j,k) - \sum_k T_r(k,i,j)]\} Y(i,j)$$

$Y(i,j) = Y(j,i)$

$Y(i,j) + Y(j,k) - Y(i,k) \leqslant 1 \quad i \neq j \neq k$

$Y(i,j) \in \{0,1\}$

## 6.1.1. Interpretation of the obtained linear criterion

## Corollary

The above linear program (6.1) can be written as

$$\begin{cases} \underset{i \neq j}{\text{Max}} \ \Sigma\Sigma \ [ \sum_{r=1}^{m} B_r(i,j) - \sum_{r=1}^{m} \bar{B}_r(i,j)] Y(i,j) \\ Y \ \text{is a partition} \end{cases}$$

$$B_r(i,j) = 2b_r(i,j) + f_r(i,j)$$

$$\bar{B}_r(i,j) = 2\bar{b}_r(i,j) + \bar{f}_r(i,j)$$

$$b_r(i,j) = \text{Card}\{k \mid k \in E \quad \text{and} \quad \{i,j\} >_r \{i,k\}, \ \{i,j\} >_r \{j,k\}\}$$

$$\bar{b}_r(i,j) = \text{Card}\{k \mid k \in E \quad \text{and} \quad \{i,j\} <_r \{i,k\}, \ \{i,j\} <_r \{j,k\}\}$$

$$f_r(i,j) = \text{Card}\{k \mid k \in E \quad \text{and} \quad [\{i,j\} =_r \{i,k\}$$
$$\text{and} \ \{i,j\} \geqslant_r \{j,k\}] \quad \text{or} \quad [\{i,j\} \geqslant_r \{i,k\}$$
$$\text{and} \ \{i,j\} =_r \{j,k\}]\}$$

$$\bar{f}_r(i,j) = \text{Card}\{k \mid k \in E \quad \text{and} \quad [\{i,j\} =_r \{i,k\}$$
$$\text{and} \ \{i,j\} \leqslant_r \{j,k\}] \quad \text{or} \quad [\{i,j\} \leqslant_r \{i,k\}$$
$$\text{and} \ \{i,j\} =_r \{j,k\}]\}$$

$b_r(i,j)$ corresponds to the number of configuration of type :

$$i*$$
$$(*k$$
$$j*$$

(i.e. the number of E element "further" from i,j , according to the $V_r$ variable). Remind that the contribution of a k element (observer) which is in this position is + 2.

$f_r(i,j)$ corresponds to the cardinal of the "boundary" of the E subset which verify the previous configuration.

$\bar{b}_r(i,j)$ corresponds to the number of configuration of type :

$$i*$$
$$(*k$$
$$j*$$

The contribution of a k element is in this position is - 2.

$\bar{f}(i,j)$ corresponds to the "boundary" of the E element which verify the previous configuration. The contribution of a k element which is in this position is - 1.

6.1.2. The corresponding paired comparison rôle

Without taking into account the linear constraint  "Y  is a Parition" the  Y*  variable which maximizes the economic function :

$$\sum_i \sum_j [\sum_r B_r(i,j) - \sum_r \bar{B}_r(i,j)] Y(i,j)$$

can be obtained using the following paired comparison rôle :

$$i,j \in E \qquad \sum_{r=1}^{m} B_r(i,j) > \sum_{r=1}^{m} \bar{B}_r(i,j) \quad \text{then} \quad Y^*(i,j) = 1.$$

Because of effects of types :

$$\sum_r B_r(i,j) > \sum_r \bar{B}_r(i,j) \quad \text{and} \quad \sum_r B_r(j,k) > \sum_r \bar{B}_r(j,k) \quad \text{but}$$

$$\sum_r B_r(i,k) < \sum_r \bar{B}_r(i,k)$$

The solution obtained without taking into account the constraint "Y is a Partition", is not a partition.

6.1.3. The case where all the variables are qualitative

Proposition

If all the description variables are qualitative then the (6.1) program can be written as :

$$\text{Max } 2n \sum_{i \neq j} \{C(i,j) - [\frac{C(i,.) + C(.,j)}{2n}]\} Y(i,j)$$

$$Y(i,j) = Y(j,i)$$
$$Y(i,j) + Y(j,k) - Y(i,k) \leq 1 \quad i \neq j \neq k$$
$$Y(i,j) \in \{0,1\}$$

$$C(i,j) = \sum_r C_r(i,j), C(i,.) = \sum_r C_r(i,.), C(.j)$$

$$= \sum_r C_r(.,j), C_r(i,.) = \sum_j C_r(i,j), C_r(.j) = \sum_i C_r(i,j)$$

$$C_r(i,j) = \begin{cases} 1 & \text{if } V_r(i) = V_r(j) \\ 0 & \text{if } V_r(i) \neq V_r(j) \end{cases}$$

In the case where all the description variables are qualitative, the (6.1) criterion, introduced at first to classify heterogeneous data, is equivalent to the "related to Jordan" classification criterion (studied by Marcotorchino, 1984b). This classification criterion can be written as the covariance between the variables $\xi$ and $T$ :

$$i,j,k \in E \qquad \xi(i,j,k) = C(i,j) - C(j,k),$$
$$T(i,j,k) = Y(i,j) - Y(j,k)$$

## 6.2. Classification criterion associated with the correlation coefficient

In the sense of the $\Psi_{Cor}$ adjustment measure, the problem of looking for the R* partition (the T* associated triordonnance) which fits best the $T_r$ (r = 1,m) triordonnances is given by :

$$\underset{P \in \Omega_e}{\text{Max}} \sum_{r=1}^{m} \Psi_{Cor}(P,P_r) \quad \Leftrightarrow \quad \underset{T \in \Phi_e}{\text{Max}} \sum_{r=1}^{m} Cor(T,T_r)$$

$$\Leftrightarrow \quad \underset{T \in \Phi_e}{\text{Max}} \sum_{r=1}^{m} \frac{\Sigma\Sigma\Sigma\, T(i,jk)T_r(i,j,k)}{\sqrt{\Sigma\Sigma\Sigma\, T^2(i,j,k)}\,\sqrt{\Sigma\Sigma\Sigma\, T_r^2(i,j,k)}}$$

## 7. Use ot the triordonnance aggregation methods for the clustering of microcomputers

Each description variable is represented by a triordonnance, the (6.1 or 6.1.1) optimization problem is solved by the approximate algorithm {8}. The obtained solution (i.e. the optimal partition) is given by :

| Class 1 | PAP | | PAP(2) | PAP(3) | | |
|---------|-----|--|--------|--------|--|--|
| Class 2 | QX 10<br>IBM PC<br>OPLITE<br>M 20<br>TANDY 2000<br>NEC PC 8000 | TI PC<br>TI PC(2)<br>HORIZON<br>PSI 80 | | APRICOT<br>T 200<br>SKS 2550<br>DESKTOP 10<br>TRS MODELE 80 | Z 150<br>AS 100<br>ZEPHYR | GOUPIL 3<br>MZ 35<br>MBC 4050<br>IPC MO-<br>DELE 15<br>SANCO 8000 |
| Class 3 | MACINTOSH | | | LISA-2 | | LISA 2-S |
| Class 4 | APPLE 3 | | | APPLE 2E | | BASIS 108 |
| Class 5 | TELE PC<br>FOXY | Z 150(2)<br>IBM PC(2) | | VICTORS S1<br>CORONA PC(2) | | EUROPE PC<br>TANDY<br>2000(2) |

References

S.Chah (1984a). Calcul des partitions optimales d'un critère d'adé-
      quation à une préordonnance. Publication de l'ISUP, vol.XXIX-
      Fascicule 1.

S.Chah (1984b). Agrégation des préordonnances, IBM-France Scientific
      Center Technical Report n° F63.

S.Chah (1985a). Comparaisons par triplets en classification automa-
      tique. IBM-France Scientific Center Technical Report n° F86.

S.Chah (1985b). Critères de classification sur des données hétéro-
      gènes. Revue de Statistique appliquée, volume XXXIII, n°2.

J.L.Chandon & S.Pinson (1981). Analyse typologique : théorie et
      applications. Masson, Paris.

I.C.Lerman (1981). Classification et analyse ordinale des données.
      Dunod, Paris.

F.Marcotorchino & P.Michaud (1981). Heuristic approach of the simila-
      rity aggregation proglem. Methods of Operations Research n°43,
      pp.395-404. Verlagsgruppe Athenaum, Scriptor, Hanstein, Gunn
      and Hain.

F.Marcotorchino (1984a)  Utilisation des comparaisons par paires en
    statistique des contingences. Part I, IBM-France Technical
    Report n° F069.

F.Marcotorchino (1984b). Utilisation des comparaisons par paires en
    statistique des contingences. Part II, IBM-France Technical
    Report n° F071.

F.Marcotorchino (1985). Utilisation des comparaisons par paires en
    statistique des contingences. Part III, IBM-France Technical
    Report n° F081.

# USING CLASSICAL TOOLS OF DATA ANALYSIS TO SOLVE THE PROBLEM

P. Van Brussel

CADEPS, Université Libre de Bruxelles

## 1. Introduction

Although the problem of analysing heterogeneous data sets has already been the source of the development of very sophisticated technics, it has seemed interesting to us to solve it by using only very classical Data Analysis methods such as Principal Component Analysis and Correspondence Analysis.
Different way of coding will make us able to use those to classify the forty Micro Computers. The objective of this approach is to compare the results of each analysis and to study the differences between them and also to confront the final solution with those obtained from more appropriate technics.
The main advantage of the use of factorial analysis is the possibility it gives to visualize the clusters and also to characterize them in the structure of the variables.

Because Factorial Analysis only gives plane representation of the variables, it would be dangerous to restrict our analysis to this approach. That is the reason why it has been completed by a hiearchical clustering analysis on the coordinates with respect to the six first factorial axes.

233

## 2. The correspondence analysis approach

In this section, a comparison of the results of two different approach of Multiple Correspondence Analysis will be performed.  In order to perform those analyses, it has been necessary to recode the quantitative variables so that they became qualitative.  That was done by dividing them in classes.

The coding was the following :

Price :

1 : 20 000 - 30 000 FF

2 : 30 001 - 40 000 FF

3 : 40 001 - 50 000 FF

4 : > 50 001 FF

RAM configuration :

1 : < 64 Kb

2 :  64 Kb

3 : 64 - 128 Kb

4 : 128 Kb

5 : 128 - 256 Kb

6 : 256 Kb

7 : > 256 Kb

RAM maximum :

1 : < 256 Kb

2 : 256 Kb

3 : 256 - 512 Kb

4 : 512 Kb

5 : 512 - 1024 Kb

6 : 1024 Kb

Mass Storage :

1 : < 320 Kb

2 : 320 Kb

```
3 :  360 Kb
4 :  360 - 720 Kb
5 :  720 Kb
6 : >720 Kb
```

Let us note that the division in classes was made according to a certain standard of the PC market.

## 2.1. The Multiple Correspondence Analysis results

The use of Multiple Correspondence Analysis seems to be the more appropriate to solve the problem.  In this case, the factorial axes are obtained by diagonalisation of the Burt table (Burt, 1950; Lebart, Morineau, Tabard, 1977).  Unfortunately, MCA does not allow to visualize individuals, so that the only way to make it possible was to associate a new variable to the data (see below) and treat it as a supplementary variable.

|  | Variables | Supplementary variables |
|---|---|---|

```
P   ┌───────────────┐ ┌──────────────────────────────────┐
C   │               │ │ 1 0 0 .......... 0 .......... 0   │
=   │  Qualitative  │ │ 0 1 0 .......... 0 .......... 0   │
    │               │ │ 0 0 1 .......... 0 .......... 0   │
I   │               │ │  . . .               .         . │
n   │       +       │ │  . . .               .         . │
d   │               │ │  . . .               .         . │
i   │               │ │                                  │
v   │  Recoded      │ │  . . .               .         . │
i   │  Quantitative │ │ 0 0 0 .......... 1 .......... 0   │
d   │  Variables    │ │  . . .               .         . │
u   │               │ │  . . .               .         . │
a   │               │ │  . . .               .         . │
l   │               │ │ 0 0 0 .......... 0 .......... 1   │
s   └───────────────┘ └──────────────────────────────────┘
```

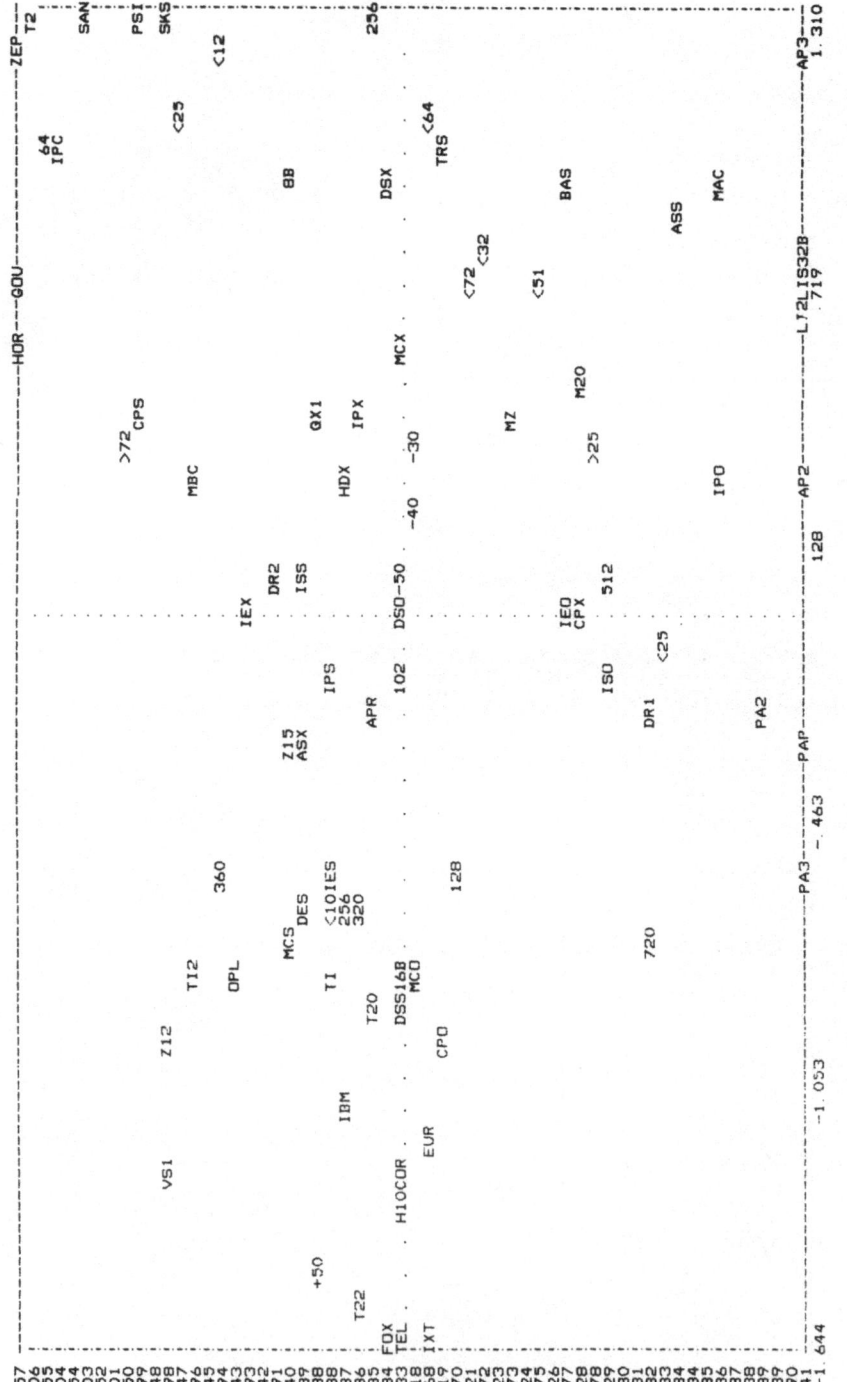

Fig.1. Contains the results of the MCA. Only the first factorial plane was represented.

Axis 1 (horizontal) opposes the PCs using a 16 bit-processor to the others. Those computers are using DOS operating system with a standard mass storage of 320 or 360 Kb. They support Color monitor and a 10 Mb hard disk. Here we have the "IBM like" family with the highest level of price.

Axis 2 (vertical) separates two groups of microcomputers. On the upper side of the axis, we found PCs using CP/M as operating system. They are characterized by having a 8 bit processor and supporting no color monitor nor hard disk. On the lower side, several microcomputers using others operating systems can be found. All "Apple like" computers are to be found in this area. Some of them use 32 bit processor and support 5 Mb hard disk.

A fourth group is localised between the "IBM like" and the "Apple like" machines. It is composed by the three PAP computers which are similar to the IBM standard according to the operating system and the processor, and close to the Apple standard because of a 720 Kb mass storage.

Now the 4 groups are defined, let us now validate them by performing others analyses.

## 2.2. The binary correspondence analysis results

Although the powerful tool it is for analysing large tables of qualitative data, MCA doesn't give any precise information concerning the projection of supplementary variables. The validation of the groups found in the first analysis can be done by using another approach of MCA. It consists to consider the total disjonctive table as a contingency table and to perform a classical Binary Correspondence Analysis. In this case, PCs are considered as row variables, and, according to the barycentric representation, their position on the factorial plane will be absolutely correct, so that we are going to be able to compare them with those obtained by the MCA.

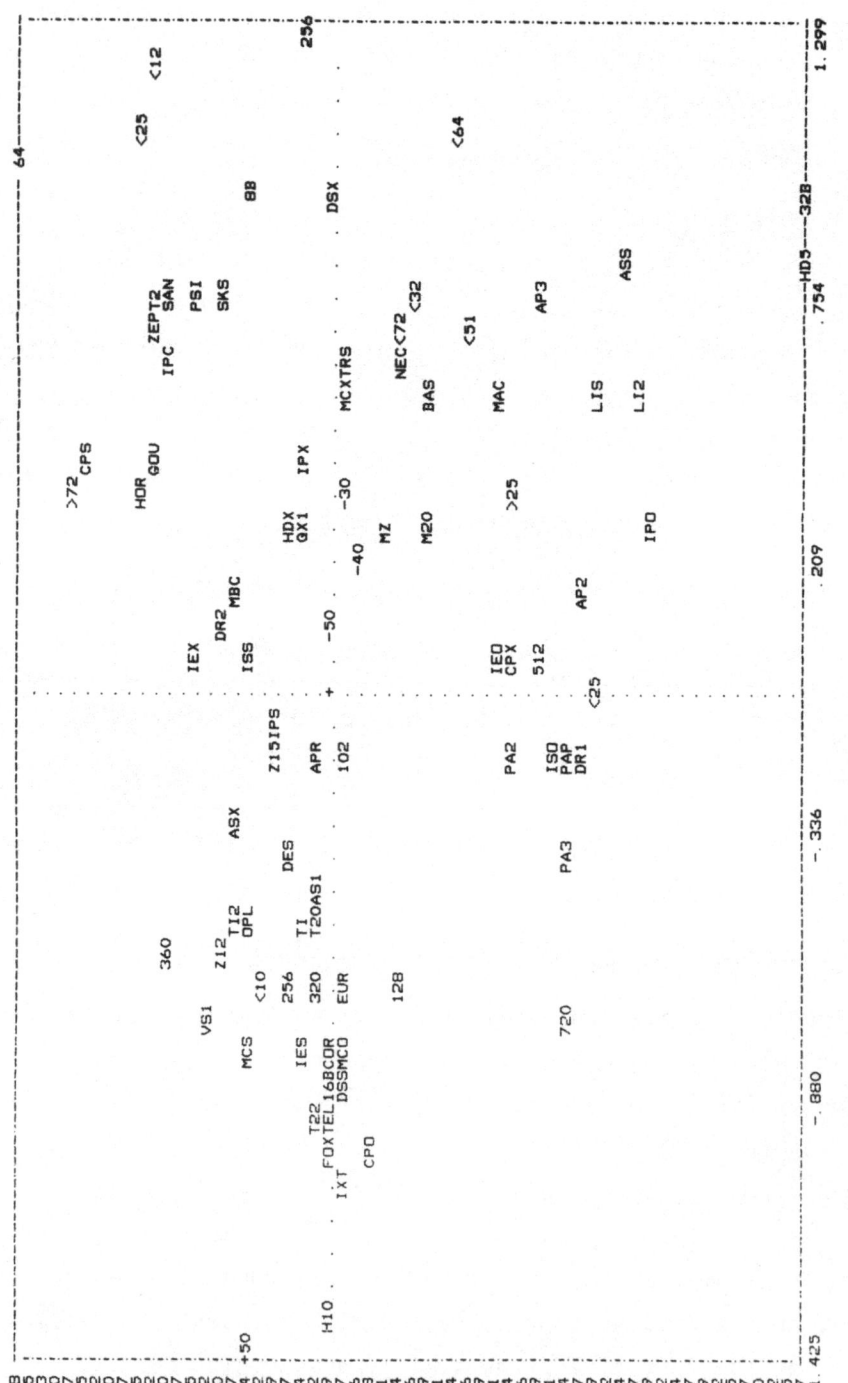

Fig.2

Figure 2 is the first factorial plane issued by the analysis. We can see that the structure is globally conserved especially if we consider the positions of the PCs.

At this stage of the study, we can consider that the 4 groups built according to the Correspondence Analysis point of view are rather stable. Let us now study what is happening if all variables are made quantitative. This will be done in the next point relating to the Principal Component Analysis approach.

## 3. The principal component analysis approach

In this section, we tried to approach the solution of the problem in another way and to compare it with our first results. This will be done by performing Principal Component Analysis on recoded variables. Two different tentative of recoding qualitative variables were made.

## 3.1. First coding tentative and results

The first coding was made in a very classical way by considering each modality of the different qualitative variables as a quantitative variable. To make it possible, a modality has to be written in a binary form (1 - 0) (Stemmelen, Pagès and Morlat, 1981). In this case, the mean of a modality represent its proportion in the sample.

Figure 3 is the first factorial plane of the analysis. Let us remember that the representation of variables and individuals on the same figure has no statistical meaning.

Concerning variables, the same structure as the one of correspondence analysis is found. Axis 1 (horizontal) still opposes 16 bit processor to 8 and 32 bit ones. The same remark is to be done concerning the operating systems, the support of Hard disk and Color monitor. The position of variable price in the 16 bit processor

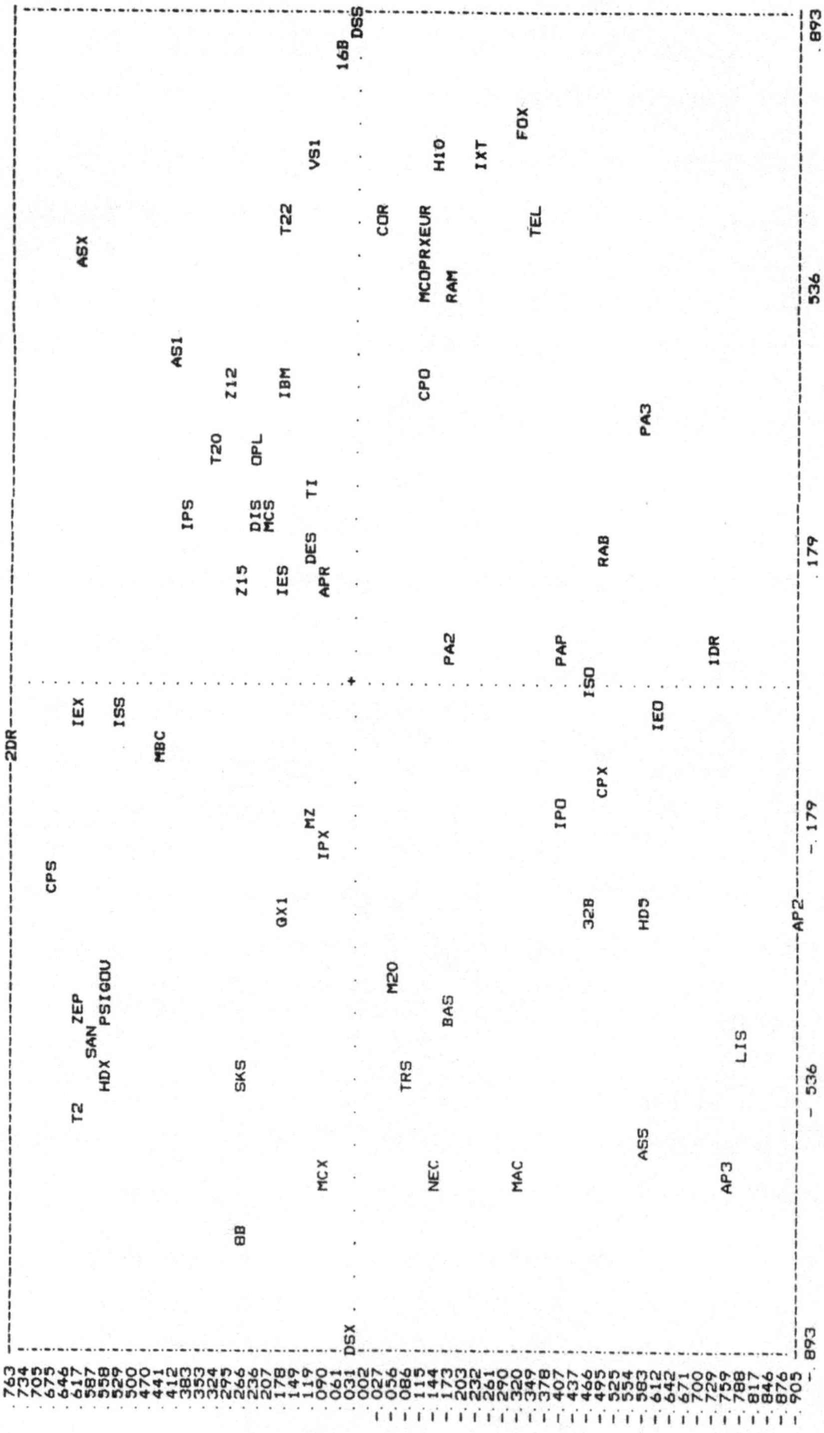
2DR
AP2
DSS
16B
DSX
VS1
FOX
H10
IXT
T22
COR
TEL
ASX
MCOPRXEUR
RAM
AS1
Z12
IBM
CPO
PA3
T20
OPL
TI
IPS
DIS
MCS
DES
RAB
Z15
IES
APR
PA2
PAP
1DR
IEX
ISS
ISO
IEO
MBC
CPX
IPO
MZ
IPX
CPS
GX1
32B
HD5
M20
ZEP
PSIGOU
SAN
HDX
BAS
SKS
LIS
T2
TRS
ASS
MCX
NEC
MAC
AP3
BB
893
536
179
-.179
-.536
-.893

Fig.3

area means that those computers are the most expensive. Axis 2 (vertical) separates 8 bit and 32 bit processor machines, and also the operating system used by them.

The position of the individuals (PCs) on the graphics shows us a rather similar distribution as found before. Let us remark that the groups are less clear, in the sense that they are less concentrated.

## 3.2. Second coding tentative and results

A great disadvantage of previous coding is that the variance of each modality is rather small, so that quantitative variables can be favoured in the analysis (Hallin and Ingenbleek, 1981). Let us now divide quantitative variables in classes like in section 2. After that, let us recode them in the following way :

$$x_i = 1 \quad \text{if value} \in i+1, i+2, \ldots$$
$$\phantom{x_i =} 0 \quad \text{otherwise}$$

For example, variable "price" would be recoded as

|             | $x_1$ | $x_2$ | $x_3$ |
|-------------|-------|-------|-------|
| 20 - 30 000 | 0     | 0     | 0     |
| 30 - 40 000 | 1     | 0     | 0     |
| 40 - 50 000 | 1     | 1     | 0     |
| > 50 000    | 1     | 1     | 1     |

Such a coding allows us to analyse only binary variable without having the scaling factor lost.

Figure 4 gives us the representation of all variables on factorial plane n°1. All except a 180° rotation around horizontal axis, the structure is still the same.

The existence of the four groups and their validity is now proved.

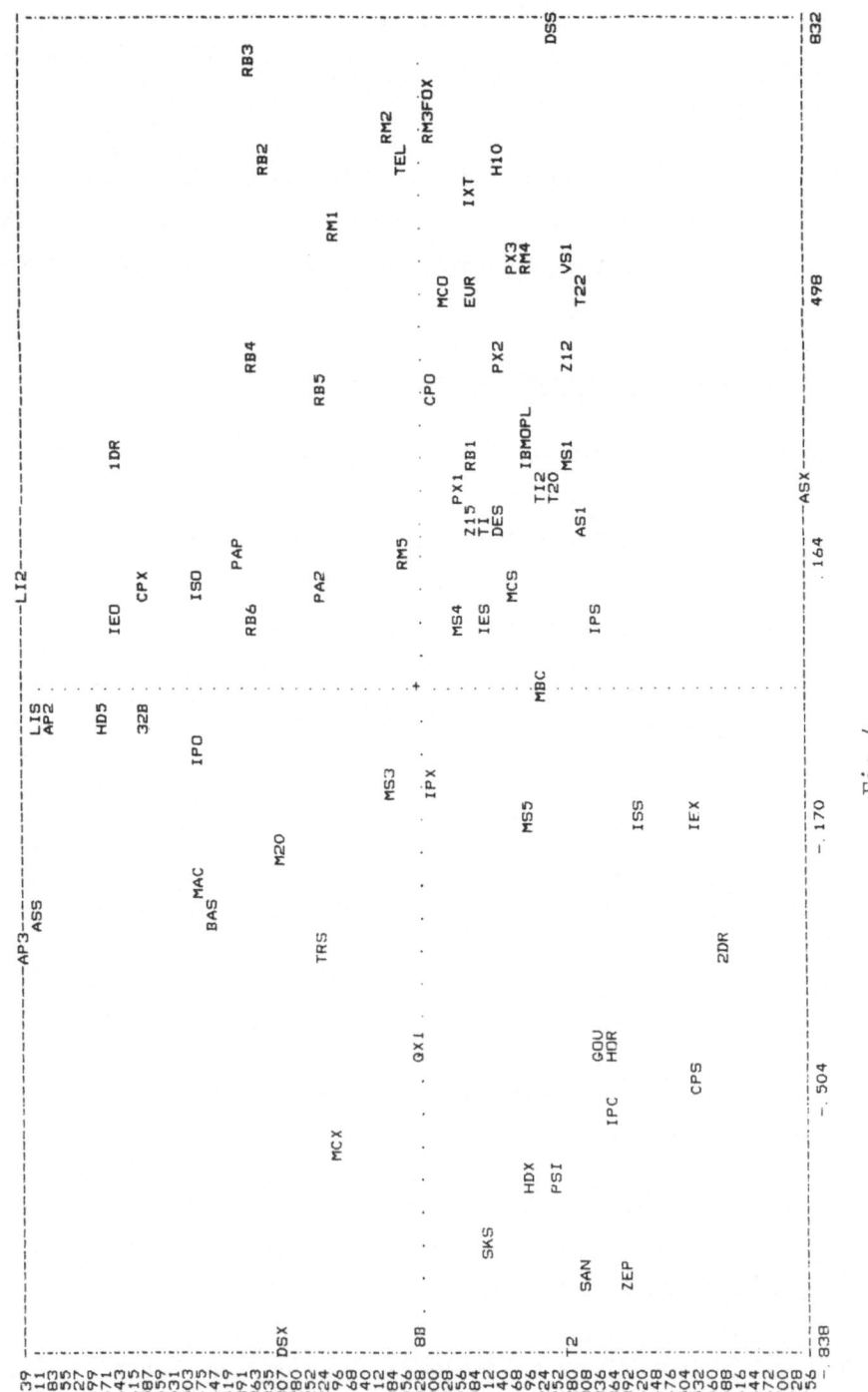

Fig. 4

4. The final solution

Factorial analysis represent variables and individuals on
planes, so that factor are only analysed by pairs.  In order to see
if the first plane is a sufficient representation of the data structure, let us a now perform a hierarchical clustering analysis
(Ward algorithm) on the coordinates with respect to the six first
factorial axes.

Table 5 is the classification of the microcomputers.  Two
solutions are possible :

|               Solution 1               |               Solution 2               |
|----------------------------------------|----------------------------------------|
| PAP                                    | PAP                                    |
| PAP(2)                                 | PAP(2)                                 |
| PAP(3)                                 | PAP(3)                                 |
| -------------------------------------- | -------------------------------------- |
| MACINTOSH                              | MACINTOSH                              |
| LISA 2-S                               | LISA 2-S                               |
| LISA 2                                 | LISA 2 _____        |
| QX 10                                  | QX 10                                  |
| NEC PC 8000                            | NEC PC 8000                            |
| TRS 80 12                              | TRS 80 12                              |
| M 20                                   | M 20                                   |
| MZ 35                                  | MZ 35                                  |
| BASIS 108                              | BASIS 108 _____         |
| APPLE 2                                | APPLE 2                                |
| APPLE 3                                | APPLE 3                                |
| -------------------------------------- | -------------------------------------- |
| SKS 2500                               | SKS 2500                               |
| PSI 80                                 | PSI 80                                 |
| IPC 15                                 | IPC 15                                 |
| GOUPIL 3                               | GOUPIL 3                               |
| HORIZON                                | HORIZON                                |
| MBC 4050                               | MBC 4050                               |

| | |
|---|---|
| ZEPHYR | ZEPHYR |
| T 200 | T 200 |
| SANCO 800 | SANCO 800 |

-----------------------------------------------------------------

| | |
|---|---|
| TI PC | TI PC |
| TI PC 2 | TI PC 2 |
| IBM PC | IBM PC |
| AS 100 | AS 100 _____ |
| Z 150 | Z 150 |
| Z 150 (2) | Z 150 (2) |
| OPLITE | OPLITE |
| APRICOT | APRICOT |
| TANDY 2000 | TANDY 2000 |
| TELE PC | TELE PC |
| FOXY | FOXY |
| VICTOR S1 | VICTOR S1 |
| DESKTOP 10 | DESKTOP 10 |
| IBM XT | IBM XT |
| EUROPE PC | EUROPE PC |
| CORONA PC 2 | CORONA PC 2 |
| TANDY 2000 (2) | TANDY 2000 |

## 5. Conclusion

Even if there exists more appropriate methods to solve such a problem, we may conclude that, one more time, the use of classical Data Analysis technics is sometimes as powerful as them. We also can say that it is the only method that allows to characterize clusters without having to make a systematic feed back to the data. In other words, using Factorial Analysis will never be a waste of time.

## References

J.P.Benzecri et collaborateurs (1973). L'Analyse des Données, tome 2 : l'Analyse des Correspondances, Dunod, Paris.

C.Burt (1950). The factorial analysis of qualitative data. Brit.J. of Stat. Psych. 3 (3).

M.Buyse (1983). Correspondence Analysis as an exploratory tool in Medical Trials in "New Trends in data Analysis and Applications". North Holland, 247-259.

R.A.Fisher (1940). The Precision of discriminant functions. Ann. Elugen Lond. 10, 422-429.

M.Hallin, J.F.Ingenbleek (1981). Etude statistique des facteurs influençant le risque automobile : Le montant cumulé des sinistres dans le portefeuille suédois en 1979; Série Actuarielle de l'U.L.B., n°10.

H.O.Hirschfeld (1935). A connection between correlation and contingency. Proc.Camb.Phil.Soc. 31, 520-524.

J.Janssen, J.L.Mentior et P.Van Brussel (1983). Young Belgian Economists in front of the Crisis in "Data Analysis in Real Life Environment: "Ins and Outs of Solving Problems," North-Holland, 301-317.

J.O.Lancaster (1957). Somer properties of the bivariate normal distribution considered in the form of a contingency table. Biometrika 44, 289-292.

L.Lebart, A.Morineau, N.Tabard (1977). Technique de la description statistique. Methode et logicels pour l'analyse des grands tableaux, Dunod, Paris.

C.Stemmelen, J.P.Pagès et G.Morlat (1981). Structures de l'opinion publique. Le progrès technique, N.24, 10-15.

SECTION 4.4

CLASSIFICATION OF MICROCOMPUTERS REPRODUCED BY SPECTRAMAP

P.J. Lewi, A. Quets and K. Van Reusel

Research Laboratories, Janssen Pharmaceutica NV

B-2340 Beerse, Belgium

## Introduction

A table of heterogeneous charateristics of microcomputers has been submitted for analysis by Chah (1985). The solution proposed here is based on spectral map analysis (SMA). A mathematical description of SMA is presented elsewhere in this volume (Lewi et al., 1985). Briefly, SMA includes the following steps : re-expression of the data, row- and column- wise centering, factorization, and simultaneous plot of factor scores and factor loadings in the plane of the first two dominant factors. Factor scores and factor loadings refer to projections of the items representing rows and columns upon the computed factor axes.

Factor scores and factor loadings are scaled such as to produce identical factor variances. Orthogonal projection of factor scores upon factor loadings (and vice-versa) reproduces the doubly centered data. In this application to heterogeneous data, identical weight coefficients have been assigned to rows and columns of the table.

Spectral map analysis (SMA) belongs to the class of methods which comprises factorial analysis of correspondences (FAC). Both

247

methods reduce the rank of the data matrix by one; they can apply individual weighting coefficients to rows and columns; and they produce a joint representation of interactions between rows and columns of the data table. Using an ordinary euclidean metric and a weighted matrix product, SMA, as defined above, is oriented toward analysis of multiple differences.

Spectramap is the name of a particular program that performs spectral map analysis. The method and the program have been developed at Janssen Pharmaceutica.

## 1. Re-expression of the original data

The original data are composed of ten categorical attributes and four quantitative measures. Categorical attributes are defined on a three-point scale (0,1,2). Quantitative measures describe price and various memory capacities of the microcomputers.

In order to render the data more homogeneous, the four quantitative measures have been re-expressed logarithmically. The results have been rescaled to the same three-point scale of the categorical attributes and the values have been rounded to the nearest integers. The re-expressed data are shown in Table 1.

## 2. Analysis of microcomputers vs attributes

In the Spectramap of Figure 1, circles refer to microcomputers, and squares represent attributes. Areas of circles are made proportional to Price. Areas of squares are kept constant, because of the heterogeneous nature of the attributes.

The positions of a circle on the map is the result of interactions with the squares. The microcomputer IBM PC XT is strongly attracted by MS-DOS, its standard operating system. The same microcomputer is also repelled by CP/M and other OS for which it has no standard provisions. This way, each microcomputer obtains a unique

Table 1. Classification of microcomputers according to ten qualitative attributes and four quantive measures.

```
A : COLOR
B : CP/M
C : MS-DOS
D : OTHER OS
E : PROCESSOR
F : PARALLEL INTERFACE
G : SERIAL INTERFACE
H : IEEE INTERFACE
I : HARD DISK
J : DISKETTES
K : LOG PRICE
L : LOG RAM SIZE
M : LOG MAX RAM SIZE
N : LOG DISKETTE SIZE
```

| | A | B | C | D | E | F | G | H | I | J | K | L | M | N | TT |
|---|---|---|---|---|---|---|---|---|---|---|---|---|---|---|---|
| 1 : PAP | 0 | 0 | 2 | 0 | 1 | 1 | 1 | 1 | 0 | 0 | 0 | 1 | 1 | 1 | 9 |
| 2 : QX 10 | 1 | 2 | 0 | 0 | 0 | 2 | 1 | 1 | 0 | 1 | 0 | 1 | 0 | 0 | 10 |
| 3 : MACINTOSH | 0 | 0 | 0 | 2 | 2 | 0 | 2 | 0 | 0 | 0 | 0 | 0 | 1 | 0 | 8 |
| 4 : TI PC | 2 | 2 | 2 | 0 | 1 | 2 | 1 | 0 | 0 | 0 | 0 | 0 | 1 | 0 | 12 |
| 5 : PAP (2) | 0 | 0 | 2 | 0 | 1 | 1 | 1 | 1 | 0 | 1 | 0 | 1 | 1 | 1 | 10 |
| 6 : APRICOT | 0 | 0 | 2 | 0 | 1 | 2 | 2 | 1 | 0 | 1 | 0 | 1 | 1 | 0 | 12 |
| 7 : Z 150 | 0 | 0 | 2 | 0 | 1 | 2 | 2 | 0 | 0 | 1 | 0 | 1 | 1 | 0 | 11 |
| 8 : GOUPIL 3 | 0 | 2 | 0 | 0 | 0 | 2 | 2 | 0 | 0 | 1 | 0 | 0 | 2 | 0 | 10 |
| 9 : APPLE 3 | 0 | 0 | 0 | 2 | 0 | 1 | 2 | 1 | 1 | 0 | 1 | 1 | 0 | 0 | 9 |
| 10 : TANDY 2000 | 1 | 0 | 2 | 0 | 1 | 2 | 2 | 0 | 0 | 1 | 0 | 0 | 1 | 1 | 12 |
| 11 : IBM PC | 1 | 1 | 2 | 0 | 1 | 2 | 1 | 0 | 0 | 1 | 1 | 0 | 1 | 0 | 12 |
| 12 : TI PC (2) | 2 | 2 | 2 | 0 | 1 | 2 | 1 | 0 | 0 | 1 | 1 | 1 | 1 | 0 | 15 |
| 13 : APPLE 2E | 1 | 1 | 0 | 2 | 0 | 1 | 1 | 1 | 1 | 0 | 1 | 0 | 2 | 0 | 11 |
| 14 : TELE PC | 1 | 0 | 2 | 0 | 1 | 2 | 2 | 1 | 2 | 0 | 2 | 1 | 1 | 0 | 15 |
| 15 : PAP (3) | 0 | 0 | 2 | 0 | 1 | 1 | 1 | 1 | 2 | 0 | 1 | 1 | 1 | 1 | 12 |
| 16 : IBM PC XT | 1 | 1 | 2 | 0 | 1 | 2 | 1 | 0 | 2 | 0 | 1 | 0 | 1 | 0 | 13 |
| 17 : Z 150 (2) | 0 | 2 | 2 | 0 | 1 | 2 | 2 | 0 | 2 | 1 | 1 | 1 | 1 | 0 | 16 |
| 18 : TANDY 2000 (2) | 1 | 0 | 2 | 0 | 1 | 2 | 2 | 0 | 2 | 1 | 1 | 0 | 1 | 1 | 15 |
| 19 : VICTOR S1 | 1 | 2 | 2 | 0 | 1 | 2 | 2 | 1 | 2 | 1 | 2 | 1 | 2 | 2 | 21 |
| 20 : T 200 | 0 | 2 | 0 | 0 | 0 | 2 | 2 | 0 | 0 | 1 | 0 | 0 | 0 | 0 | 7 |
| 21 : AS 100 | 2 | 1 | 2 | 0 | 1 | 2 | 2 | 2 | 0 | 1 | 1 | 0 | 1 | 1 | 16 |
| 22 : MZ 35 | 1 | 0 | 2 | 0 | 2 | 2 | 2 | 0 | 0 | 1 | 1 | 0 | 1 | 0 | 12 |
| 23 : BASIS 108 | 1 | 0 | 0 | 2 | 0 | 2 | 2 | 1 | 0 | 1 | 0 | 1 | 1 | 0 | 12 |
| 24 : LISA 2 | 0 | 0 | 0 | 2 | 2 | 2 | 1 | 1 | 0 | 0 | 1 | 2 | 1 | 0 | 12 |
| 25 : EUROPE PC | 1 | 0 | 2 | 0 | 1 | 2 | 2 | 0 | 2 | 0 | 1 | 0 | 2 | 0 | 14 |
| 26 : PSI 80 | 0 | 2 | 0 | 0 | 0 | 2 | 2 | 0 | 0 | 1 | 1 | 0 | 0 | 0 | 9 |
| 27 : CORONA PC 2 | 1 | 1 | 2 | 0 | 1 | 2 | 2 | 1 | 2 | 1 | 1 | 1 | 1 | 0 | 16 |
| 28 : OPLITE | 1 | 0 | 2 | 0 | 1 | 2 | 2 | 0 | 0 | 1 | 1 | 1 | 1 | 0 | 12 |
| 29 : HORIZON | 0 | 2 | 0 | 0 | 0 | 2 | 2 | 0 | 0 | 1 | 1 | 0 | 1 | 0 | 9 |
| 30 : FOXY | 1 | 1 | 2 | 0 | 1 | 2 | 2 | 1 | 2 | 0 | 1 | 1 | 2 | 0 | 17 |
| 31 : SKS 2500 | 0 | 2 | 0 | 0 | 0 | 1 | 2 | 1 | 0 | 1 | 1 | 0 | 0 | 1 | 9 |
| 32 : ZEPHYR | 0 | 2 | 0 | 0 | 0 | 2 | 2 | 0 | 0 | 1 | 1 | 0 | 0 | 1 | 9 |
| 33 : MBC 4050 | 0 | 2 | 0 | 0 | 1 | 2 | 2 | 0 | 0 | 1 | 1 | 1 | 2 | 1 | 13 |
| 34 : SANCO 8000 | 0 | 2 | 0 | 0 | 0 | 2 | 2 | 0 | 0 | 1 | 0 | 0 | 0 | 0 | 8 |
| 35 : IPC MODEL 15 | 0 | 2 | 0 | 0 | 0 | 0 | 2 | 0 | 0 | 1 | 1 | 0 | 1 | 1 | 8 |
| 36 : DESKTOP 10 | 0 | 1 | 2 | 0 | 1 | 0 | 2 | 1 | 0 | 1 | 1 | 0 | 1 | 0 | 11 |
| 37 : LISA 2-S | 0 | 0 | 0 | 2 | 2 | 2 | 1 | 1 | 0 | 1 | 0 | 2 | 2 | 0 | 15 |
| 38 : NEC PC 8000 | 0 | 0 | 0 | 2 | 0 | 2 | 1 | 1 | 0 | 1 | 1 | 0 | 0 | 0 | 8 |
| 39 : M 20 | 0 | 0 | 0 | 2 | 1 | 2 | 2 | 1 | 0 | 1 | 0 | 0 | 1 | 0 | 11 |
| 40 : TRS 80 MODEL 12 | 0 | 0 | 0 | 2 | 0 | 2 | 2 | 0 | 0 | 0 | 1 | 0 | 1 | 1 | 9 |
| TT : TABLE TOTAL | 20 | 35 | 42 | 20 | 28 | 68 | 68 | 20 | 21 | 27 | 30 | 25 | 43 | 22 | 470 |

position on the map. Interactions between circles and squares are reciprocal. (Newton's third law of the forces may serve as an analogy here). As a result, the attribute MS-DOS is pulled toward those microcomputers that provide it as the standard or optional operating system (e.g. : IBM PC XT, TANDY 2000 and others). The same MS-DOS is driven away on the map from those microcomputers that operate under the control of Other OS (e.g. : LISA2, ZEPHYR and others).

Circles and squares appearing near the barycenter of the map possess little or no contrasting power (e.g. : Price, Diskette Size and Maximal RAM Size). The barycenter is represented by a small cross on the map.

The plane of the map in figure 1 is defined by the two most important factors of interaction between microcomputers and their

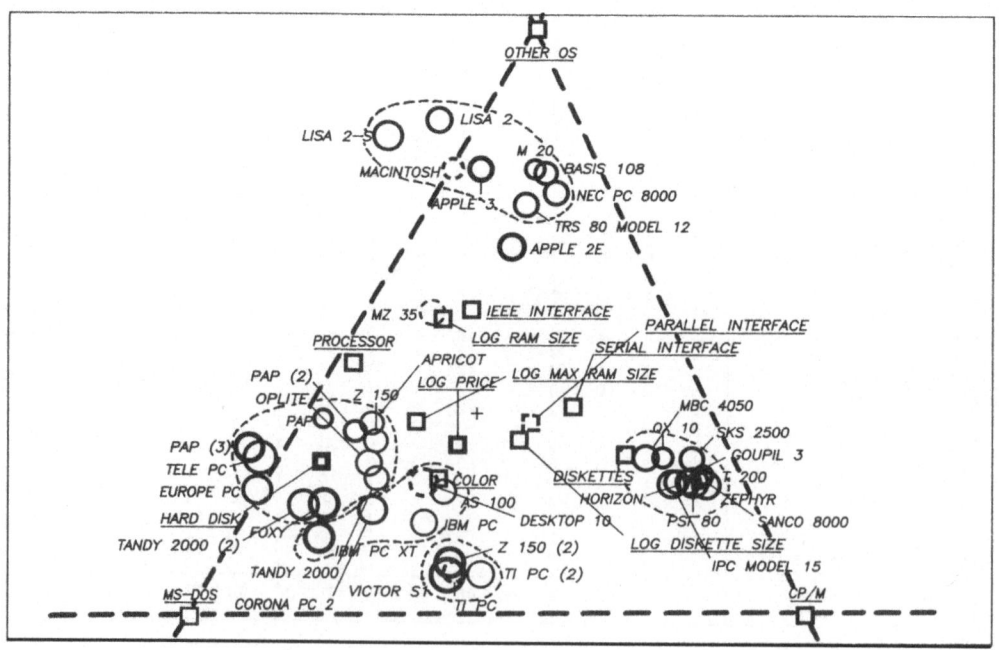

Fig.1. Spectramap produced from Table 1

attributes. A third factor is encoded by means of the thickness of
the contours of the symbols. For example, IBM PC is located somewhat below the plane and is shown, therefore, with a thinner than
average outline. Apple 2E, on the other hand, lies above the plane
of the map, and is rendered with a thick contour. The same graphic
convention applies to the mapping of attributes. Finally, some
symbols appear with a broken outline (e.g. : MZ35 and Parallel
Interface). These items cannot be fully represented by the three-
factor model. Broken contours indicate that part of the interaction
is contained in a fourth or higher-order factor.

The three-factor model of figure 1 reproduces 76% of the variance
in the data of table 1. About 64% of the variance produced by the
interactions between microcomputers and attributes is rendered graphically by the map.

## 3. Interpretation

Three attributes stand out very clearly on the map, namely the
three types of operating system (OS) that are described in the table
(MS-DOS, CP/M, Other OS). These appear as the polarizing forces that
draw the various microcomputers into clusters.

On the map, microcomputers have been clustered according to
their standard and optional OS. Because of the large polarizing forces
of the latter, all clusters are non-overlapping. Note the distinct
clustering of the APPLE, IBM, PAP, TANDY and TI families. According
to the table, APPLE 2E and MZ35 offer unique combinations of standard
and optional OS. As a result, they appear as singletons on the map.

Some attributes seem to be more associated with one type of OS
rather than with another. Hard disks come more often with MS-DOS;
IEEE-Interface is more frequent with Other OS; and double diskettes
seem to be more favored by CP/M. None of these associations, however,
appear to be very strong.

Several attributes are associated to two types of OS, at the exclusion of the remaining one.  This seems to be the case with Color and Processor Architecture.  But, once more, the associations are weak.

On the map we have added (broken) line segments in order to highlight the contrasts produced by the three types of OS;  Each of these line segments defines a particular contrast.  For example, projection of the circles upon the horizontal line segment, joining MS-DOS to CP/M, reproduces the contrast between these two types of OS.

Typical single-standard computers are found on the left (MS-DOS) and at the right (CP/M).  Microcomputers, which offer both of these OS or none of them, obtain their projections around the middle of the contrasting line segment.

## 4. Discussion and conclusion

Analysis of the heterogeneous description of microcomputers by means of Spectramap is straightforward, once the date have been suitably re-expressed.

Using spectral map analysis, we obtained a clear and almost obvious classification of the microcomputers.  Operating systems seem to polarize the microcomputers into families that are more or less specialized with respect to some of the other attributes (especially Hard Disk, Color, Processor Architecture, Interfaces and RAM Size).

Price does not appear to possess great contrasting power. Computers of various price levels are found in most of the larger clusters.

References

J.-P.Benzécri (1973). L'analyse des données. Vol.II, L'analyse des correspondances, Dunod, Paris.

S.Chah (1985). Classification of microcomputers. Problem submitted to the 4th Symposium on Data Analysis, Brussels.

L.J.Cronbach and G.C.Gleser (1953). Assessing similarities between profiles, Psychol.Bull., 50, 456-473.

P.J.Lewi (1976). Spectral Mapping, a technique for classifying biological activity profiles of chemical compounds. Drug Res. (Arzneim.Forsch.), 26, 1295-1300.

P.J.Lewi (1979). Datascope, An interactive program for multivariate data analysis. IBM Program 5787 FAE(F).

P.J.Lewi (1982). Multivariate analysis in industrial practice. Research Studies Press (J.Wiley), Chichester, Engl.

P.J.Lewi (1984). Multivariate data representation in medicinal chemistry. In : Chemometrics. Mathematics and Statistics in Chemistry. B.R.Kowalski, Ed., Reidel Publ., Dordrecht, The Neth.; 351-376.

P.J.Lewi, G.Calomme and J.Van Hoof (1985). Multivariate and longitudinal data on growing children, analyzed by Spectramap. Proceedings 4th Symposium on Data Analysis, J.Janssen e.a., Eds. Plenum, London, pp. in this volume.

SECTION 4.5

PARTIAL-ORDER CLASSIFICATION OF MICROCOMPUTERS

S. Levy and L. Guttman

Hebrew University of Jerusalem and the Israel Institute

of Applied Social Research

The problem is to classify forty given microcomputers according
to several given equipment characteristics.  Ten of the characteristics are called "qualitative" by the researchers.  These are
features which, if present, make for better functioning of a microcomputer.  The item categories deal with the sheer existence of the
features, being ordered from "not exist" (1), to "exist" (3), the
intermediate category (2) being "optional".  (The original data code
"0" for our "1", and "2" for our "3").  Only two out of the ten
features listed below, namely 5 and 10, are numerically ordered.
Following are listed the ten items for classifying the microcomputers (category ranks are in parentheses) :

1. Color monitor (1,2,3)

2. Disk Operating System CP/M (1,2,3)

3. Disk Operating System MS-DOS (1,2,3)

4. Disk Operating System "Other" (1,2,3)

5. Processor (1 = 8 bits, 2 = 16 bits, 3 = 32 bits)

6. Parallel Interface (1,2,3)

7. Serial Interface (1,2,3)

8. IEEE 488 Interface (1,2,3)

9. Hard Disk (1 = not exist, 2 = 5 MB, 3 = 10 MB)

10. Number of diskette drives (1,2).

The original data contain four additional classification variables, namely, price, access memory configuration, access memory
maximum, and mass storage. These four differ from the above ten in
meaning of internal order of categories, and hence cannot act as an
extension of the previous partial order. The first ten variables
have a common range, in the sense of "level of existence" in the
given microcomputer. The four further variables concern features
that exist in every computer. Their assessment cannot be made by
looking at the computer itself in a "presence-absence" fashion. An
example is "price". We shall explore to what extent this is determined by the ten traits of the presence-absence classifications.

The major problem to be faced here, then, is the dimensionality
and structure of the typology of the microcomputers from the ten
qualitatively ordered variables. The analysis will take into account
not only the existence of each feature but also its type. Price will
be examined as an external variable with respect to the partial-order
determined by the analysis.

## 1. The concept of partial-order

A partial-order structuple analysis (POSA) begins with some
different criteria for stratification of the population (in the
present case, microcomputers). Each microcomputer has an observed
structuple (profile) composed of fourteen structs, specified by
Chah's table of data. As already noted, just the ten "qualitative"
variables listed above have a common notion of order for the structs.
Therefore, for the classification purpose of POSA, each of the forty
microcomputers is assigned a structuple consisting of ten structs,
one for each feature. The external variable "price" will be used as
a criterion for external discrimination.

The overall partial-order is an automatic consequence of the
simple orders on each item separtely, namely, the extent of existence of each structural feature.  By definition, the least equipped
microcomputer will be assigned the structuple "1111111111", indicating absence of all ten features.  In practice, we cannot of course
expect to find such a machine.  Any microcomputer must have at least
one of the ten features.  Indeed, it turns out that all the microcomputers share one feature to some degree (optional or existing),
namely, Serial Interface.

Actually, among the forty microcomputers under consideration,
the least equipped have at least some three out of the ten listed
structural features.  But such three features differ with the kind
of microcomputer, as can be clearly observed for the IPC Model 15,
TRS 80 Model, and MacIntosh.

A microcomputer with maximum equipment would have the structuple "3333333332".  The highest possible structuple also cannot be
expected to occur in practice : some features can be substituted for
by others.  Two of the microcomputers turn out to be relatively
highly equipped, each lacking only one feature.  These two are
Victor SI and Corona PC 2, Victor being the better equipped, with
more actually existing features than just optional.  It should also
be noted, that while the least equipped microcomputers vary in the
kinds of feature they lack, the best equipped lack the identical
structural feature, namely "other" disk operating system.  This is
not a real lack, because they have CP/M and MS-DOS disk operating
systems, specified in items 2 and 3.  Item 4 cannot be properly interpreted without reference to items 2 and 3.  De facto, the two
best equipped microcomputers do have all the kinds of structural
features.

All other microcomputer structuples are partly ordered between
the extreme lowest and highest possible.  Some microcomputers share

the same structuple;   the forty micro-computers yield only 34 different structuples.   The distinct structuples are listed below.
The structs in the structuples are written from left (1) to right
(10), according to the above classification list.

| Name of Microcomputer | ID | Profile (Structuple) | Sum of category ranks |
|---|---|---|---|
| (Non-existent) | 1* | 3333333332 | 29 |
| VICTOR SI | 2 | 2331233232 | 24 |
| AS 100 | 3 | 3231233312 | 23 |
| CORONA PC 2 | 4 | 2231233232 | 23 |
| Z 150 (2) | 5 | 1331233132 | 22 |
| FOXY | 6 | 2231233231 | 22 |
| TI PC(2) | 7 | 3331232112 | 21 |
| TELE PC | 8 | 2131233231 | 21 |
| TANDY 2000 (2) | 9 | 2131233132 | 21 |
| IBM PC XT | 10 | 2231232131 | 20 |
| MZ 35 | 11 | 2133133112 | 20 |
| EUROPE PC | 12 | 2131233131 | 20 |
| TI PC 13 | 13 | 3331232111 | 20 |
| TANDY 2000 ) OPLITE ) | 14 | 2131233112 | 19 |
| IBM PC | 15 | 2231232112 | 19 |
| BASIS 108 | 16 | 2113133212 | 19 |
| APRICOT | 17 | 1131233212 | 19 |
| LISA 2-S | 18 | 1113332221 | 19 |
| M 20 | 19 | 1113233212 | 19 |
| PAP (3) | 20 | 1131222231 | 18 |
| LISA 2 | 21 | 1113332211 | 18 |
| Z 150 | 22 | 1131233112 | 18 |
| APPLE 2E | 23 | 2213122221 | 18 |
| MBC 4050 | 24 | 1311233112 | 18 |
| DESKTOP 10 | 25 | 1231213212 | 18 |
| QX 10 | 26 | 2311132212 | 18 |
| GOUPIL 3 ) T 200 ) PSI 80 ) HORIZON ) ZEPHYR ) SANCO 8000 ) | 27 | 1311133112 | 17 |
| PAP (2) | 28 | 1131222212 | 17 |
| SKS 2500 | 29 | 1311123212 | 17 |

* Extreme profile added by program

| Name of Microcomputer | ID | Profile (Structuple) | Sum of category ranks |
|---|---|---|---|
| APPLE 3 | 30 | 1113123221 | 17 |
| NEC PC 8000 | 31 | 1113132212 | 17 |
| MACINTOSH | 32 | 1113313111 | 16 |
| TRS 80 MODEL | 33 | 1113133111 | 16 |
| PAP | 34 | 1131222211 | 16 |
| IPC MODEL 15 | 35 | 1311113112 | 15 |
| (Non-existent) | 36* | 1111112111 | 11 |

By definition, one structuple is <u>higher</u> than another if and only if it is higher on at least one item (structural feature), and not lower on any other item. For example, consider the structuple "1331233132" for Z 150 (2) (ID 5 in the above list), and the structuple "1131233112" for Z 150 (ID 22). These two structuples are <u>comparable</u>, with Z 150 (2) being the better equipped. Z 150 (2) has a CP/M operating system and hard disk; Z 150 lacks both these features, and is equal to Z 150 (2) on the rest.

Again by definition, two structuples are <u>not comparable</u> if and only if one structuple is the higher on at least one struct while the other structuple is also the higher on at least one struct. Many, if not most, of the pairs of microcomputer structuples are noncomparable.

A systematic and detailed study of the similarities and differences among the structuples is facilitated by viewing them in the space of smallest dimensionality that can preserve the partial-order. Unidimensionality is the case where <u>all</u> structuples are comparable. Noncomparability requires two or more dimensions for its portrayal. To determine the (smallest) dimensionality of the partial-order, the technique of Partial Order Structuple (Scalogram) Analysis with Base Coordinates (POSAC-I) is employed (Levy and Guttman, 1985).

---

* Extreme profile added by program

POSA is one of the families of techniques for intrinsic ("non-metric") data analysis introduced by Guttman. It is a special case of Multidimensional Structuple (Scalogram) Analysis (MSA), when partial-order is specified. MSA-I does not take advantage of the powerful order feature when it exists; POSAC-I does, and its special algorithm is therefore more effective than MSA-I for partly ordered data. For details on MSA in general see Zvulun (1978) and Lingoes (1968). A discussion of the mathematical and technical aspects of POSA in general is presented in Shye (1978, 1985). Published examples of uses of partial orders can be found in Guttman and Bat-Miriam (1961), Brown and Sime (1982), Yalan et al. (1972), Shye and Elizur (1976) and Levy (1984, 1985).

## 2. The empirical partly ordered structure

POSAC shows that the partial order of the forty microcomputers according to their structural features is essentially two-dimensional. The space diagram of the computer output is portrayed in Figure 1.

The reference axes of Figure 1 are the base coordinates calculated by POSAC-I. Such base coordinates have an intrinsic meaning for POSA, as will be illustrated later below.

In Figure 1, each structuple appears as a point, the serial numbers being those in the above list. Actually, as already noted, each of the structuples but two - 14 and 27 - is associated with but one microcomputer. In Figure 1, each structuple serial number is accompanied by the name of the corresponding microcomputer. Structuple 1*, at the northeast corner of Figure 1, is the highest possible structuple, representing the best equipped (but non-existent) microcomputer. The lowest possible equipment level (again non-existent) for the microcomputers is represented by structuple 36*, at the southwest corner of the Figure.

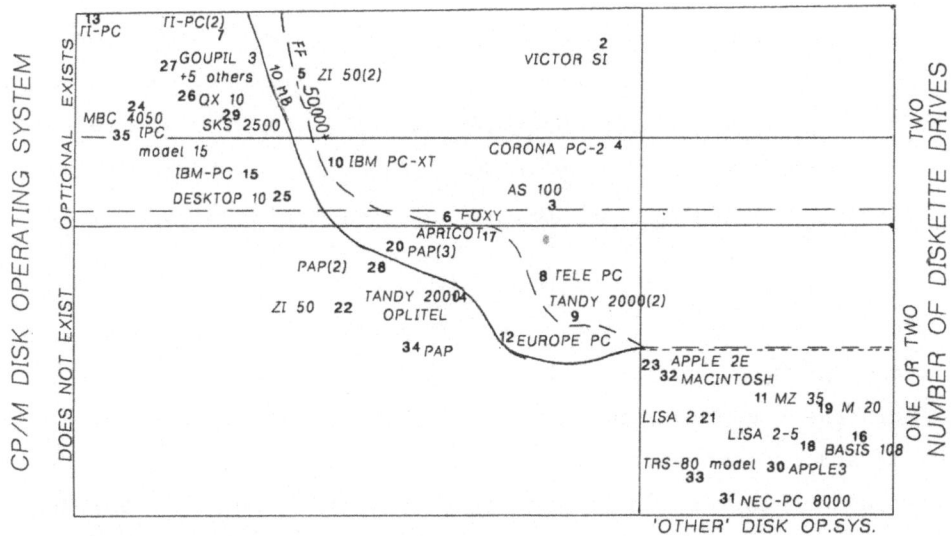

Fig.1. Two-Dimensional Partial Order of Microcomputers by Ten Qualitative Items

The diagonal direction going from the northeast corner of figure
1 to the southwest corner defines the intermediate levels of equipment of microcomputers, and is called the joint direction of the
partial order.  Any two structuples that are comparable will be represented by two points whose common line has a positive slope, the
one closer to the upper right having the better equipment.  The extreme corner regions are empty here.  Most of the points are distributed around a negatively sloped line in a rather narrow belt.  Only
VICTOR SI - and to some extent CORONA PC 2 - tend towards the upper
right corner of superior equipment.  This means that the majority of
the forty microcomputers have a rather similar overall level of equipment.  They differ largely in the kind of features they possess,
rather than overall level of features.  This is what makes the structuples noncomparable.  Two noncomparable structuples have their two
points aligned with negative slope, or in the lateral direction (as
presented in figure 1).  The joint and lateral directions are essentially 45° rotations of the two base coordinates.  All four kinds of

directions in the 2-space have a rôle in interpreting the results.

Detailed analysis of the systematic differences among the structural features is made in terms of ten POSAC diagrams, one for each
item.  We shall not present here each of the item diagram separately,
but rather indicate in the space diagram (figure 1) the boundaries
of regions corresponding to certain categories of the ten items.

POSAC-I calculates a mathematically optimal pair of base axes
(X  and  Y) for the empirical partial-order.  There need not always
be items that correspond to the directions of these basic coordinates.  In the present example, item (feature) 2 (CP/M disk operating
system) and to some extent also item 10 (number of diskette drives)
were found to correspond to the  Y  direction.  The solid horizontal
lines (cutting the  Y  direction) in figure 1 indicate the partitions
according to CP/M disk operating system (item 2).

All microcomputers in the upper part of the figure, above the
solid line, do have a CP/M disk operating system, while the microcomputers at the extreme bottom region do not have it.  Between these
two regions are located, in a region of their own, micrcomputers that
have the CP/M disk operating system as an optional feature.  The
broken line in the horizontal direction of figure 1 indicates the
partition according to number of diskette drives.  Microcomputers
located in the region above the broken line have 2 diskette drives,
while those below it may have either 1 or 2 diskette drives.  The
fact that this boundary is within the second region of the partition
according to the PC/M disk operating system, means that microcomputers
which actually have a CP/M disk operating system (and most of those
which have it as an option) have two diskette drives.  Thus, the
two features constitute a "scale" among themselves.

The horizontal axis  X  is partitioned by item 4, namely "other"
disk operating system.  This partition is indicated by the solid

vertical line. Microcomputers which have "other" disk operating systems are located to the right of the vertical line, while those who do not have "other" disk operating system are to its left. There is of course no "optional" category, as this feature is by itself an optional one for those microcomputers which do have the disk operating systems mentioned explicitly (CPM or MS DOS). Having such items corresponding to the base coordinates means that the partly ordered space is essentially spanned by these three structural features (two for the Y axis, and one for the X).

The difference between levels on the base items determines the differentiation among the structuples in the lateral direction of the partial order. This is portrayed in the lower right corner region and upper left corner region. The boundaries of the lower right region are indicated in figure 1 by the vertical solid line partitioning the X direction, and by the dotted line partitioning the Y direction. The uniqueness of this region stems from the fact that there are no microcomputers above the dotted line in the region of "other" disk operating system. Microcomputers which have "other" disk operating system are located solely in this region.

According to the difference between levels on the base items, this region contains microcomputers that do not have CP/M or MS-DOS operating system but have "other" disk operating systems. It should also be noted that most of the microcomputers in this region have no hard disk, and only few have 5 MB on this feature. As will be shown later, the Hard Disk feature plays an important rôle in refining the base regions according to the joint direction, and in discriminating among microcomputers according to their prices.

Computers which have a CP/M disk operating system and do not have "other" disk operating system, are of course located in the upper left corner region. Remaining computers, which do not have the CP/M or offer it only as optional, have MS-DOS. This is not shown explicitly in the present drawing.  Thus, the lateral direction distinguishes among <u>kinds</u> of microcomputers according to the difference between their levels on the base items.

Each of the other features behave like a combination of the base items, essentially partitioning their categorical regions into finer regions.

As already mentioned, the joint direction is rather weak in this space, most of the microcomputers being located in a quite narrow band along the negative diagonal.  But it turns out, that the joint direction plays an important rôle, not only by further refining the regions derived according to the base items, but also in discriminating among microcomputers.  This is especially outstanding for feature no.9, namely, Hard Disk. Microcomputers which are located at the upper right corner of the solid diagonal line, in figure 1, are the highest on "Hard Disk" having 10 MB.  The microcomputers left to this line either do not have hard disk or only 5 MB.  Being high on this feature implies being high on most of the other features as well (for example see "VICTOR SI"), except of course for "other" disk operating system.  For details see discussion above.

Features 6 and 7 (Parallel and Serial Interface), and to some extent also feautre 3 (MS-DOS disk operating system), partition the POSAC space in a fashion similar to that of Hard Disk, though with less sharp regionality.  The remaining three features do not play any special rôles in partitioning the space.

## 3. Discriminant analysis for prediction of price

The above POSAC technique revealed that it is possible to classify the forty microcomputers in a two dimensional space with a lawfulness according to the ten structural features suggested.  We shall now go on to see how this structure helps provide a prediction for

price, by treating this variable as external to the partial order
structure.

The price criterion was projected onto the partial-order. Such
a superimposition helps reveal the relationship - if any - between
the partial order of the microcomputers and the criteria to be predicted. The same printout device is used to relate the partial-
order with traits extrenal to the POSAC traits such as price. The
relationships can be summarized in part by the (monotone) correlations with one or more directions of the POSA : base, joint and/or
lateral. The microcomputers were grouped into four price categories
from the least to the most expensive :

1) 20,000 - 29,999 F.F.

2) 30,000 - 39,999 F.F.

3) 40,000 - 49,999 F.F.

4) 50,000+ F.F.

The relationships between the partial order of the microcomputers
and their prices can be summarized in part by correlations with two
directions of the POSAC : joint and lateral. (We say "in part",
because the prediction is clear for only part of the price categories).

It turns out that an almost perfect discriminant function is
provided by the joint direction for the most expensive microcomputers.
This is shown by the diagonal broken line in Figure 1. The microcomputers to the upper right corner of this line are the most expensive (over 50,000 F.F.), and no microcomputers of over 50,000 F.F.
are to be found in the region below the diagonal broken line. This
partition of the space according to price categories almost coincides
with the partition of the space according to feature 9, namely Hard
Disk. All the expensive computers have 10 MB Hard Disk. Only the
microcomputers [PAP(3) and Europe PC] with Hard Disk (10 MB) are not
included in the most expensive region. But they are relatively expensive too, their price being over 47,000 F.F. Thus, a perfect

discriminant function is provided by the joint direction for the most
expensive microcomputers, and the discrimination is also according to
the partition by the Hard Disk feature.

As already discussed above, the distribution of the forty microcomputers is mainly along the lateral direction band.  This direction
helps somewhat in predicting prices.  The lower right region of microcomputers – for those which do not have the CP/M disk operating system
but have some "other" – is characterized by inexpensive prices.  All
the microcomputers in this region cost less than 39,000 F.F. (which
is the second price level according to the above grouping).

Thus we have two discriminant functions according to price.  One,
an almost perfect discriminant function distinguishes among the most
expensive microcomputers and the others.  This distinction is in the
joint direction, mostly according to the Hard Disk feature.  The
other discriminant function distinguishes among microcomputers by
the lateral direction.  All those which have "other" disk operating
systems (and not a CP/M) are inexpensive.  This discriminant function
is imperfect because other inexpensive microcomputers may be found in
other regions of the POSAC space as well, except for the joint region
of 10 MB Hard Disk.

## 4. Afterword

The data and analysis presented above are very relevant to market
research.  However, they are too limited to give practical advice to
manufacturers and distributors of microcomputers.  Successful sales
depend not only on objective features of the product but also on how
these are perceived by the potential consumer.  In this regard, the
list of traits as given may not be sufficiently inclusive from the
consumer's point of view.  A full-blown market research program requires a facet design which will help ensure that all relevant features are studied, and the reactions of potential customers be studied.  This will enlarge the data base, but the POSAC approach to

the data analysis will remain a powerful way for helping to arrive
at practical conclusions.

References

J.M.Brown and J.D.Sime (1982). Multidimensional Scaling Analysis of
        Qualitative Data.  In E.Shephard and J.P.Watson (Editors) :
        Personal Meanings. John Wiley & Sons.

L.Guttman and M.Bat-Mirian (1961). A New Approach to Fingerprint
        Analysis in Population Studies. In Proceedings of the Second
        International Congress of Human Genetics, Rome.

S.Levy (1984). Partial Orders of Israeli Settlements by Adjustive
        Behaviors. Israel Social Science Research 2, 44-65.

S.Levy (1985). Partial Order Analysis of Crime Indicators. Social
        Indicators Research 16, 195-199.

S.Levy and L.Guttman (1985). The Partial Order of Severity of Thyroid
        Cancer with the Prognosis of Survival. In J.F.Marcotorchino,
        J.M.Proth and J.Janssen (Eds). Data Analysis in Real Life
        Environment : Ins and Outs of Solving Problems. Amsterdam,
        Elsevier Science Publisher B.V. (North-Holland), 111-119.

J.C.Lingoes (1968). The Multivariate Analysis of Qualitative Data.
        Multivariate Behavioral Research 3, 61-94. Also in J.C.Lingoes,
        E.E.Roskam, I.Borg (Editors) Geometric Representations of
        Relational Data. Ann Arbor : Mathesis Press, 1979, 575-608.

S.Shye (1978). Partial Order Scalogram Analysis. In S.Shye (Ed.)
        Theory Construction and Data Analysis in the Behavioral Sciences.
        San Francisco, Jossey-Bass, 265-279.

S.Shye (1985). Multiple Scaling. Amsterdam, North-Holland.

S.Shye and D.Elizur (1976). Worries about Deprivation of Job Rewards
        Following Computerization : A Partial Order Scalogram Analysis.
        Human Relations 29, 63-71.

Yalan et al. (1972). The Modernization of Traditional Agricultural
        Villages. Rehovot, Israel, Settlement Study Center.

E.Zvulun (1978). Multidimensional Scalogram Analysis : The Method
    and Its Application.  In S.Shye (Ed.) Theory Construction and
    Data Analysis in the Behavioral Sciences. San Francisco :
    Jossey-Bass, 237-264.

SECTION 4.6

CONCLUSION

P. Cullus

CADEPS, Brussels, Belgium

Forty Microcomputers, described by either qualitative and quantitative data had to be classified. Several approaches were presented during the Workshop, that gave quite similar results. Four of them had been accepted for publication.

The "Triple Comparison" approach of Chah's solution seems to be a very powerful one for solving this type of problem. His paper is rather theoretical, due to the fact that the author had first to present the methodology of his method.

Using "Spectramap" in Lewi's paper shows us one more time the simplicity and the power of this method that appears as a three-dimensional way of performing Factor Analysis.

The "Partial-Order Classification" presented by Levy and Guttman appears as a quite attractive one. Its originality was to work separately on the qualitative data first and then to include the quantitative variables. The power of the so-called "structuples" is one more time demonstrated and the graphic results are quite easy to read.

Although it is not an original method for solving problems, the use of Factorial Analysis in Van Brussel's presentation make us conclude that it is as powerful as more appropriate technics for solving this problem.  In this case the characteristics of the clusters are quite easy to determine due to the fact of the graphic presentation of the results.

In conclusion, the fact that the structure of the Microcomputers set was very strong made us able to perform a real comparison of the approaches without having to take into account a large diversity of results.

CHAPTER 5. GROUP TECHNOLOGY IN PRODUCTION MANAGEMENT

SECTION 5.1

THE SHORT HORIZON PLANNING LEVEL-GENERALIZED ISSUE

H. Garcia and J.M. Proth

## 1. The problem

### 1.1. Introduction

We mean by working process a sequence of tasks to be performed in order to manufacture a part. We say that two parts belong to the same part type if their working processes are identical. We suppose that a machine is able to perform only one task, and that only one machine is available to perform a task. These hypotheses do not restrict the following results, but simplify the explanations.

The general problem consists in finding a partition of the set of part types into a given number of subsets called "part families", a partition of the set of machines into the same number of subsets called "production subsystems", and a biunivocal correspondence between these partitions in such a way that :

1. the total amount of time spent by the parts in non corresponding production subsystems is as small as possible (the correspondence has to be taken in the sense of the above biunivocal correspondence).

2. in addition, the tasks performed in a non corresponding pro-

duction subsystem are, as far as possible, those which need the smallest amount of time.

We first give an algorithm which leads to a "good" solution of this problem.

The problem we have to solve is slightly different from the previous one : we are looking for part families and production sub-systems in such a way that the number of travels between subsystems is as small as possible.

This problem is equivalent to the previous one if we consider that the processing time of a part which visit a machine is equal to 1.

## 1.2. Setting of the general problem

X is the set of n parts types and Y the set of m tasks. We also consider the matrix :

$$T = [t_{i,j}]; \quad i = 1,2,\ldots,n; \quad j = 1,2,\ldots,m; \quad (1)$$

where $t_{i,j}$ is the time needed to perform the task j on a part belonging to the part type i.
$k_i$ (i = 1,...,n) is the number of parts belonging to the part type i.
We define the matrix :

$$T^* = [t^*_{i,j}]; \quad i = 1,\ldots,n; \quad j = 1,\ldots,m;$$

as follows :

$$t^*_{i,j} = t_{i,j}/U; \quad U = \text{Max } t_{i,j} \quad i = 1,\ldots,n; \quad j = 1,\ldots,m \quad (2)$$

Let p be the number of part families (or production subsystems) chosen by the user.
If :

$$P_X = \{X_1,\ldots,X_p\} \quad \text{and} \quad P_Y = \{Y_1,\ldots,Y_p\}$$

are respectively a partition of X and a partition of Y, we define $W_{r,s}$ (r = 1,...,p; s = 1,...,p) as follows :

$$W_{r,s} = \{(i,j) \in (1,\dots,n) \times (1,\dots,m)/i \in X_r \text{ and } j \in Y_s\} \quad (3)$$

Let us now consider the partitions $P_X^* = \{X_1^*,\dots,X_p^*\}$ and $P_Y^* = \{Y_1^*,\dots,Y_p^*\}$, and $\varphi^*$ which is a biunivocal correspondence between $P_X^*$ and $P_Y^*$.

In order to simplify the notation, we suppose that the biunivocal correspondence always associates to an element of the partition of X the element of same rank in the partition of Y. For instance :

$$\varphi^*(X_i^*) = Y_i^*; \quad i = 1,\dots,p$$

As we said in the introduction, $(P_X^*, P_Y^*, \varphi^*)$ is a "good" solution of the problem if :

1. $\displaystyle\sum_{\substack{r=1 \\ }}^{p} \sum_{\substack{s=1 \\ s \neq r}}^{p} \sum_{(i,j)\in W_{r,s}^*} k_i t_{i,j}$ is "as small as possible"

2. If $(i,j) \in W_{r,s}^*$ and $r \neq s$, then "$t_{i,j}$ is small".

Obviously, the optimal solution of this problem is :

and $\begin{cases} X_1^* = X, X_2^* = \dots = X_p^* = 0 \\ \\ Y_1^* = Y, Y_2^* = \dots = Y_p^* = 0 \end{cases}$

But we are looking for a solution $(P_X^*, P_Y^*, \varphi^*)$ in which the elements of $P_X^*$ and $P_Y^*$ are non-empty. We then propose to solve the following problem : find $P_X^*$ and $P_Y^*$ such that :

$$h \sum_{r=1}^{p} \sum_{(i,j)\in W_{r,r}^*} k_i (t_{i,j}^*)^{\alpha}$$

$$+ (1 - h) \sum_{\substack{r=1 \\ }}^{p} \sum_{\substack{s=1 \\ s \neq r}}^{p} \sum_{(i,j)\in W_{r,s}^*} k_i (1 - t_{i,j}^*)^{1/\alpha}$$

$$(4)$$

$$= \underset{w\in\mathcal{D}}{\text{Max}} \left\{ h \sum_{r=1}^{p} \sum_{(i,j)\in W_{r,r}} k_i (t_{i,j}^*)^{\alpha} \right.$$

$$+ (1 - h) \sum_{\substack{r=1 \\ }}^{p} \sum_{\substack{s=1 \\ s \neq r}}^{p} \sum_{(i,j) \in W_{r,s}} k_i (1 - t^*_{i,j})^{1/\alpha} \Bigg\}$$

where

- $W^*_{r,s}$ is obtained using (3) by replacing $X_r$ by $X^*_r$ and $Y_s$ by $Y^*_s$
- $\mathbb{W} = \{W_{r,s}\}$; $r = 1,\ldots,p$; $s = 1,\ldots,p$
- $\mathbb{D}$ is the set of admissible partitions of $\{1,\ldots,n\} \times \{1,\ldots,m\}$
- $\alpha \in R^{*+}$. The value of $\alpha$ is chosen by the user ($\alpha \leqslant 1$)
- $h \in (0,1)$ is also chosen by the user.

The more value of $\alpha$ is small, the more the tasks which are not performed using the corresponding subsystem have a small processing time.

The more the value of $h$ is great, the more the tasks will be performed using the corresponding subsystem.

In both the cases, the risk to obtain some empty elements in $P^*_X$ and $P^*_Y$ increases, but not in the same way.

The problem given by (4) allows to take into account more or less the various parameters of the problem. An example will be given below.

The following paragraph is devoted to the search of a "good" solution of this problem.

## 1.3. An example

The following matrix is the example we want to solve in order to illustrate the following algorithm. Each line represents a part type and each column represents a machine, except the last one which gives the weights of the part types.

| | | | |
|---|---|---|---|
| 1 | 0.1.0.1.0.0.0.0.0.0.0.0.1.3. | 31 | 0.0.0.1.0.0.0.1.0.0.0.0.1. |
| 2 | 1.0.0.0.1.0.0.0.0.1.1.0.3. | 32 | 0.0.0.0.0.1.0.0.1.0.0.1.3. |
| 3 | 0.0.0.1.0.1.0.0.1.0.0.0.1. | 33 | 0.1.0.1.0.0.0.1.0.0.0.0.1. |
| 4 | 1.0.0.0.0.1.0.0.1.1.0.0.2. | 34 | 0.0.0.0.0.0.1.0.0.1.1.1.2. |
| 5 | 1.1.0.1.0.0.0.1.0.0.0.1.2. | 35 | 0.0.1.0.1.0.0.0.0.0.0.0.1. |
| 6 | 0.0.0.0.0.0.0.1.0.0.0.0.2. | 36 | 1.0.1.0.0.0.1.0.0.1.0.1.2. |
| 7 | 0.1.0.0.0.0.1.0.1.1.0.1.2. | 37 | 0.0.0.0.0.0.1.0.1.0.0.4. |
| 8 | 0.0.0.0.0.0.0.0.0.1.0.1.4. | 38 | 0.0.0.0.1.0.0.0.0.0.0.0.5. |
| 9 | 1.0.0.0.0.0.0.0.0.0.1.1.1. | 39 | 0.0.0.0.0.1.0.0.0.0.0.1. |
| 10 | 0.0.0.0.0.0.0.0.0.1.0.0.1. | 40 | 1.0.1.0.1.0.0.0.0.0.0.3. |
| 11 | 0.0.0.0.0.0.0.0.1.0.0.0.3. | 41 | 1.0.0.0.0.1.0.0.0.1.0.3. |
| 12 | 0.0.0.0.0.1.0.0.1.1.0.0.1. | 42 | 0.1.0.1.0.0.0.1.0.0.0.0.1. |
| 13 | 0.1.0.1.0.0.0.1.0.0.0.0.1. | 43 | 0.0.1.0.1.0.0.0.0.1.0.0.2. |
| 14 | 0.0.0.0.0.1.1.0.1.0.1.0.2. | 44 | 0.1.0.0.0.0.0.1.0.1.0.0.1. |
| 15 | 0.1.0.0.0.0.0.1.0.0.0.0.1. | 45 | 0.0.0.1.0.1.0.1.0.0.0.0.1. |
| 16 | 0.0.0.1.0.0.0.1.0.0.0.1.2. | 46 | 0.0.1.0.1.0.0.0.0.1.0.0.3. |
| 17 | 0.1.0.0.0.0.1.0.0.1.0.1.1. | 47 | 0.1.0.0.0.0.0.1.0.0.0.0.1. |
| 18 | 0.1.0.0.1.1.0.0.1.0.0.0.1. | 48 | 0.1.0.1.0.0.0.1.1.0.0.0.4. |
| 19 | 0.1.0.0.0.1.0.1.0.0.0.0.5. | 49 | 0.0.0.0.0.0.1.0.0.1.0.0.1. |
| 20 | 0.0.0.0.0.1.1.1.1.0.0.0.2. | 50 | 0.0.0.0.0.1.0.0.1.0.0.0.1. |
| 21 | 0.0.0.0.1.0.1.1.0.1.0.1.4 | 51 | 0.1.0.1.0.0.0.0.0.0.0.1.3. |
| 22 | 0.0.0.0.0.1.0.0.1.0.1.0.3. | 52 | 1.0.0.0.1.0.0.0.0.1.1.0.3. |
| 23 | 1.0.1.0.0.0.1.0.1.0.0.1.3. | 53 | 0.0.0.1.0. .0.0.1.0.0.0.1. |
| 24 | 0.1.0.0.0.0.0.0.1.0.0.0.3. | 54 | 1.0.0.0.0.1.0.0.1..1.0.0.2. |
| 25 | 0.0.1.0.1.0.1.0.0.0.0.2. | 55 | 1.1.0.1.0.0.0.1.0.0.0.1.2. |
| 26 | 1.0.0.0.0.0.0.0.0.1.0.4. | 56 | 0.0.0.0.0.0.0.1.0.0.0.0.2. |
| 27 | 0.1.1.0.1.0.1.0.0.0.0.2. | 57 | 0.1.0.0.0.0.1.0.1.1.0.1.2. |
| 28 | 0.0.0.0.0.0.0.0.0.1.0.1.3. | 58 | 0.0.0.0.0.0.0.0.0.1.0.1.4. |
| 29 | 0.0.1.0.0.0.1.0.1.1.1.0.3. | 59 | 1.0.0.0.0.0.0.0.0.0.1.1.1. |
| 30 | 0.1.0.1.0.0.0.1.0.0.0.1.2. | 60 | 0.0.0.0.0.0.0.0.0.1.0.0.1. |

## 2. An heuristic algorithm

We suppose that a partition of the part types is given. We will show that it is possible to obtain a good solution of (4) starting from this initial partition. We first prove a fundamental result.

## 2.1. A fundamental result

### Theorem 1

Let $P_X^K = \{X_1^K, \ldots, X_p^K\}$ and $P_Y^K = \{Y_1^K, \ldots, Y_p^K\}$ be partitions of X and Y respectively.

1.  For  $j = 1,2,\ldots,m$, we compute :

$$a_r^K(j) = h \sum_{i \in X_r^K} k_i (t_{i,j}^*)^\alpha + (1 - h) \sum_{i \notin X_r^K} k_i (1 - t_{i,j}^*)^{1/\alpha}$$

   for  $r = 1,\ldots,p$

and we denote by  $r^K(j)$  the integer which verifies :

$$a_{r^K(j)}^K = \underset{r=1,\ldots,p}{\text{Max}} \; a_r^K(j)$$

We assign the task  $j$  to the production subsystem  $Y_{r^K(j)}^{K+1}$

Finally, we obtain the partition :

$$P_Y^{K+1} = \{Y_1^{K+1}, Y_2^{K+1}, \ldots, Y_p^{K+1}\}$$

in which some elements may be empty.

2.  For  $i = 1,2,\ldots,n$, we compute :

$$b_r^K(i) = h \sum_{j \in Y_r^{K+1}} (t_{i,j}^*)^\alpha + (1 - h) \sum_{j \notin Y_r^{K+1}} (1 - t_{i,j}^*)^{1/\alpha}$$

   for  $r = 1,\ldots,p$

and we denote by  $r^K(i)$  the integer which verifies :

$$b_{r^K(i)}^K = \underset{r=1,\ldots,p}{\text{Max}} \; b_r^K(i)$$

We assign the part type  $i$  to the part family  $X_{r^K(i)}^{K+1}$

Finally, we obtain the partition :

$$P_X^{K+1} = \{X_1^{K+1}, X_2^{K+1}, \ldots, X_p^{K+1}\}$$

Then, either the partitions  $(P_X^{K+1}, P_Y^{K+1})$  are better than the partitions  $(P_X^K, P_Y^K)$, or the value of the criterion is the same for both solutions.

Proof

1. Let us consider $j \in \{1,2,\ldots,m\}$ and suppose that

$$j \in Y^K_{s^K}, \quad s^K \in \{1,2,\ldots,p\}$$

If $s^K = r^K(j)$, then the contribution of the column $j$ to the criterion (4), knowing $P^K_X$, remains the same.

If $s^K \neq r^K(j)$, then this contribution increases.

Finally, $P^{K+1}_Y$ leads to a greater value for the criterion (4) than $P^K_Y$ if $P^K_X$ is known.

To conclude, the solution $(P^K_X, P^{K+1}_Y)$ is either better than $(P^K_X, P^K_Y)$, or leads to the same value for the criterion (4).                    (5)

2. Using the same arguments, it is easy to show that the solution $(P^{K+1}_X, P^{K+1}_Y)$ is either better than $(P^K_X, P^{K+1}_Y)$, or leads to the same value for the criterion (4).                    (6)

3. From (5) and (6), we conclude that either the solution $(P^{K+1}_X, P^{K+1}_Y)$ is better than the solution $(P^K_X, P^K_Y)$, or the value of the criterion (4) is the same for both solutions.

  The proof is then finished.                    □

## 2.2. The algorithm

  We suppose that we know $P^0_X$, initial partition of $X$ (i.e. initial set of part families).

  The algorithm is then the following :

1. Set $K = 1$
2. Compute $P^K_Y$ starting from $P^{K-1}_X$ (see the first part of theorem 1)
3. Compute $P^K_X$ starting from $P^K_Y$ (see the second part of theorem 1)
4. If $(K = 1)$ or if (the solution $(P^K_X, P^K_Y)$ is better than the solution $(P^{K-1}_X, P^{K-1}_Y)$ acoording to the criterion (4)), go to 2 else $(P^K_X, P^K_Y)$ is a "good" solution.

Theorem 2

The previous algorithm converges.

Proof

Obvious, if we consider that :

1. the number of partitions of $X$ and $Y$ is finite.

2. the value of the criterion (4) increases at each step of the algorithm or the algorithm stops.                                    □

Remarks

1. This algorithm leads to a solution which depends on the initial set of part families. This solution may be not optimal. A consequence of this remark is that we have to execute a number of trials to be sure to obtain a solution close to the optimal one.

2. The initial set of part families can be obtained using the "nuées dynamiques" method, which is well known in data analysis. In this method, the number of part families has to be chosen by the user (see Diday, Lemaire, Pouget & Testu, 1982).

3. The solution depends on $h$ and $\alpha$.

4. If $p$ is the number of part families in the initial partition of $X$, it is possible to arrive at a solution which contains less than $p$ part families and production subsystems. We observe that there exists, for given values of $h$ and $\alpha$, an equilibrium value $p^*$ such that :

a. if $p \leqslant p^*$, then the solution is composed of $p$ part families and $p$ production subsystems,

b. if $p > p^*$, the solution is composed of $p^*$ part families and $p^*$ production subsystems.

2.3. The example

For the example we have to solve, the result is the same whatever the value of $\alpha$ may be. We choose $h = 1/2$. We obtain the following result. Some other trials could be done using some other values for $h$. In the case of various trials, the final choice is the responsibility of the user.

| | | 1 | 1 | 2 | 2 | 4 | 4 | 5 | 8 | 8 | 11 | 15 | 15 | |
| - | - | - | - | - | - | - | - | - | - | - | - | - | - | - |
| | | 12 | 10 | 8 | 4 | 11 | 1 | 7 | 9 | 6 | 2 | 5 | 3 | |
| 1 | 7 | 1. | 1. | 0. | 0. | 0. | 0. | 1. | 1. | 0 | 1. | 0. | 0. | 2. |
| 1 | 8 | 1. | 1. | 0. | 0. | 0. | 0. | 0. | 0. | 0. | 0. | 0. | 0. | 4. |
| 1 | 10 | 0. | 1. | 0. | 0. | 0. | 0. | 0. | 0. | 0. | 0. | 0. | 0. | 1. |
| 1 | 17 | 1. | 1. | 0. | 0. | 0. | 0. | 1. | 0. | 0. | 1. | 0. | 0. | 1. |
| 1 | 21 | 1. | 1. | 1. | 0. | 0. | 0. | 1. | 0. | 0. | 0. | 1. | 0. | 4. |
| 1 | 28 | 1. | 1. | 0. | 0. | 0. | 0. | 0. | 0. | 0. | 0. | 0. | 0. | 3. |
| 1 | 34 | 1. | 1. | 0. | 0. | 1. | 0. | 1. | 0. | 0. | 0. | 0. | 0. | 2. |
| 1 | 36 | 1. | 1. | 0. | 0. | 0. | 1. | 1. | 0. | 0. | 0. | 0. | 1. | 2. |
| 1 | 57 | 1. | 1. | 0. | 0. | 0. | 0. | 1. | 1. | 0. | 1. | 0. | 0. | 2. |
| 1 | 58 | 1. | 1. | 0. | 0. | 0. | 0. | 0. | 0. | 0. | 0. | 0. | 0. | 4. |
| 1 | 60 | 0. | 1. | 0. | 0. | 0. | 0. | 0. | 0. | 0. | 0. | 0. | 0. | 1. |
| 2 | 31 | 0. | 0. | 1. | 1. | 0. | 0. | 0. | 0. | 0. | 0. | 0. | 0. | 1. |
| 2 | 33 | 0. | 0. | 1. | 1. | 0. | 0. | 0. | 0. | 1. | 0. | 0. | | 1. |
| 2 | 16 | 1. | 0. | 1. | 1. | 0. | 0. | 0. | 0. | 0. | 0. | 0. | | 2. |
| 2 | 6 | 0. | 0. | 1. | 0. | 0. | 0. | 0. | 0. | 0. | 0. | 0. | | 2. |
| 2 | 37 | 0. | 1. | 1. | 0. | 0. | 0. | 0. | 0. | 0. | 0. | 0. | | 4. |
| 2 | 42 | 0. | 0. | 1. | 1. | 0. | 0. | 0. | 0. | 1. | 0. | 0. | | 1. |
| 2 | 45 | 0. | 0. | 1. | 1. | 0. | 0. | 0. | 1. | 0. | 0. | 0. | | 1. |
| 2 | 48 | 0. | 0. | 1. | 1. | 0. | 0. | 0. | 1. | 0. | 1. | 0. | | 4. |
| 2 | 55 | 1. | 0. | 1. | 1. | 0. | 1. | 0. | 0. | 0. | 1. | 0. | | 2. |
| 2 | 56 | 0. | 0. | 1. | 0. | 0. | 0. | 0. | 0. | 0. | 0. | 0. | | 2. |
| 2 | 5 | 1. | 0. | 1. | 1. | 0. | 1. | 0. | 0. | 0. | 1. | 0. | | 2. |
| 2 | 13 | 0. | 0. | 1. | 1. | 0. | 0. | 0. | 0. | 0. | 1. | 0. | | 1. |
| 2 | 30 | 1. | 0. | 1. | 1. | 0. | 0. | 0. | 0. | 0. | 1. | 0. | | 2. |
| 4 | 41 | 0. | 0. | 0. | 0. | 1. | 1. | 0. | 0. | 1. | 0. | 0. | 0. | 3. |
| 4 | 2 | 0. | 1. | 0. | 0. | 1. | 1. | 0. | 0. | 0. | 0. | 1. | 0. | 3. |
| 4 | 26 | 0. | 0. | 0. | 0. | 1. | 1. | 0. | 0. | 0. | 0. | 0. | 0. | 4. |
| 4 | 9 | 1. | 0. | 0. | 0. | 1. | 1. | 0. | 0. | 0. | 0. | 0. | 0. | 1. |
| 4 | 59 | 1. | 0. | 0. | 0. | 1. | 1. | 0. | 0. | 0. | 0. | 0. | 0. | 1. |
| 4 | 52 | 0. | 1. | 0. | 0. | 1. | 1. | 0. | 0. | 0. | 0. | 1. | 0. | 3. |
| 5 | 29 | 0. | 1. | 0. | 0. | 1. | 0. | 1. | 1. | 0. | 0. | 0. | 1. | 3. |
| 5 | 49 | 0. | 1. | 0. | 0. | 0. | 0. | 1. | 0. | 0. | 0. | 0. | 0. | 1. |
| 5 | 23 | 1. | 1. | 0. | 0. | 0. | 1. | 1. | 1. | 0. | 0. | 0. | 1. | 3. |
| 8 | 18 | 0. | 0. | 0. | 0. | 0. | 0. | 0. | 1. | 1. | 1. | 1. | 0. | 1. |
| 8 | 20 | 0. | 0. | 1. | 0. | 0. | 0. | 1. | 1. | 1. | 0. | 0. | 0. | 2. |
| 8 | 32 | 1. | 0. | 0. | 0. | 0. | 0. | 0. | 1. | 1. | 0. | 0. | 0. | 3. |
| 8 | 50 | 0. | 0. | 0. | 0. | 0. | 0. | 0. | 1. | 1. | 0. | 0. | 0. | 1. |
| 8 | 12 | 0. | 1. | 0. | 0. | 0. | 0. | 0. | 1. | 1. | 0. | 0. | 0. | 1. |
| 8 | 53 | 0. | 0. | 0. | 1. | 0. | 0. | 0. | 1. | 1. | 0. | 0. | 0. | 1. |
| 8 | 54 | 0. | 1. | 0. | 0. | 0. | 1. | 0. | 1. | 1. | 0. | 0. | 0. | 2. |
| 8 | 22 | 0. | 0. | 0. | 0. | 1. | 0. | 0. | 1. | 1. | 0. | 0. | 0. | 3. |
| 8 | 3 | 0. | 0. | 0. | 1. | 0. | 0. | 0. | 1. | 1. | 0. | 0. | 0. | 1. |
| 8 | 14 | 0. | 0. | 0. | 0. | 1. | 0. | 1. | 1. | 1. | 0. | 0. | 0. | 2. |
| 8 | 39 | 0. | 0. | 0. | 0. | 0. | 0. | 0. | 0. | 1. | 0. | 0. | 0. | 1. |
| 8 | 4 | 0. | 1. | 0. | 0. | 0. | 1. | 0. | 1. | 1. | 0. | 0. | 0. | 2. |

| | | | | | | | | | | | | | | |
|---|---|---|---|---|---|---|---|---|---|---|---|---|---|---|
| 8  | 11 | 0. | 0. | 0. | 0. | 0. | 0. | 0. | 1. | 0. | 0. | 0. | 0. | 3. |
| 11 | 47 | 0. | 0. | 1. | 0. | 0. | 0. | 0. | 0. | 0. | 1. | 0. | 0. | 1. |
| 11 | 15 | 0. | 0. | 1. | 0. | 0. | 0. | 0. | 0. | 0. | 1. | 0. | 0. | 1. |
| 11 | 24 | 0. | 0. | 0. | 0. | 0. | 0. | 0. | 1. | 0. | 1. | 0. | 0. | 3. |
| 11 | 1  | 1. | 0. | 0. | 1. | 0. | 0. | 0. | 0. | 0. | 1. | 0. | 0. | 3. |
| 11 | 51 | 1. | 0. | 0. | 1. | 0. | 0. | 0. | 0. | 0. | 1. | 0. | 0. | 3. |
| 11 | 44 | 0. | 1. | 1. | 0. | 0. | 0. | 0. | 0. | 0. | 1. | 0. | 0. | 1. |
| 11 | 19 | 0. | 0. | 1. | 0. | 0. | 0. | 0. | 0. | 1. | 1. | 0. | 0. | 5. |
| 15 | 40 | 0. | 0. | 0. | 0. | 0. | 1. | 0. | 0. | 0. | 0. | 1. | 1. | 3. |
| 15 | 27 | 0. | 0. | 0. | 0. | 0. | 0. | 1. | 0. | 0. | 1. | 1. | 1. | 2. |
| 15 | 25 | 0. | 0. | 0. | 0. | 0. | 0. | 1. | 0. | 0. | 0. | 1. | 1. | 2. |
| 15 | 43 | 0. | 1. | 0. | 0. | 0. | 0. | 0. | 0. | 0. | 0. | 1. | 1. | 2. |
| 15 | 38 | 0. | 0. | 0. | 0. | 0. | 0. | 0. | 0. | 0. | 0. | 1. | 0. | 5. |
| 15 | 35 | 0. | 0. | 0. | 0. | 0. | 0. | 0. | 0. | 0. | 0. | 1. | 1. | 1. |
| 15 | 46 | 0. | 1. | 0. | 0. | 0. | 0. | 0. | 0. | 0. | 0. | 1. | 1. | 3. |

## Conclusion

The method given in this paper is robust and fast. The initial set of part families can be choose at random. In that case, the algorithm becomes particularly fast, and we can made a great number of trials.

## References

M.R.Anderberg (1973). Cluster analysis for applications. Academic Press, New York.

E.Diday, J.Lemaire, J.Pouget, F.Testu (1982). Eléments d'analyse des données. Dunod.

N.Dridi. Optimisation d'ordonnancements de tâches contraintes. Thèse, Université Paris-Dauphine, UER Mathématiques de la Décision, France.

J.Erschler (1976). Analyse sous contraintes et aide à la décision pour certains problèmes d'ordonnancements. Thèse de doctorat d'Etat, Toulouse, France.

G.Fontan (1980). Notion de dominance et son application à l'étude de certains problèmes d'ordonnancement. Thèse de doctorat d'Etat, Toulouse, France.

SECTION 5.2


SPATIAL ORGANIZATION OF MACHINE TOOLS IN A WORKSHOP

G. Caraux

Department of Biometrics, Ecole Nationale Supérieure

Agronomique, Place Viala, F-34060 Montpellier, France

## 1. Introduction to the problem

The manufacturing of several types of mechanical parts by a machine tool system creates a problem of workshop organization. In fact, if the machine system which plays a rôle in manufacturing a part is different from one part to another, it is then desirable, in order to reduce the unnecessary transfers of the parts being worked on, to group machines that work together.

In this particular problem we need to find a spatial organization of work in the form of workshops. Each workshop will group together as much as possible a set of machines that have a similar production type, so that the number of mechanical parts to be transferred from one workshop to another to complete manufacturing, is as low as possible. In an ideal case, one hopes to define separate workshops in such a way that each part might be entirely manufactured in one workshop.

In order to solve this work organizational problem, Garcia and Proth have propose a data set in a rectangular matrix of indicative variables (see fig.1). The list of machines intervening in the

283

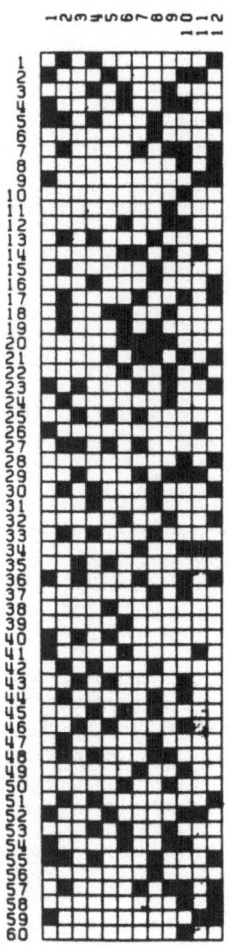

Fig.1. Initial data matrix

manufacturing of a set of parts (here  n = 60) is herein given encoded in it.  That matrix, which will known as "A", is indexed in rows
by the  I  set of mechanical parts and in columns by the  J  set of
machines.  It contains the values 1 or 0, according to the following
code :

$A_{ij} = 0$  the  i  part is not processed by the  j  machine

$A_{ij} = 1$  the  j  machine processes the  i  part

In the following paragraphs, the set of parts to be machined

will be known as  I  (A lines), and the set of machines will be
known as  J  (A columns).

Once these notation codes have been defined, the optimization
problem was stated by the authors in the following terms :

"Starting from the  A  matrix which has just been defined, draw
a  B  matrix, by permuting lines and colums of  A, so that in  B  a
set of  p  $E_1, \ldots, E_p$  diagonal blocks is obtained and that the number
of 1 without this blocks is minimal".

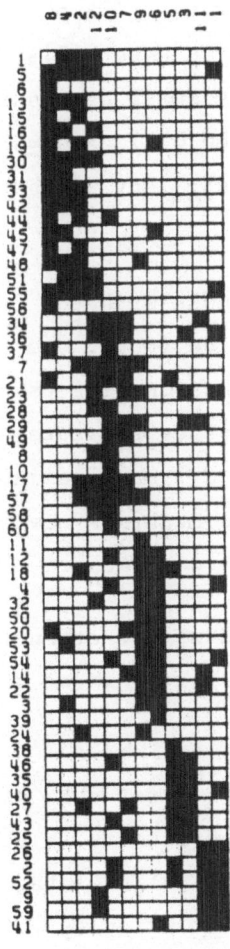

Fig.2. Solution proposed by the authors

Once the problem thus has been formulated, it was illustrated by an example of reorganization (see fig.2).

At least three preliminary remarks are brought to light by these terms  of the problem.

* To perfectly define the problem, the  p  number of diagonal blocks to be found must be set from the very beginning.  In other words the number of workshops in which machines will be grouped must be determined a priori.  In fact, any partition in  p  classes of machines will give a higher value of the criterion than a rougher partition in  p' classes with  p' < p.  Consequently if  p  was not a constant value given in hypothesis, it would always be better to assign to  p  a value as small as possible.  In the extreme case where  p' = 1  the criterion would always be zero !  The problem would then be of no interest.  Any reorganization would then be optimal.

* The second remark deals with the fact that number  0  appearing in diagonal blocks to be found, does not intervene in the criterion to be optimized.  This enables us to state that the authors desiring to define workshops as separate as possible, accept parts which are nor processed by all machines available.  However, they hope to find almost all the machines needed during the manufacturing process of the parts in the workshops.

* A last remark must draw our attention on the fact that in the data supplied to us the 5 first rows of the matrix are identical to the last 5.

As a conclusion of this first data study, we may notice that the solution proposed by the authors is far from being the ideal situation sought.  Perfectly defined diagonal blocks should be present in it.  Although this solution is relevant (as stressed further on), we may wonder if the data to be processes are likely to result in a regrouping of the machines in autonomous and independent work-shops.

## 2. Reorganizing data array

These remarks and notably the last one encouraged us to make a graphic and descriptive study of the problem. In effect, it seemed useful to have a descriptive and exploratory approach of data before building an optimized heurism as far the criterion proposed by the authors is concerned.

For that we used a number of known methods each of them ena-bling us to rearrange the data matrix. We will now review them and try to draw conclusions we could use for our study.

These techniques have in common the aim of grouping together rows as well as columns according to their similarities in the ini-tial array. Except for the last method, a distance criterion will be explained in each case, and rows and columns will be grouped together by affinity according to this distance criterion.

If there is a diagonal block structure in the array submitted to our analysis, then these different approach types will give an optimum solution for the criterion the authors desire to optimize. Otherwise, we think these methods, without precisely meeting the optimization sought, will supply pertinent reorganizations for the solution to the problem.

### 2.1. Approach using Correspondence Analysis

It has been demonstrated (Benzécri et al., 1973; Schriever, 1982) that with an array showing an order structure for rows and another for columns, so that row and column profiles are shifted sideways, a first factor which determines an increasing function defined on the set of indexes and meets the initial order can be found by the method of analysis of correspondences. A famous example

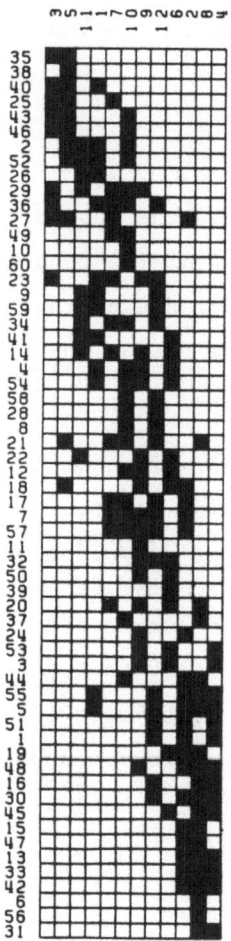

Fig.3. Reorganization obtained with Correspondence Analysis

of this property is known in literature under the name of Guttman effect. This well-known case is found when studying a matrix of 0 and 1 values in which the 1 values are distributed along the main diagonal under a parallelogram shape. The property above mentioned applies to more general cases that may have certain similarities with our problem.

Rearranging our array starting from the order defined by the first factorial axis of correspondences yields the result shown in figure 3.

In this reorganization, the absence of diagonal blocks as well as a rather strongly diagonal aspect, which recalls a Guttman matrix, are immediately noticed. This result should not be surprising. In this reorganization the opposition derived by the first factorial axis is expressed. Each mechanical part and each machine tool is positioned in function of the two extremities. When browsing through the 60 rows of the array we progressively pass from parts requiring processing of machines 3, 5, 11,... to those to be processed by machines ..., 2, 8, 4.

This result prompts us to think of the optimum organization to be adopted for the spatial installation of machine tools. Indeed the authors of this experiment are looking for a machine organization in separate workshops, but with this result we may wonder whether it would not be preferable to build a single workshop longitudinally organized. There machines would be arranged according to the column order given in this solution. Thus, mechanical parts tabulated in the first rows of the solution would undergo processing in the first sector of the workshop. Conversely, parts appearing in the last rows would be machined in the last sector of the workshop.

It is not our rôle to settle a debate likely to arise from the above proposal. Let us simply state that our approach, deliberately descriptive, can suggest original solutions, the relevance of which was not initially percieved.

## 2.2. Approach through Hierarchical Clustering

Having asserted above our desire to reorganize the data matrix through grouping together rows which look alike as well as columns with similar profiles, we naturally adopt a taxonomic method.

With this aim in mind, we used a hierarchical ascending clustering method based on Jaccard index of distance (see section 2.3).

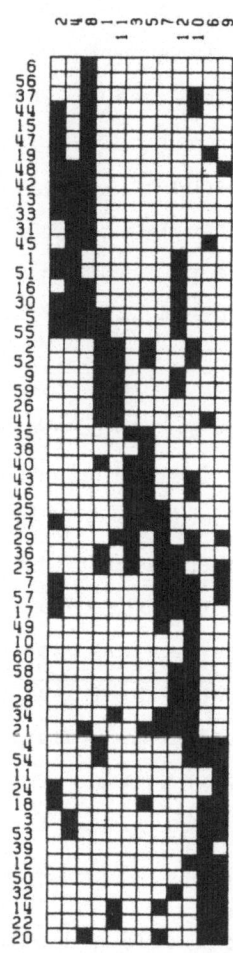

Fig.4. Reorganization obtained through a Hierarchical Cluster Analysis

The reorganization shown here (fig.4) resulted from listing rows and columns in the order stated by the dendrogram obtained through hierarchical upward classification of rows then columns. Class agglomeration was carried out according to maximization criterion of order-2 centered partition moment.

The order induced by this reorganization on rows or columns is not completely defined. In fact, this methodology does not enable us to determine an order on the items it handles. One solution pro-

posed through ascending hierarchical clustering allows us to distinguish separate groups of rows and columns.  Once the number of
retained groups defined, rows belonging to a same group can be per-
muted without refuting classification, similarly, the listing order
is not specific to the solution.

The reorganization found through this typological approach gives
an allocation of values very clearly distributed along the diagonal.
While having a closer look at this distribution, 5 diagonal blocks
can be distinguished, which when examined are very similar to those
proposed by the authors.

## 2.3. Approach through the structuration of the distance matrix

The approach to be introduced was recently developed by the
author (Caraux, 1984).  Its dual target is to group together by
permutation rows that have similar profiles, and separate those
which are dissimilar.

The reorganization of rows and columns is performed through an
identical technique but on a separate mode.  In order to simplify
our explanation, we will here only deal with the case of row reorganization.  That of columns will be deducted by a symmetrical
process.

Let  D  be a matrix with a defined distance on  I  where
$d(i,i')$  is a value for the distance allocated to the space between
the pair of rows  i  and  i'.

Any  P  permutation of  I  rows in table  A  will induce an
organization change of values in the symmetrical matrix of  D
distances.

The method considered here tries to determine the permutation
of the items of  I  in such way that in the  PDP'  matrix (the
matrix of distances between items of  I  after reorganization) the

lowest values are distributed as close as possible to the diagonal and conversely that the high values are far from it. This realization replies to our expectation of grouping together similar rows and spacing apart distant rows.

The criterion to be minimized in the search of P is expressed as a moment of inertia with the following formula:

$$M = \sum_{i=1}^{n} \sum_{i'=1}^{n} (i - i')^2 d(i,i')$$

Mathematical properties of this quadratic form allow determining algorithmic simplifications in the search of the optimum P permutation to be performed. This permutation is worked out through successive transpositions. For each step, the two lines, which when transposed induce a maximum increase of M criterion, are sought.

Obtaining a global optimum is not guaranteed by the working-out step by step of the P permutation. Generally, the process converges rather to a local optimum which is function of the initial matrix configuration.

The Jaccard distance index was adopted here as a distance measurement to solve the problem. This index, seeming well adapted to the problem, represents the machine proportion working together in the manufacturing process of i and i'.
This is written :

$$d(i,i') = \frac{nii'}{ni. + ni'. - nii'}$$

The reorganizations achieved by this method (see fig.5) varied according to the initial matrix configuration. The matrices thus found were all very disappointing as far as our problem is concerned. This method designed both to get similar rows together and to distance opposite rows cannot simultaneously reach both objectives. Certain rows, in order to be apart from others, are positioned in

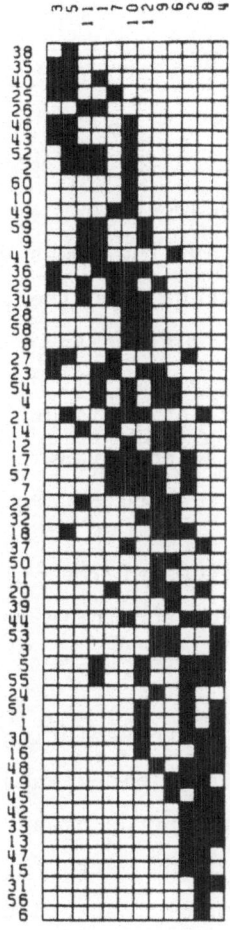

Fig.5

a set of rows which, otherwise, would be homogeneous.  This can also
be understood by the fact that rows are vertically indexed, following therefore a one-dimension axis.  This axis cannot alone sum
up the complexity of similarities and dissimilarities between rows.

From that remark we defined a less global criterion allowing a
row to be located in its immediate environment and not in its global
one.  We defined this criterion with the sum of the items from the
first overdiagonal of the  D  distance matrix.

The criterion can be written :

$$S = \Sigma \, d(i, i + 1) \qquad i \in I$$

We drew inspiration from the method explained above in order to reorganize the initial matrix. The P permutation, which induces in the matrix of distances between rows a distribution of values minimizing S, is sought. Thus the lowes values of D will tend to be distributed along the diagonal and therefore rows from the initial matrix will tend to regroup by affinity.

As before, the P permutation is built with successive transpositions of rows. These transpositions were selected step by step by minimizing for each step the sum of PDP' overdiagonal items.

From several configurations of the initial matrix, several solutions were found among which the three matrices shown in figure 6 seemed to be the most pertinent.

The first one is close to the solution proposed by the authors, in which there is a very similar block subdivision. However, this reorganization because of the approach used, is visually more satisfying, without for all that being better as far as the optimum defined by the authors is concerned. Note that in this reorganization the first rows associated with columns 9 and 6 draw a block comparable to that defined by the last rows.

In this reorganization, the fact that machines 12, 10, and 7 have very different uses can be noted. The operation of these three machines is found again in several workshops that might be designed after this reorganization. In the same way, mechanical parts 34, 29, 23, and 36 require a very large range of processing. From the very nature of these machines and parts, the problem we are to solve cannot be satisfactorily solved. If these three machines and these few mechanical parts could be dealt with separetely, then the solution likely to be achieved would be much closer to an ideal situation.

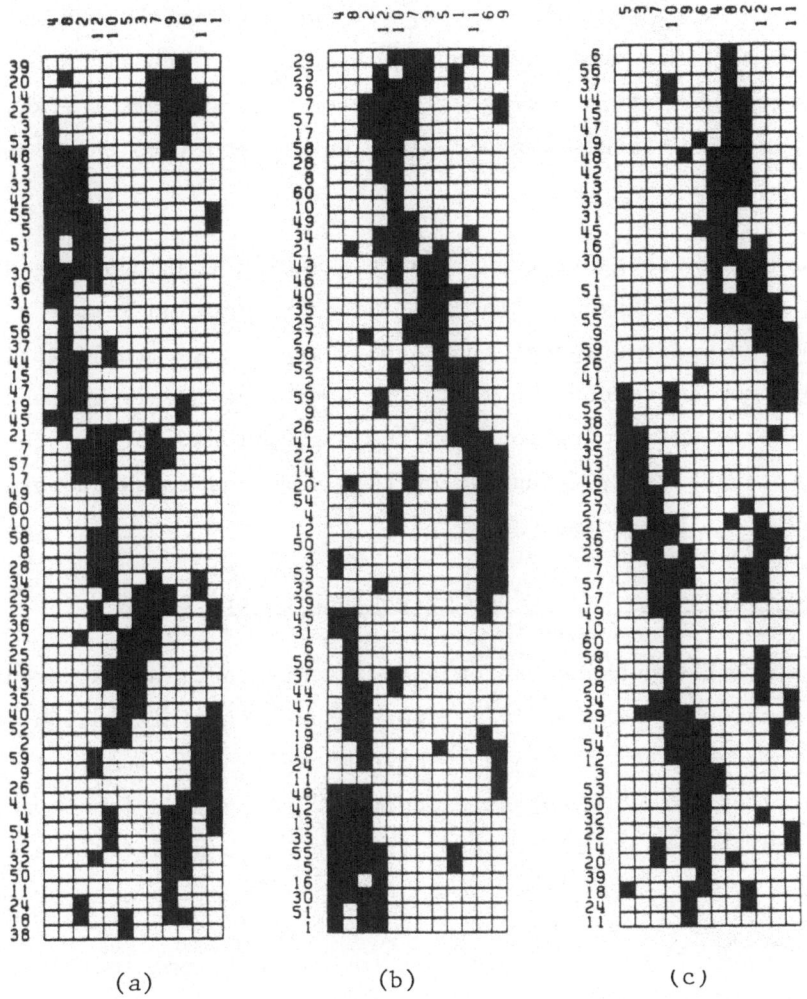

(a)                    (b)                    (c)

Fig.6. Reorganization based on the overdiagonal minimization

It is preferable to sacrifice machining optimization of a few parts
if globally a better work organization is gained for others.  It is
up to the authors of this experiment to decide if the analysis should
be carried on that way.

The second reorganization obtained with this method is significantly different.  It brings to the fore a first block located in

the left lower part of the matrix. It isolates in the median part
the three machines mentioned above. Finally, in the right part of
the reorganized matrix a diagonal-shaped block appears thanks to the
median lines. This block could still be divided into three under-
blocks.

The third reorganization reveals a matrix that could be sub-
divided into two diagonal blocks. This distribution leaves a number
of values in under-diagonal blocks but respects seemingly well the
original problem requirements. After having seen this representation and drawing inspiration from the suggestions made for the correspondence analysis, grouping together machines in two separate
workshop might be recommended.

## 2.4. Approach following Leduc reorganization criterion

The graphic representation used in this document was proposed
by the geographer Bertin (1971), for structuring information contained in a statistical table. While manually and empirically hand-
ling matrices formed by plastic dominoes he revealed the power of a
visual representation for interpreting values contained in a matrix
of data.

A number of authors tried to think of an automation for the
empirical process of reorganization proposed by Bertin (refer to
bibliographic references in Caraux, 1984).

Among these authors, Leduc proposes a reorganization methodology based on a minimization of the length of the border between 0
and 1 values in the data matrix. With this criterion, homogeneous,
compact and well defined spots tend to appear. It seemed to us
interesting to examine the results of our problem through this
method.

The heurism borrowed from the graph theory that this author proposes does not give, as for the former method, a global optimum.

According to the initial matrix, several solutions can be obtained.

Starting each time from a different solution we several times submitted our matrix to this heurism.  The results are shown in figure 7.

The three solutions proposed by this algorithm are very different as far as our problem is concerned.  At first, a graphism with well defined shapes is noticed.

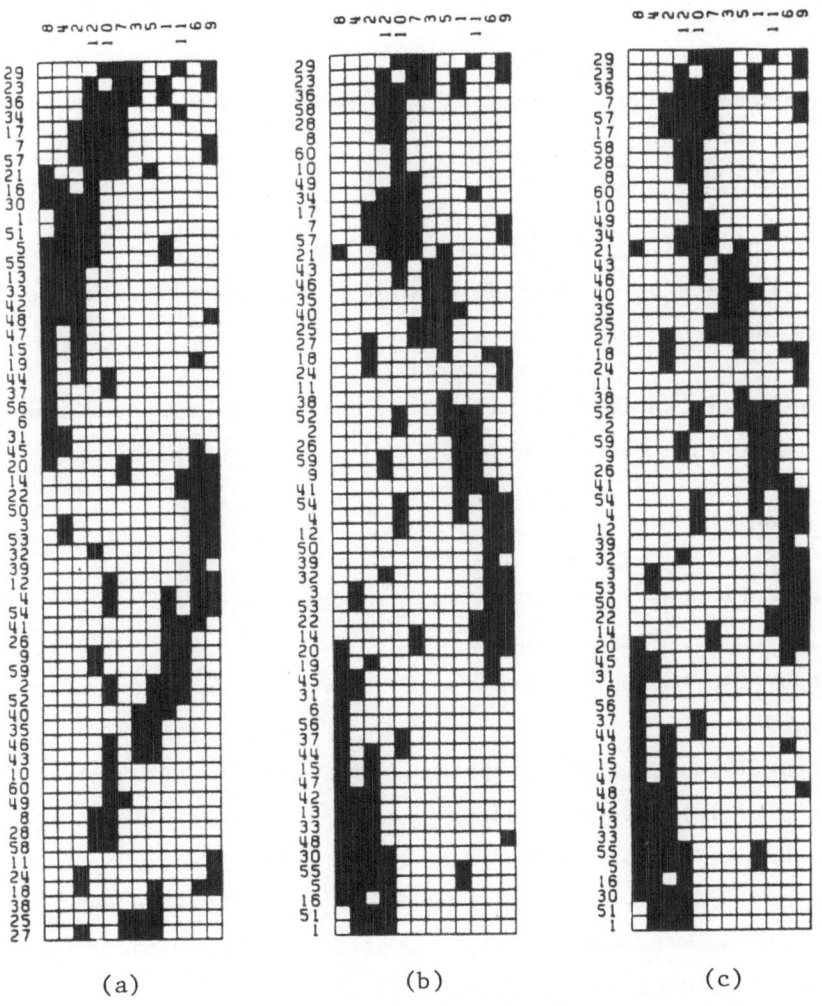

(a)                    (b)                    (c)

Fig.7. Reorganisations based on Leduc criterion

In the first graph, a two diagonal block subdivision is found
again, as in the former section, each block having an elongated
value distribution.  This subdivision of rows and columns into two
groups is not identical to that found above, although it has the
same structure.

In the two other reorganizations, as seen in the previous section, a subdivision of the matrix in 5 well distinct blocks can be
found.  Indeed these blocks are not separate but they are visually
more distinct than in figure 2.  When considering in detail the
block constitution defined by this reorganization, the solution initially presented by the authors is almost retrieved.

## 3. Conclusion

Having thus approached this problem from various different
angles, we have shown various solutions which may interest the
authors who proposed the data.

The optimization initially sought has not been achieved but
every angle gave us elements which clarified the problem.

## References

J.P.Benzécri et al. (1973). L'analyse des données. Tome 1 : La
      Taxinomie. Paris, Dunod, 2ème éd. 1976.

J.Bertin (1967). Sémiologie graphique. Mouton-Gauthier Villars,
      Paris.

J.Bertin (1971). Article graphique. Encyclopedia Universalis.

J.Bertin (1977). Le graphique et le traitement graphique de l'infor-
      mation. Flammarion, Paris.

J.Bertin (1980). Traitements graphiques et Mathématiques.
      Différence fondamentale et complémentarité. Mathématiques et
      Sciences Humaines, n°72, 60-71.

G.Caraux (1984). Réorganisation et représentation visuelle d'une
    matrice de données numériques : Un algorithme itératif. Revue
    de Statistique Appliquée, 1984, Vol.XXXII, n°4, 5-23.

A.Leduc (1982). Chaînage automatique des matrices ordonnables.
    2ème Colloque de Micro-Info-Graphique, Université de Rouen
    Haute Normandie, septembre 1982.

B.F.Schriever (1982). Scaling of order dependent categorical
    variables with correspondence analysis. Stichting matematisch
    centrum. Amsterdam.

SECTION 5.3

CONCLUSIONS

J.M. Proth

The goal of this work is not to compare various algorithms, but only to outline a central problem in Production Management and to propose several approaches to solve it.

Garcia and Proth's approach is an iterative one. It starts from a set of part families to compute a set of production subsystems, then it uses this set in order to define a new set of part families, and so on. The process converges very fast.

Lerman's approach consists in finding independently a set of part families and a set of production subsystems. Some statistical information are useful to define the number of elements in these sets. The results are comparable to those obtained using the previous algorithm.

Caraux reorganizes the matrix using several techniques. He first uses the Factorial Analysis of Correspondences. He then classifies independently the rows and the columns of the matrix : it is, in some way, comparable to Lerman's approach. Finally, Caraux proposes an algorithm which consists in grouping the similar rows (resp. column) and in separating the rows (resp. columns) which

are different.   It seems that these approachs are not well adapted
to our specific problem.

   To summarize, Garcia and Proth's algorithm seems to be interesting to solve the specific type of problems presented here.   Some
extensions of the method are expected.   In addition, a more detailed
comparison with Lerman's method would be appreciated, specially concerning the computation times.

CHAPTER 6. JUVENILE DELINQUENCY

SECTION 6.1

BACKGROUNDS OF DELINQUENT BEHAVIOR

M. Junger and J. Junger-Tas

Research and Documentation Centre, Ministry of Justice
Postbus 20301, 2500 EH The Hague, The Netherlands

## 1. The data

The RDC (Research and Documentation Centre, Ministry of Justice
in The Hague) conducted a study on juvenile delinquency (Junger &
Junger-Tas, 1984), from which three concepts were issued :

1. Social integration.
Questions were asked in an interview on social integration in family
life, in school, leisure time activities and relationships with
friends.

2. Selfreported deliquency.
Youngsters (12-17 year old) were asked if they ever commited an
offense. The interview included 4 property offenses (shoplifting,
joy-riding, bicycle theft and theft at school) and 3 aggressive
offenses (malicious damage, soccer hooliganism, violence against
persons).

3. Judicial contacts.
Information was gathered concerning recorded contacts of the youngsters with police and prosecutor.

In total 332 youngsters were interviewed in 1981 as well as in
1983.

## 2. Social control or social integration theory

We used Hirschi's social control theory in order to explain delinquency.  The theory states that when youngsters are well integrated in society they won't commit offenses and thus will not have contacts with the judicial system.  When they are badly integrated the opposite will of course be the case.

### 2.1. Interfering variables on delinquency

In criminology some socio-demographic variables are known to have an impact on social integration (which in turn influences delinquency) :

1. Sex : girls are better integrated than boys.

2. Age : older children are less integrated.

3. Education level : pupils in schools of low educational level are less integrated.

4. Degree of urbanisation of municipality : this variable is related to social control;  less urbanisation results in better social control.

### 2.2. Variables influencing the decision of the police and the prosecutor

Given the schoollevel, social economic status (SES), sex and father's employment, police officers and prosecutors react differently to youngsters.  This means that boys whose father is unemployed or has a low SES will have a greater chance to be prosecuted.  Furthermore, boys with a lower education level will also be prosecuted more frequently.

## 3. The key problem

Social control theory doesn't predict that the evolution of social integration at one point will be strongly related to social integration two years later.  Based on social control theory we can make the following scheme of relationships (scheme 1) :

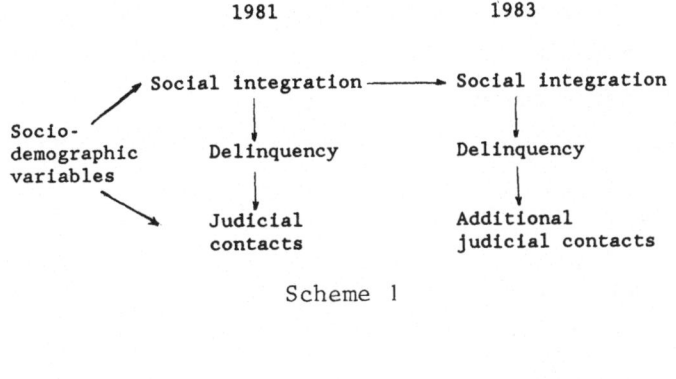

Scheme 1

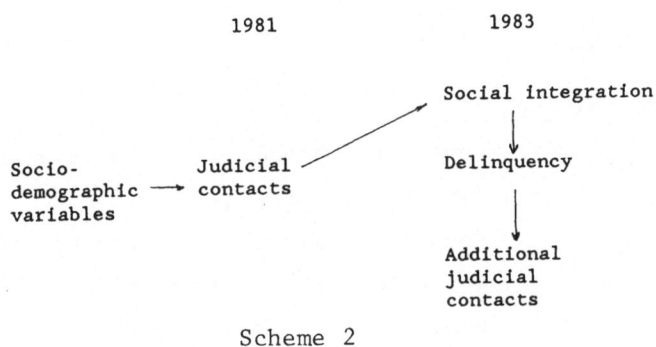

Scheme 2

## 3.1. Labelling theory

The labelling theory emphasizes other relations. There is a strong influence of former judicial contacts on delinquency and on judicial contacts according to this theory. This leads us to a quite different scheme where social integration is no longer of much importance (scheme 2).

Within the judicial system socio-demographic variables are important elements which weigh heavily on the decision whether to prosecute.

An additional problem is the influence of age. Age affects both social integration and the decision of the prosecuter whether to pro-

secute or not.  This means that, if nothing else changes, the effect
of age alone (the youngsters being two years older in 1983) will have
a negative influence on integration and a positive influence on judicial contacts.

## 3.2. The question

Our question is : which of both theories is right, or are they
both predicting existing relationships ?  Which model fits the data
best ?

## 4. The variables and the data

### 1. Judicial contacts

| | | |
|---|---|---|
| 1981 | JC81 | Counts the number of contacts with the juvenile justice system |
| 1983 | AC83 | Counts the additional contacts with the juvenile justice system since 1981 |
| codes | 0 | none |
| | ↓ | |
| | n | number of contacts |

### 2. Delinquency

| | | |
|---|---|---|
| 1981 | DEL81 | Concerns the offenses of the past year and is weighed for frequency |
| 1983 | DEL83 | idem DEL81 |
| codes | 0 | no deliquency |
| | ↓ | |
| | n | number of contacts |

### 3. Social integration, 1981

Different indices were made to measure integration in family
life, school and leisure time activities.  Sometimes indices were
put together in more comprehensive indices.  Finally one big index
was made of 4 variables and of seven indices measuring integration.

SI81 and SI83 measure social integration of the youngsters in 1981 and 1983. Scheme 3 shows the variables and indices composing SI81 and SI83

1981        <u>SI81</u>
1983        <u>SI83</u>
codes       0  : low
            ↓
            10 : high

All the indices were made in the same way by counting the number of positive answers on several variables (excepting Z3 which is based on a crosstable).

(variables)

    -Running away from home
    -Leisure time spent away from home
    -Number of delinquent friends
    -Truancy

(indices)

| | | |
|---|---|---|
| Family integration | -Functioning in the family | -Direct control of parents<br>-Communication with parents |
| | -Attachment to parents | -Number of activities spent with the family<br>-Rows/conflicts |
| School integration | -Attachment and commitment to school | -Attachment to school<br>-Commitment to school |
| | -Involvement in school | -School record<br>-Social behaviour in school |
| Leisure time activities | -Nature of leisure time activities<br>-Bravado behaviour<br><br>-Commitment in leisure time activities | |

Scheme 3

## 4. Socio-demographic variables

4.1.  <u>Sex</u>        <u>V5</u>

      codes     1        boy

                2        girl


4.2.  <u>Age</u>

      1981      <u>V9</u>

      1983      <u>XV9</u>

      codes     2        12 years old

                3        13  "      "

                4        14  "      "

                5        15  "      "

                6        16  "      "

                7        17  "      "

                8        18  "      "

                9        19  "      "


4.3.  <u>Urbanisation level</u> <u>v300</u>

      codes     1   The Hague  (ca. 450.000 inhabitants)

                2   Venlo      (ca.  63.000 inhabitants)


4.4.  <u>Education</u>

      1981      <u>V35</u>

      1983      <u>XV35</u>

      codes     1        extension primary school

                2        lower vocational/junior high school

                3        professional training/high school

                4        grammarschool preparing for university


4.5.  <u>Social economic status</u> <u>V300</u>

      codes     1   low

                ↓

                6   high

4.6.  Father's employment V303

     codes      1  father works

                   2  father out of work

                   3  father is handicapped

5. The data

The data are presented in table 1 in the form of a correlation matrix.

Table 1

| | V5 | V9 | XV9 | V300 | V35 | XV35 | V303 | V302 | SI81 | SI83 | DEL81 | DEL83 | JC81 | AC83 |
|---|---|---|---|---|---|---|---|---|---|---|---|---|---|---|
| V5 | 1. | | | | | | | | | | | | | |
| V9 | -.066 | 1. | | | | | | | | | | | | |
| XV9 | -.065 | .936 | 1. | | | | | | | | | | | |
| V300 | -.054 | .210 | .099 | 1. | | | | | | | | | | |
| V35 | -.014 | .059 | .106 | -.140 | 1. | | | | | | | | | |
| XV35 | -.016 | -.049 | -.043 | -.098 | .674 | 1. | | | | | | | | |
| V303 | .067 | .049 | .079 | -.165 | .301 | .259 | 1. | | | | | | | |
| V302 | -.052 | .148 | .124 | .204 | -.150 | -.143 | -.202 | 1. | | | | | | |
| SI81 | .059 | -.375 | -.325 | -.276 | .186 | .255 | .029 | -.152 | 1. | | | | | |
| SI83 | .092 | -.245 | -.220 | -.223 | .156 | .230 | .097 | -.107 | .566 | 1. | | | | |
| DEL81 | -.080 | .077 | .048 | .177 | -.088 | -.079 | -.113 | .178 | -.487 | -.360 | 1. | | | |
| DEL83 | -.144 | -.054 | -.092 | .026 | -.109 | -.147 | .025 | .010 | -.221 | -.410 | .340 | 1. | | |
| JC81 | -.053 | .113 | .060 | .293 | -.138 | -.132 | -.140 | .088 | -.362 | -.305 | .407 | .130 | 1. | |
| AC83 | -.048 | -.026 | -.040 | .062 | -.135 | -.173 | -.140 | .030 | -.195 | -.264 | .288 | .161 | .398 | 1. |

* n may vary due to missing values.

SECTION 6.2

SOCIAL INTEGRATION OR LABELLING

M. Junger

Research and Documentation Centre, Ministry of Justice
Postbox 20301, 2500 EH The Hague, The Netherlands

## 1. Introduction

This study was undertaken by the research department of the
Ministry of Justice at the request of a prosecutor in youth cases
in The Hague, one of the four largest cities of the Netherlands.
His question was whether the policy of minimal punitive intervention
by the prosecutor and the police, is in any way effective in pre-
venting youngsters to come into contact with the juvenile justice
system again.

The leading idea in this policy is the labelling theory :
according to this theory too much intervention by the juvenile
justice system contributes to delinquent behaviour and therefore
leads to additional (so called secundary) deviant behaviour.  The
expectation is that a policy of reprimanding and dismissing cases
leaves juveniles to the normal processes of maturation so that they
will outgrow their delinquent behaviour.  However, in practice
certain youngsters do not respond favourably to this treatment, but
keep coming back instead.

Judicial intervention is probably only one of the factors that could influence the behaviour of juveniles.  Other factors are perhaps more important in determining which youngsters come into contact with the juvenile justice system and which ones keep coming back.  More specifically the influence of social integration on the delinquent behaviour of youngsters, as social control theory predicts (Hirschi, 1969), can be mentioned here.

This leads to two theories which can be tested :
1. labelling theory, which predicts additional contacts with the judicial system from previous contacts;
2. social control theory, which explains deviance and judicial contacts from social integration.

## 2. The research design

In 1981 a representative sample of youngsters was interviewed (n = 1980) by means of a structured questionnaire.  They were selected randomly from the population of youngsters in a large city (The Hague) and a smaller town (Venlo) in the Netherlands.  Information was gathered about their social integration and selfreported delinquency.  Information was also collected about their judicial contacts. After two years a group of 332 youngsters was interviewed a second time and again information was collected using the same variables as in the 1981 questionnaire.  The 332 subjects were selected from the 1980 subjects in such a way that youngsters of different levels of contacts with the juvenile justice system were represented.  For a presentation of the research refer to Junger-Tas and Junger (1983).

## 3. The theoretical framework

## 3.1. Social control theory

Social control theory states (Hirschi, 1969; Kornhauser, 1978) that the more a person is integrated in conventional society the less

he will transgress legal norms.  This theory makes use of four important concepts :
- attachment to significant others,
- commitment to conventional subsystems, such as school or work,
- involvement in conventional activities,
- beliefs in social norms and respect for legal norms.
Information on these concepts was gathered in fields important for
the youngsters, such as family, school, friends and leisure time.
According to the theory a good integration in these fields implies
that the youngster is attached to conventional society and will not
commit offenses.  A low integration in these different fields will
weaken the youngster's ties to society and society's norms, and
delinquent behaviour is more likely to occur.

On the basis of this theory one might expect that one or more
contacts with the judicial system will not affect the level of social
integration very much and will be of relatively minor importance to
the youngster's level of delinquency and his future contacts with
judicial authorities.  Social integration alone determines delinquency.

Besides this relatively minor importance of the effect of judicial intervention, social control theory doesn't predict anything
specific about the evolution of integration, delinquency or judicial
contacts through time.  However, the following does not seem to be in
contradiction with the theory.

- Integration can change but we may expect a strong correlation
between integration in 1981 and 1983.
- Delinquency in 1981 will influence delinquency in 1983 through
integration.
- Judicial contacts will have only a minor influence on social
integration;  therefore it seems probable that the correlation between
these variables will be approximately zero.

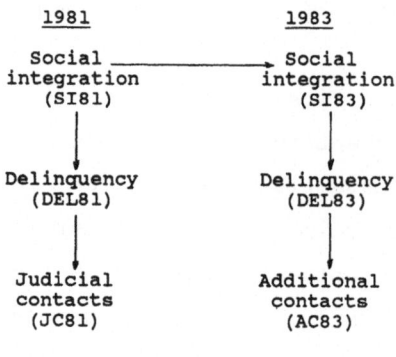

Fig.1

The scheme above could summarize the relations we may expect if social control theory is correct (arrows are representing the relations between variables and their direction).

## 3.2. Labelling theory

Labelling theory doesn't explore the origins of the first delinquent acts. For an introduction to labelling theory we refer to Knutsson (1977). In fact it assumes that every youngster commits offenses and therefore one can not differentiate among them. The difference between delinquents and non-delinquents is that delinquent youngsters have been officially labelled as 'delinquents' : during their passage through the judicial system they have undergone a stigmatization process. During this passage they develop a negative, delinquent self-concept.

We can interpret this in two ways :

1) judicial contacts have a direct influence on delinquency;

2) judicial contacts lead to a lowered self-esteem which in turn will lead to less social integration. Less self-esteem and less social integration will affect delinquency and additional contacts with the judicial authorities positively. It seems possible that within this framework there remains a direct influence of former contacts on delinquency.

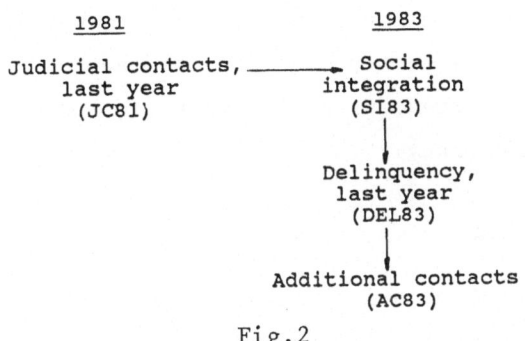

1981                          1983

Judicial contacts, ————————▸ Social
   last year                  integration
    (JC81)                      (SI83)
                                  │
                                  ▾
                              Delinquency,
                               last year
                                (DEL83)
                                  │
                                  ▾
                           Additional contacts
                                (AC83)

Fig.2

This leads to another scheme, given in figure 2.

## 4. The concepts and their operationalisation

### 4.1. Social integration

Social integration was measured by interviewing the youngsters
about their ties with significant others, their commitment to
and their involvement in the important domains of their life :
family, school and leisure time.  A large number of variables was
obtained.  From these variables the following indices were constructed : family functioning;  attachment to parents;  attachment to
teachers and commitment to school;  involvement in school;  nature of
leisure time activities;  bravado behaviour;  commitment in leisure
time activities.  One large index was constructed by summing up all
the indices.  To this large index the scores on four other variables
were added : running away from home,  leisure time spent outside the
home, number of delinquent friends, truancy.  For more details we
refer to Junger-Tas and Junger (1983).

### 4.2. Delinquency

Delinquency was measured by asking the youngsters whether they
had committed delinquent acts.  Seven offenses were included in the
questionnaire : shop-lifting,  malicious damage,  violence againts
persons,  theft at school,  bicycle theft,  soccer hooliganism,

joy-riding. This information about delinquency was assembled in an index by counting - for each youngster - the number of different type of offenses committed during the last year before the interview and weighing for frequency.

## 4.3. Judicial contacts

Information about recorded contacts with the juvenile justice system was collected by the researchers at the police and judicial administrations.

We counted all the recorded judicial contacts of juveniles during the last year (for 1981) and the additional contacts with the authorities (from 1981 to 1983).

## 5. Path analysis

The data were analyzed according to the method of path analysis as it was described by Bohrnstedt and Knoke (1982), Goldberg (1976), Blalock (1972) and Segers and Hagenaars (1980).

Firstly it is necessary to construct a scheme in which the concepts are related to each other by arrows. The direction of the arrows shows the causal ordering of the variables. Secondly, correlations and regression coefficients (standardized) are computed for the existing arrows. The importance of the relations is visualised by placing these numbers next to the arrows. When a correlation or regression coefficient is small we can conclude that this relation is of little importance and that the corresponding arrow should not remain in the scheme. Thirdly, for the arrows that could have been drawn but are not, possible direct relations are supposed to be non-existent. After computing the coefficients which belong to the arrows in the scheme, one should check whether the remaining possible relations are equal to zero. The purpose of path analysis is to check whether the model fits the data : all the arrows present

in the scheme should be important enough and the arrows which are
not present must be of no importance.  To check this, one can compute the partial correlations.  If some partial correlations differ
significantly from zero they can not be neglected and the scheme
should be revised by placing more arrows.  The revised scheme should
then be tested again the same way.

A comment should be made here on the use of .  Whenever we
use tests to see if correlation or regression coefficients differ
significantly from zero we assume the data to follow a multinormal
distribution.  However, this assumption is not met in all cases.
The counts of the number of contacts (for the data of 1981 and 1983)
have a rather asymmetric distribution because the majority of the
youngsters have no judicial contacts at all, some have one and very
few youngsters have more than one judicial contact.  This skewed
distribution holds also – but to a lesser extent – for the delinquency indices (1981 and 1983).  Therefore the significance levels should
be treated with caution.

The computation of partial correlations is not unequivocal.  The
general rule is that one should control for all relevant variables.
More precise guides for action are needed.  Blalock (1972) gives two
additional specifications :
1) one should never check for dependent variables :
   "The phrase 'controlling for all variables' does not mean
   that one should automatically control for all variables that
   might be available to the researcher, including possible dependent variables. It means that one should control only for variables that are either prior to one or both of the variables
   being related, or that may be intervening between them.  Even
   this rule can give misleading results depending upon the purposes of one's investigation.  But at the very least, we can
   indicate that one should not control for dependent variables".

2) In addition one needs not check for variables operating on only
one endogenous variable :

> "Whenever a variable operates on only one member of a set of
> other variables we do not need to control for this former va-
> riable in order to have the appropriate partials among the
> other variables disappear.  As long as a factor is a causal
> determinant of only one endogenous variable, we need not be
> concerned about its disturbing effects".

A few words of explanation about causality : the fact that we
test models by means of path-analysis does not mean  we pretend to
prove causality.  Correlational data are not suited for this purpose.
What we can do is eliminate models which do not fit the data and
give the best possible empirical support for the model which emerges
from the analysis and from theoretical considerations.

### 6. Analyses for the labelling-scheme

Labelling theory predicts a change in self-concept and social
integration as a reaction to contacts with the juvenile justice
system.  Because of the influence of judicial contacts on social
integration it seemed correct to exclude from the analyses the subjects with contacts older than one year.  Youngsters with old judicial contacts would already have been stigmatized - according to
labelling theory - and have therefore a lowered social integration.
This would interfere in the analysis of the theory.  This is why only
youngsters with recent contacts were kept in the analysis (n = 298).
The correlation matrix is given in Table 1.

It is possible to form $n!/2(n-2)!$ pairs among  n  variables.
Model 2 comprises four variables, this means that 6 arrows can be
drawn  $(4!/2(4-2)!)$.  As only three arrows connect the variables in
model 2, three partial correlations should be computed.

Table 1

|        | SI83   | DEL83 | AC83  | JC81  |
|--------|--------|-------|-------|-------|
| SI83   | 1.000  |       |       |       |
| DEL83  | -0.394 | 1.000 |       |       |
| AC83   | -0.259 | 0.146 | 1.000 |       |
| JC81   | -0.310 | 0.140 | 0.400 | 1.000 |

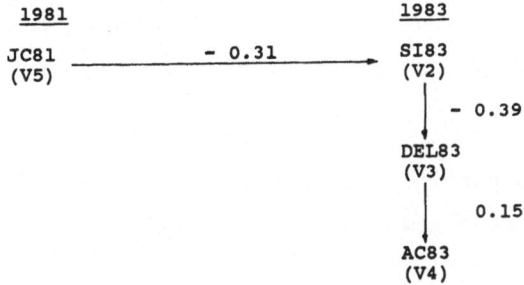

**Partial correlations**

The partial correlations are:
r 24.3  = - 0.22      ( p < 0.00 )
r 35.2  =   0.02      ( p = 0.37 )
r 45.23 =   0.35      ( p < 0.00 )

Fig.3

A first sight model 2 (figure 3) seems to be reasonably in accordance with labelling theory :
  - a higher number of judicial contacts (1981) correlates with a
lowered level of social integration (1983)  (r = - 0.31;  p < 0.001),
  - a lowered level of social integration is connected with more self-
reported delinquency (r = - 0.39;  p < 0.001),
  - more delinquency is hardly related to additional judicial contacts.
The correlation is quite low (r = 0.15;  p < 0.006).

The following remarks can be made on the basis of these first
results : Firstly the relation between JC81 and SI83 and the relations between SI83 and DEL83 both seem strong enough to keep them in
the model.  Secondly the relation between DEL83 and AC83 is signifi-

cant although not very important.  Finally, when we look at the
partial correlations we see that only  r 35.2 is close to zero.

    In 3.2 we explained that there could be two paths through which
the judicial contacts (1981) could influence the youngsters : firstly
by influencing their social integration, secondly by influencing
their delinquency level directly.  We can see that the partial correlation r35.2  relating JC81 and DEL83 (controlling for SI83) is
almost zero (r = 0.02;  $p < 0.37$).  On this basis we may conclude
that judicial contacts are not related to delinquency in any direct
way.

    The other two partial correlations are rather high (r 24.3 and
r 45.23).  Therefore a new scheme is constructed in which four arrows
are presented.

    In figure 4 we present a revised version of model 2.  The number
of contacts of the last year still has a negative influence on the
social integration level and the social integration level is related

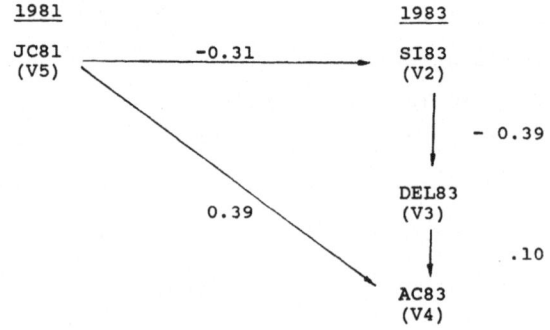

Partial correlations:

r 24.35  = -0.12    (p =  0.02)
r 35.2   =  0.02    (p =  0.37)

Fig.4

- also negatively - to delinquency. A new arrow is added : JC81 is
related directly to AC83 (Beta = 0.39;  p < 0.001).  When AC83 is
related to JC81 and to DEL83 the coefficient for the relation DEL83-
AC83 drops from .15 to .10 (p = 0.06) which is on the limit of significance.  We decided to keep the arrow in the model because it seems
more logical to relate AC83 to DEL83 as well as JC81.  Judicial intervention generally requires the precedence of a delinquent act.

All the arrows in the scheme represent relatively strong correlations (or Beta's).  The most important prediction of labelling
theory seems to find empirical support in this model : JC81 is related to DEL83 through SI83.  In addition AC83 is predicted by JC81.
This seems to underscore the influence of recidivism on the decision-
making process within the judicial system.  Apparently the police-
officers are reacting strongly to the fact that they are encountering a
recidivist whenever they catch a youngster in an offense and have to
decide if they will record him officially or dismiss him.  Surprising
is the result that given JC81 and SI81, DEL83 appears to relate only
weakly to additional judicial contacts.  We will come back to the
influence of both JC81 and social integration and delinquency on AC83
later.


## 7. Analyses for the social integration theory

We will now analyse the data with the social integration model
as our starting point.  The correlation matrix is shown in table 2.
The analyses are performed on the total sample.  Social integration
is the key concept in this model.  Therefore we do not need a selection to exclude a part of the sample from the analyses as we did to
test labelling theory (where we excluded from the analysis the youngsters with judicial contacts older than one year).

Firstly we test the relations represented in figure 1 when social
control theory was introduced.  All the arrows (figure 5) are accom-

Table 2

|        | SI81   | SI83   | DEL83 | AC83  | JC81  | DEL81 |
|--------|--------|--------|-------|-------|-------|-------|
| SI81   | 1.000  |        |       |       |       |       |
| SI83   | 0.565  | 1.000  |       |       |       |       |
| DEL83  | -0.218 | -0.405 | 1.000 |       |       |       |
| AC83   | -0.196 | -0.265 | 0.160 | 1.000 |       |       |
| JC81   | -0.361 | -0.302 | 0.128 | 0.399 | 1.000 |       |
| DEL81  | -0.491 | -0.363 | 0.346 | 0.293 | 0.409 | 1.000 |

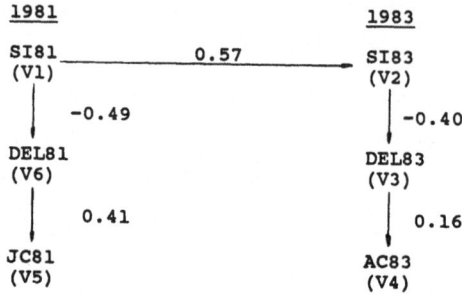

Partial correlations (significance in parentheses)

```
r15.6    = -0.20     (p = 0.00)
r26.15   = -0.09     (p = 0.06)
r25.16   = -0.10     (p = 0.03)
r13.256  =  0.10     (p = 0.03)
r35.126  = -0.06     (p = 0.13)
r36.125  =  0.26     (p = 0.00)
r24.1356 = -0.13     (p = 0.01)
r14.2356 =  0.06     (p = 0.12)
r45.1236 =  0.30     (p = 0.00)
r46.1235 =  0.12     (p = 0.02)
```

Fig.5

panied by significant coefficients. For the indices measured in
1981 we can see that a low level of 'social integration (1981)' is
related to more self-reported delinquency (1981 ($r = -0.49$;
$p < 0.001$) which is leading to more recorded judical contacts (1981)
($r = 0.41$; $p < 0.001$). SI81 is strongly connected to SI83 ($r =
0.57$; $p < 0.001$). SI83 is related negatively to DEL83 ($r = -0.40$;

p $<$ 0.001).  DEL83 in its turn does not relate very strongly with
AC83 (r = 0.16;  p $<$ 0.003).

When we take the importance of the partial correlations into
consideration, this model does not fit the data very well.  Four
partial correlations are too high to keep these relations out of
the model.  The four relations are those between :
- SI81 and JC81
- SI83 and AC83
- DEL81 and DEL83
- JC81 and AC83
Therefore a new model will be presented (figure 6).

In figure 6 we can find the same relations as were described in
figure 5 as well as four arrows which were added.  Because several

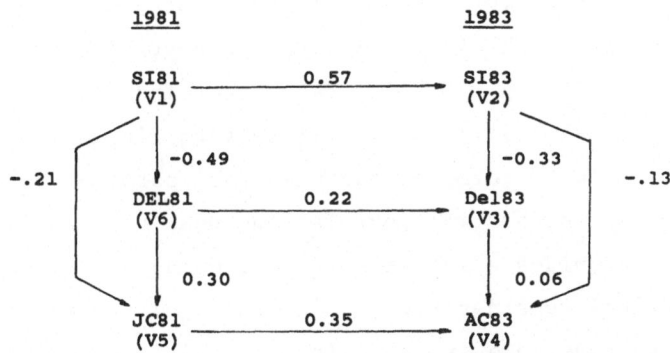

**1981**                          **1983**

Partial correlations  (significance within parentheses)

```
r25.16   =   -0.10     (p = 0.03)
r26.15   =   -0.09     (p = 0.06)
r13.256  =    0.10     (p = 0.03)
r35.216  =   -0.06     (p = 0.13)
r14.2356 =    0.06     (p = 0.12)
r46.1235 =    0.12     (p = 0.02)
```

Fig.6

indices are influenced by more than one arrow (JC81, DEL83 and AC83) we need to compute the regression coefficients. The four arrows added to the model concern the following relations :

  - Both in 1981 and 1983 social integration is related directly to judicial contacts (1981 : Beta = - 0.21;   p < 0.001;   1983 : Beta = - 0.13;   p < 0.002).

  - DEL81 is influencing DEL83 directly (Beta = 0.22;   p < 0.001).

  - There is a direct influence from JC81 to AC83 (Beta = 0.36; p < 0.001), AC83 is also related to DEL83 (Beta = 0.06; p = 0.23).

    In this model all the partial correlations are rather low. However the Beta coefficients of the relations DEL83-AC83 and of SI83-AC83 are rather low too. Please note that some coefficients change when Beta's are computed in stead of correlations. For instance the relation between DEL81-AC83 goes from .16 to .06. Because of these low values we tested a third model in which the relations DEL83-AC83 to SI83-AC83 are removed. In doing this we keep only relatively high coefficients in the model but a disadvantage is that one of them comes back as relatively high partial correlation (r 24.1356 = - .13;   p = 0.01). Because of lack of space we do not present this model. It remains difficult to find a model which is relatively simple in terms of the number of paths it contains and still keeps all the partial correlations near to zero. It was decided figure 6 provides a sufficient model because it is compatible with theoretical considerations. In addition it seems logically preferable to have some symmetry in the relations between the variables of '81 and '83. This would not be the case if we removed the relation SI83-AC83 while the relation SI81-JC81 is present. It appears that for decisions of this kind we have to rely on theory as well as on statistical tests.

    When we consider the indices measured in 1981 only, we may conclude that social integration is an important predictor for delin-

quency.  The number of contacts of juveniles with the judicial system
is related both to delinquency and rather unexpectedly, to social
integration.  We may interpret the impact that social integration
appears to have on the number of recorded judicial contacts in two
ways :

1) Probably the police and other officials within the judicial
system react to characteristics of youngsters which they interpret
as unfavourable and a 'bad prediction' for the future.  These 'bad
signs' - which we did measure partly by means of social integration -
leads them to be less tolerant towards some youngsters and less in-
clined to drop their case, which results in more recorded contacts
with the juvenile justice system.  An example of this would be
truancy.  When a boy is not doing very well at school and does not
attend to school very often, he will have a low score on the social
integration index.  If the police finds out he is a truant, they may
consider whether he needs a more severe treatment and send him to
the prosecutor instead of dismissing him.

2) Spending a lot of time outside the home (which is part of
the social integration index) gives a youngster a greater probability
of getting caught by the police and thus have recorded contacts.

We will now take a closer look at the variables from <u>1983</u>.  SI83
has a strong relation with SI81.  Thus, although social integration
is relatively stable, it is not something immutable.  DEL83 is related not only to SI83 as social control theory predicts, but also to
DEL81.  A plausible explanation is that youngsters, if they committed a few offenses and were not caught (keeping in mind that this
must be quite usual because the risk of being caught is low) might
find in this impunity an encouragement to continue their behaviour.

Finally AC83 is predicted best by JC81 - as we saw in the model
of labelling theory - and this confirms the weight judicial authorities apparently give to recidivism.  Delinquency has a weak relation

to AC83.  To understand this, one should keep in mind that DEL83 is
per definition a self-reported measure of delinquency known to the
investigators only, and not visible to the judicial authorities.
Because of this - natural - lack of knowledge the police and prosecutor have to base their decisions about whether or not to dismiss
a case on information accessible to them;  for instance the existence
of prior arrests and eventually information pointing at problems in
the sphere of social integration.

## 8. Discussion of the results

We choose figure 6 as the final model which fits the data best.
The most important difference between the final 'labelling model'
(figure 4) and the final 'social control model' (figure 6) is the
absence of the relation JC81-SI83.  This has two important implications :

1. The fact of having judicial contacts does not have an important negative influence on the development of the social integration
of youngsters.  This is why we may conclude labelling theory is in-
correct.

2. There seems to be a mechanism within the judicial system
which results in a selective treatment of youngsters.  The most important selective factor for additional judicial contacts is the
presence of former recorded contacts with the judicial system.
Taking former judicial contacts into account means that the influence
of frequency of delinquency is almost zero.

When we compare the first social control model (figure 5) with
the second model (figure 6) the following remarks can be made

1. On essential points social control seems right : social integration is related to delinquency both for the data of 1981 and
1983 and SI81 can be predicted from SI83.

The point seems to be that - while the relations predicted by
social control theory are confirmed by the data - more arrows are
needed in the model to predict delinquency and additional judicial
contacts.


2. Some features appear to characterize the judicial system and
to influence judicial contacts :
    - Elements belonging to the social integration concept seem to
influence the decision making process of police and prosecutors;
    - When one is considering additional judicial contacts of young-
sters, the most important factor influencing the decision to prosecute or drop the case is the presence of former judicial contacts.
A recidivist gets a harsher treatment and has a greater chance of
getting an additional judicial contact.

To summarize one can say that social control theory seems right
but incomplete in predicting delinquency because past delinquent
behavior is related directly to renewed self-reported delinquency.

Labelling theory is wrong in predicting lower social integration
as a result of former judicial contacts but there are features within
the juvenile justice system which are responsible for selecting youngsters for prosecution otherwise than by considering their delinquency
level alone.  Elements of social integration and especially recidivism play an important role in influencing the decision making of
police and prosecutors.

We conclude by making some remarks about the method used : as
we stated before, correlational data cannot prove causality.  It is
helpful in eliminating models which predict relations which find
no empirical support.  But, given the data, several models remain
possible and theoretical considerations should help researchers
choosing between them.

The statement that some relations can be considered as spurious is also based on theory.  The presence of a clean theoretical framework seems necessary to guide analyses.

Finally a problem of the method is that it does not provide us with an overall test.

However, path-analysis - being a relatively simple technique - proved to be a useful technique to help us choose between two contradictory criminological models.  More substantial information on this subject can be found in J.Junger-Tas and Junger (1984).

## References

H.M.Blalock Jr (1972). Causal Inferences in Non-experimental Research. New York, Norton and Company.

G.W.Bohrnstedt and D.Knoke (1982). Statistics for Social Sciences Data-Analysis. New York, Peacock Publications.

A.S.Goldberg (1976). Discerning a Causal Pattern among Data on Voting Behaviour. In : Causal models in the Social Sciences; ed. by H.M.Blalock; 4 th. ed. Chicago, Aldine Publishing Company, 33-48.

T.Hirschi (1969). Causes of Delinquency. Berkeley, University of California press.

J.Junger-Tas and M.Junger (1983). Juvenile Delinquency, Backgrounds of Delinquent Behaviour. The Hague, Ministry of Justice, Research and Documentation Centre.

J.Knutsson (1977). Labelling theory : a critical examination. Stockholm, National Council for Crime Prevention, Scientific Reference Group Report n°3.

R.R.Kornhauser (1978). Social Sources of Delinquency, An appraisel of Analytic Models. Chicago, University of Chicago press.

J.M.G.Segers, J.A.P.Hagenaars (eds.) (1980). Sociological Research Methods, part II (in dutch). Assen, Van Gorcum.

SECTION 6.3

ESTABLISHING CAUSAL ORDER AMONG VARIABLES IN DELINQUENCY RESEARCH
FAILURE OF PANEL ANALYSIS

J.A. Hagenaars

Tilburg University, Department of Sociology

P.O.Box 90153, 5000 LE Tilburg, The Netherlands

## 1. Introduction

Junger-Tas and Junger of the Research and Documentation Centre
of the Dutch Ministry of Justice conducted a panel study on juvenile
delinquency (Junger-Tas and Junger, 1984). In order to explain delinquency and the evolution of delinquency over time, Junger-Tas and
Junger put forward two theories, viz. Hirschi's social control theory
and the labelling theory.

Social integration is the key concept in the social control
theory. The amount of social integration determines the probabilities of committing delinquent acts and hence the chances of judicial
contacts; changes in delinquent behaviour have their roots in changes in the amount of social integration.

The key concepts in the social labelling theory are the amount
and the nature of the contacts with the police and the prosecutor.
The judicial contacts at one point in time determine to a large
extent the probabilities of delinquent behaviour in the future,
through a process of social labelling.

Junger-Tas and Junger pose the question whether the panel study mentioned above is capable of deciding between the two theories.

Because these two theories are neither mutually exclusive nor exhaustive explanations of delinquent behaviour, it is not possible to give a general univocal answer to this question. Therefore, we direct our attention to one main point of difference between the two theories, viz. the causal order of the crucial variables. The labelling theory regards the judicial contacts as the source of all evil; once one gets into the clutches of the judicial apparatus, social labelling begins to work, and future delinquent behaviour is unavoidable. Social integration as such is not a part of this theory, but it seems safe to assume that many judicial contacts lead to social disintegration. On the other hand the social control theory implies a different causal order viz. social integration - delinquency - judicial contacts.

We will investigate whether it is possible to determine the causal order of these key variables using the data from the panel study. At the same time attention will be paid to the importance of several background variables for the evolution of delinquency.

## 2. The data

In the Junger panel study on juvenile delinquency 332 youngsters were interviewed in 1981 as well as in 1983. Because of missing data our analyses are based on 284 respondents. The variables used, as well as their intercorrelations, means and standard deviations are presented in Table 1.

We took the variables 1 through 10 in the way they were constructed by the Jungers. Delinquency (variables 7 and 8; in the original notation D2 and XD2) measures the number of different delinquent acts committed during the last year, weighted by the fre-

Table 1. Product-moment correlations, means and standard deviations (N = 284)

| | 1 | 2 | 3 | 4 | 5 | 6 | 7 | 8 | 9 | 10 |
|---|---|---|---|---|---|---|---|---|---|---|
| 1 | 1 | | | | | | | | | |
| 2 | -.074 | 1 | | | | | | | | |
| 3 | -.000 | .039 | 1 | | | | | | | |
| 4 | .056 | .033 | .290 | 1 | | | | | | |
| 5 | .046 | -.397 | .173 | .005 | 1 | | | | | |
| 6 | .086 | -.297 | .169 | .072 | .556 | 1 | | | | |
| 7 | -.093 | .115 | -.062 | -.082 | -.497 | -.365 | 1 | | | |
| 8 | -.129 | .006 | -.117 | .038 | -.230 | -.428 | .377 | 1 | | |
| 9 | -.082 | .143 | -.115 | -.128 | -.371 | -.301 | .401 | .166 | 1 | |
| 10 | -.032 | -.001 | -.130 | -.123 | -.195 | -.263 | .256 | .171 | .349 | 1 |
| Mean | 1.130 | 4.592 | 2.352 | 3.218 | 5.965 | 4.820 | 1.342 | 1.447 | .317 | .451 |
| s.d. | .337 | 1.537 | .808 | 1.495 | 3.018 | 3.004 | 2.215 | 2.484 | .864 | 1.210 |

1 = (V5) Sex (1. male 2. female)

2 = (V9) Age in 1981 (2. 12 year old... 7. 17 years old)

3 = (V35) Educational level 1981 (1. low... 4. high)

4 = (V303) SES 1981 (1. low... 6. high)

5 = (Total2) Social Integration 1981 (1. low... 10. high)

6 = (XTotal2) Social integration 1983 (1. low... 10. high)

7 = (D2) Delinquency 1981 (0. no... 11. highest)

8 = (XF2) Delinquency 1983 (0. no... 11. highest)

9 = (V542) Judicial Contacts 1981 (0. none... 9. contacts last year)

10 = (Dif535) Additional Contacts since 1981 (0. none... 8. contacts)

quency with which they were committed. The variables 9 and 10 measure the number of contacts with the police or public prosecutor of juvenile court. The variables 5 and 6 are summated ratings on more than 20 variables. We did not investigate the dimensionality, reliability or valdity of these two overall indices (although the correlations between the indices and the separate indicators pointed to the need of such an investigation).

Although the questions concerning age, educational level and social status were asked twice, only the data from the first wave have been used here. The true scores on ages-1981 and age-1983, are of course, perfectly linearly related to each other and would be in-distinguishable from each other in the (regression) analyses that will be carried out. Moreover, the correlation between age-1981 and age-1983 is .934, and not much can be gained from combining the two. The correlations between the variables status-81/status-83 and education-81/education-83 are much lower, .668 and .683 respec-tively. However, using the information from both waves is rather complicated because it would then be necessary to separate real changes from changes due to unreliability and to establish the causal order of changes in educational or status level and changes in delinquency. Given the purposes of this article, these difficul-ties have been avoided.

Throughout this paper only regression techniques have been em-ployed, assuming interval measurements for all variables and linear relations among them. (The LISREL program has been used to obtain estimates of the parameters of the regression equations). Again, given the purposes of this article the validity of these assumptions is taken at face value.

## 3. Two-wave, two-variable models

Cross-sectional surveys are notorious for their incapacity to establish causal order among the variables. Sociologists have been led to believe that repeated surveys, panel studies, provide the solution. "Our feeling toward panels ... approaches superstitious reverence' (Davis, 1978).

Of course, panel studies do present a clear causal order from wave 1 to wave 2 to wave 3 etc., but that does not mean that all problems are solved. The so-called cross-lagged panel correlation technique is the most explicit attempt to establish empirically the

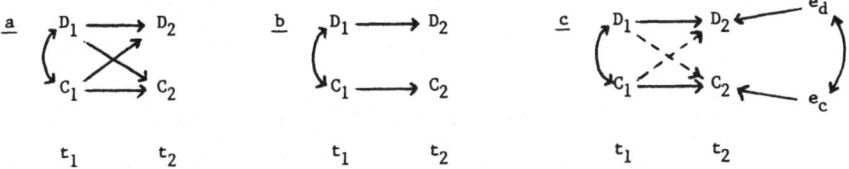

Fig.1. Two-wave, two-variable models

causal order of the variables.  The basic ideas of this technique were developed by Lazarsfeld in 1946, and later on by Campbell and Pelz and Andrew (Lazarsfeld, 1972; Campbell, 1963; Pelz and Andrews, 1964).  The scheme underlying this technique is presented in fig.1a.

In figure 1 we assume that a two-wave panel study has been conducted with two characteristics D(elinquency) and C(ontacts) with the judicial system, both measured at time 1 and time 2.  The double-headed arrows between $D_1$ and $C_1$ point to unanalysed correlations generated by previous causal processes.  The one-headed arrows represent direct causal effects.  The effects $D_1-D_2$ and $C_1-C_2$ estimate the stability of the characteristics Delinquency and Contacts respectively.  Comparison of the cross-lagged effects $D_1-C_2$ and $C_1-D_2$ in figure 1a ought to provide information about the causal order of the characteristics delinquency and contacts. For instance, if the arrow $C_1-D_2$ is absent or its effect much smaller than $D_1-C_2$ one concludes that delinquency causes contacts, rather than vice versa.

However, interpreting the model in figure 1a along these lines one encounters several practical and theoretical difficulties (see among others, Davis, 1978; Duncan, 1975, Hagenaars, 1985).  In the first place, the model in figure 1a very seldom gives an adequate description of the data;  mostly, there are too many or too few arrows in this diagram.  In many studies the cross-lagged effects turn out to be insignificant or very small, leaving the investigator with the model in figure 1b.  This model, of course, presents

no clue whatsoever as to the causal order of delinquency and
contacts.

In many other studies, model a (and b) has to be rejected because
the model does not sufficiently explain the observed correlation
between $D_2$ and $C_2$. Drawing a direct arrow between $D_2$ and $C_2$
might solve this difficulty, but then one has to choose between
$D_2 \rightarrow C_2$ and $C_2 \rightarrow D_2$. However, this choice implies a causal order
of the characteristics D and C which we try to establish empiri-
cally. Some investigators have sought a way out by postulating
model 1c. Instead of a direct path between $D_2$ and $C_2$ a correlation between the error terms belonging to these two variables is
assumed. Because the source of this correlation is unknown (and
might even represent a direct effect between $D_2$ and $C_2$) this is
a mere algebraic solution without any substantial meaning.

But even if model 1a is acceptable, inferring the causal order
of D and C from it may be unjustified on theoretical grounds.
In the first place one has to assume that all variables in figure
1a are perfectly measured. Different reliabilities of the various
measurements may lead to different cross-lagged effects which have
nothing to do with the causal order.

Secondly the assumption has to be made that the causal system
in 1a is closed. There are no other (background) variables or variables autonomously evolving in time which influence D and C
at each point in time.

And, last but not least, the causal lag ought to be in correspondence with the time lag between the waves. If the effects of
one characteristic work 'slower' than the effects of the other characteristic the time lag between the measurements should adequately
capture these different causal lags. Otherwise wrong conclusions
will be drawn.

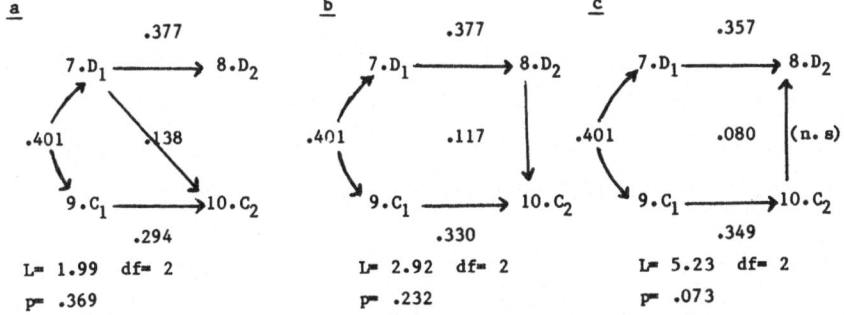

Fig.2. Delinquency and Contacts; Data – table 1; standardized
solution

We tried to use the cross-lagged panel correlation technique
to establish the causal order of delinquency and contacts with the
judicial system (the variables 7 through 10 from table 1). First,
we tested the model in figure 1a (setting up the necessary regression equations and estimating the parameters through the LISREL V
program). This model fitted very well : L = 1.91, df = 1, p = .167.
However, the cross-lagged effect $C_1$-$D_2$ was not significant, and
very small indeed (standardized regression coefficient $\beta_{89.7}$ = .017).
This leads to the model in figure 2a. Model 2a seems to imply the
failure of the social labelling theory : contacts with the judicial
system do not lead to delinquent behaviour in the future. However,
there is something strange going on. In figure 2a there exists no
influence whatsoever from $D_2$ on $C_2$. It is highly improbable that
the police and the presecution will only act upon delinquent acts in
the past, and take no action with regard to present criminal activities.

Models b and c in figure 2 represent more realistic models.
Both are acceptable, but each one makes a different a priori assumption with regard to the causal order of delinquency and contacts.

The strange result in figure 2a may be the consequence of a
lack of correspondence between the causal lag and the time lag used

(1981-1983) or of the influences of other variables.

To check this last possibility the direct effects in model 2a were estimated controlling the variables 1 through 6 (table 1).   No major changes occurred.   The cross-lagged effect $D_1$-$C_2$ dropped slightly from .138 to .102 (and was no longer significant).

This leaves us with an outcome of the cross-lagged panel correlation technique in which the figures seem to point to the causal order Delinquency-Contacts;   however, within the context of a highly improbable model.

## 4. Social control theory

In the previous section an attempt has been made to compare the social control theory and social labelling theory and to decide between the two in a most elementary fashion by focusing on the causal order of delinquency and contacts, using the cross-lagged panel correlation technique.   Obviously the attempt failed.   Therefore, we now turn to the question whether the full model derived from social control theory (as depicted by Junger) is acceptable.

In the first instance a system of recursive equations was set up in which the background variables 1 through 4 (table 1) performed the role of exogeneous variables influencing all other variables (5 through 6).   The causal order of the other variables was : 5, 7, 9, 6, 8, 10, in such a way that each variable exercises direct influences on all variables of 'lower' causal order.

After estimating the parameters of this recursive system, non-significant effects were left out, all pertaining to effects (much) lower than .11.

Testing this more parsimonious model yields the following results : L = 38.03, df = 25, p = .046.   (Putting in a few border-line effects from the saturated model resulted in a more satisfac-

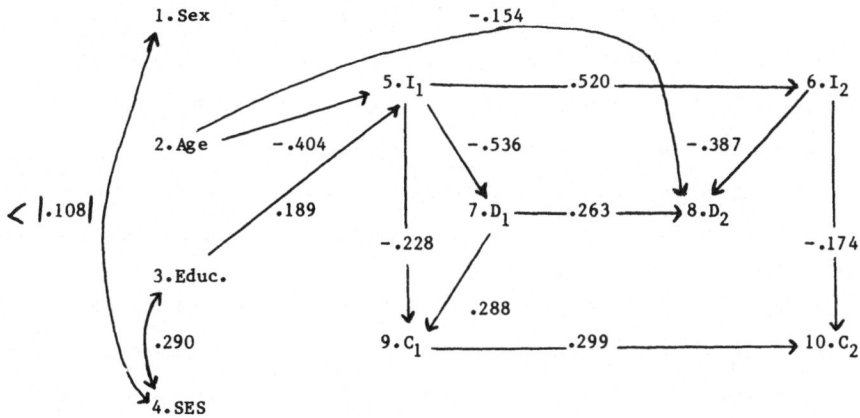

additionally: Age – $D_1$ : –.098 (n. s.)

               Age – $I_2$ : –.090 (n. s.)

               SES – $D_2$ :   .093 (n. s.)

Fig.3. The social control theory; standardized effects

tory fit in terms of  L.  However, the sizes of these extra effects
were still negligible).

In figure 3 the results of the restricted model are shown.
Effects that were significant in the original, saturated model, but
not in this restricted model, are not drawn in this diagram, but
mentioned beneath the figure.

Junger-Tas and Junger (1984) state some hypotheses concerning
the influence of the background variables, some of which are corrobo-
rated and some are not.

Contrary to the expectations sex has no direct influence on
social integration and girls are not better integrated than boys.
Age, indeed, has a negative direct influence on integration : older
children are relatively less integrated.  A smaller, but expected
influence is exerted by education : the higher the educational
level, the higher the social integration.  The small negative influence of age on delinquency was not predicted, but found.  *Most*

striking is the absence of any direct effects of social status on judicial contacts. It was predicted (esp. according to the labelling theory) that SES would have a large negative influence on judicial contacts, that youngsters of low SES will be prosecuted sooner than their high SES counterparts. Only very low, insignificant effects in this direction were found in the saturated model.

Most central to the social control theory is the assumption that social integration is an autonomously developing characteristic; the evolution of delinquency over time mirrors the changes and stabilities of this fundamental characteristic, and the scores on judicial contacts, in turn, are derived from the delinquency scores at each point in time.

Looking at figure 3, we find that social integration appears to be the most stable characteristic (not necessarily meaning the most autonomous), more so than delinquency or judicial contacts. But the direct relation between $D_1$ and $D_2$ cannot be explained away by social integration. Delinquency does depend on social integration, but at the same time the evolution of delinquency over time is (partly) independent of social integration.

Although it was not expected, social integration has a direct influence on judicial contacts, not mediated by the delinquency score. And also : judicial contacts have an evolution of their own, not dependent on either integration or delinquency. And again, most surprisingly : at time 3 there is no direct influence of delinquency on judicial contacts.

The strong effects exercised by social integration point to the adequacy of the social control theory. On the other hand, the direct influence of $C_1$ on $C_2$ and the absence of a direct effect between $D_2$ and $C_2$ are more easily explained in terms of the labelling theory. To give this last theory a fairer chance, we have, in the next section, reanalysed the data according to figure 3, this

time assuming a different causal order between delinquency and judicial contacts.

## 5. Judicial contacts causing delinquency

In setting up the full recursive system of equations the causal
order of the variables applied in the previous section was maintained, with the exception of 7-9 and 8-10, whose order was reversed.
In the next step all insignificant effects were removed, resulting
in a well fitting model : L = 32.64, df = 24, p = .112.

This model looked very much like the model in figure 3.  The
direct effect between $I_1$ and $D_1$ was now estimated to be -.442;
the effect of $I_1$ on $C_1$ was -.370;  the relation between $D_1$
and $C_1$ was of course reversed and estimated to be : .251;  a significant relation between SES and $C_1$ appeared, albeit a small
one : -.126.  All other relations were very close to the effects
presented in figure 3.

What is the meaning of these results ?  They show in the first
place that there is no empirical way to establish the causal order
of delinquency and judicial contacts with these data, a result we
have encountered before in section 3.  Furthermore, it is clear that
the labelling theory also does not give a perfect description of the
data : there is no direct link between $C_1$ and $D_2$ or between $C_1$
and $I_2$;  delinquency has a stability of its own, not explained by
judicial contacts;  in wave two, there is no direct effect of contacts on delinquency.

## 6. Social integration; a dependent variable ?

The autonomy of the variable social integration is essential to
the social control theory.  But imagine someone studying the phenomenon of social integration and looking for its causes.  It is not
impossible that this person will point to delinquency and judicial

contacts as causes of social (dis)integration.  If we set up a model
like figure 3, but now with the causal order : delinquency → judicial
contacts → social integration, we obtain a well fitting model in
which the effects are very much like the effects in figure 3, but
with several arrows pointing to different directions.  (The largest
difference is the stability coefficient of social integration, which
drops from .520 to .394).

Conclusions based on these outcomes would be diametrically opposed to the basic idea underlying the social control theory, which
was partially 'supported' by the results in section 4.

7. Conclusions

What we have arrived at, in a way, is just the truism that the
same set of empirical data can be explained by several, but not by
all theories.

Several parts of the labelling and the social control theories
were discussed and, in a way, tested.  However, on very crucial
points a definite choice between the models could not be made.

Contrary to popular belief,panel analysis does not establish a
clear causal order among all the variables concerned.  One reason
for this failure of the cross-lagged panel correlation technique
could be the omission of important variables and the lact of sufficiently developed theories (see Saris and Stronkhorst, 1984).
More plausible in this case seems to me the lack of correspondence
between the causal lag and the time lag.

In order to unravel the many intricacies and complexities of
the causal processes concerning the effects of background variables,
delinquent acts, judicial contacts and social integration, one needs
almost continuous observations.  These can be obtained by studying
case histories and recording all the major changes in 'status';

time series and event analyses might then provide more definite
answers to the problem of causal order.

References

D.T.Campbell (1963). From description to experimentation : inter-
    preting trends as quasi-experiments, in : C.W.Harris (ed.),
    Problems in Measuring Change, Univ.of Wisconsin Press, Madison,
    212-242.

J.A.Davis (1978). Studying categorical data over time. Social
    Science Research, vol.7, 151-179.

O.D.Duncan (1975). Some linear models for two-wave, two-variable
    panel analysis  with one-way causation and measurement error,
    in : J.M.Blalock (ed.), Quantitative Sociology, Academic Press,
    New York, 285-306.

J.A.P.Hagenaars (1985). Loglineaire analyse van herhaalde surveys;
    panel-, trend- en cohortonderzoek, diss.Tilburg.

J.Junger-Tas and M.Junger (1984). Juvenile delinquency, backgrounds
    of delinquent behavior. Ministry of Justice, R.D.C., The Hague.

P.F.Lazarsfeld (1972). Mutual effects of statistical variables,
    in : P.F.Lazarsfeld, A.K.Pasanella and M.Rosenberg (eds).
    Continuities in the language of social research, Free Press,
    New York, 388-398.

D.C.Pelz and F.M.Andrew (1964). Detecting causal priorities in panel
    study data. American Sociological Review, vol.29, 836-848.

W.Saris en H. Stronkhorst (1984). Causal modelling in nonexperimen-
    tal research. Sociometric Research foundation, Amsterdam.

SECTION 6.4


USING LISREL TO CHOOSE BETWEEN CONTRADICTORY CRIMINOLOGICAL THEORIES

P.G.M. van der Heijden, F. Meijerink and A. Mooijaart

University of Leiden, Dept. of Research
Methods and Psychometrics, Hooigracht 15
2300 RA Leiden, The Netherlands

1. Introduction

   In Junger-Tas and Junger (1984) the problem was formulated how
to answer the question which of two criminological theories explains
the relationships in a given data-set best.  The two theories are the
social control theory and the labelling theory.  Both theories differ
in the emphasis they give to different causes of delinquent behavior.
We try to shed some light on this problem using structural equation
models, as implemented in the program LISREL, an acronym for LInear
Structural RELations.  In this approach population covariances (or
correlations), following some prespecified model, are estimated and
compared with the matrix of observed covariances (or correlations).
When the difference between the observed and the estimated matrix is
small, the hypothesized model is thought to give an acceptable explanation for the data.

   It is possible to state the social control theory in terms of
a structural equation model.  For the labelling theory some extra
causal relations have to be specified.

In the following we first give a short description of LISREL. Then an outline of the variables and the problem is given. In section 4 we discuss some preliminary problems that have to be solved before using LISREL. In section 5 our results are discussed.


## 2. LISREL and structural equation models

LISREL, the name of a computer program, has become a label for an approach to model covariance structures. This approach can be used to fit models such as confirmatory factor analysis models, structural equation models, and combinations of these two. This model fitting is done by minimizing a loss-function of the difference between the sample covariance matrix $S$ and a hypothesized cova- riance matrix $\hat{\Sigma}$ being a function of the unknown parameters. The population covariance matrix is estimated taking into account the restrictions on some parameters that follow from the hypothesized model. We will now describe LISREL in more detail. Good introduc- tory texts are Bentler (1980), Long (1983a, b) and Saris and Stronkhorst (1984).

Long describes LISREL as a program unifying two different fields originating from psychology and econometrics. From psychology the concept of errors in variables is taken, which is prominent in con- firmatory factor analysis. The idea is that we try to measure latent variables with observed variables. This measurement is necessarily contaminated with error. From econometrics stems the emphasis on errors in equations, which has as a starting point the idea that var- iables never can be predicted completely, because not all relevant predictors are taken into account. It is possible that errors of some equations are correlated. LISREL joins these two approaches by allowing to define both errors in variables and errors in equations: it is possible to define equations between latent variables, which are measured with error. This combined approach is known under different names, one of which is 'covariance structure models' (Long, 1983b).

In the following we restrict ourselves to the 'errors-in-equations' approach (the word 'structural' denotes the fact that an assumed causal structure of the process is modelled). This self imposed restriction can also be described as that we assume that our latent variables are measured without error. This restriction is made because it was possible to find structural equation models that fitted the data well. Therefore we did not have to introduce the 'weaker' error-in-variables concept.

In structural equations a distinction is made between <u>endogenous</u> variables, i.e. variables that are to be explained by the model, and <u>exogenous</u> variables, variables that are not explained by the model. Endogenous variables can also be explained by other endogenous variables. This is reflected in the general linear structural equation (1)

$$\eta = B\eta + \Gamma\xi + \zeta \tag{1}$$

where $\eta$ is a $(r \times 1)$ vector of endogenous variable, $\xi$ a $(s \times 1)$ vector of exogenous variables, $\zeta$ a $(r \times 1)$ vector of errors, $B$ a $(r \times r)$ matrix of coefficients, and $\Gamma$ a $(r \times s)$ matrix of coefficients. From (1) it can be seen that, for the structural model to hold, the <u>effects</u> should be <u>additive</u>. Another assumption is that the <u>relations</u> between the variables are <u>linear</u>, i.e. the variables should be measured on an interval or ratio scale level. Sometimes a suitable transformation is necessary to obtain this property.

Structural equations can be graphically represented in <u>diagrams</u> of variables, in which <u>arrows</u> indicate which variables are to be predicted directly by which variables. $\beta$- and $\gamma$-values are usually placed near arrows, to indicate the relative weight of the effects. Positive values of $B$ and $\Gamma$ indicate positive relations, negative

values negative relations. The absence of certain arrows is also reflected
in values of B and $\Gamma$, which are constraint to be zero in this case. Furthermore, correlations between errors are collected in the matrix $\Psi$,
and covariances between exogenous variables in the matrix $\Phi$. It is
possible to define the population covariance matrix between all
(r + s) variables $\Sigma$, in terms of B, $\Gamma$, $\Phi$ and $\Psi$ (see Long, 1983b).
Constraints on B, $\Gamma$, and $\Psi$ restrict values in $\Sigma$. The problem is
to find values for B, $\Gamma$, $\Phi$ and $\Psi$, such that the difference between
the observed covariance matrix S and the estimated population covariance matrix $\hat{\Sigma}$ is a small as possible.

In the following we do not consider the case of reciprocal
causal relations : we restrict ourselves to the case that B is a
triangular matrix. Often the matrix $\Psi$ is restricted to be diagonal, i.e. the errors of different equations are not correlated, but
since we are going to deal with panel data, it makes sense to allow
for correlation of errors. In this context correlation of errors
can be interpreted as being caused by the same (perhaps unknown)
variable that is left out of both structural equations.

So far, we have discussed some issues regarding the specifica-
tion of models. Now we come to discuss identification, estimation,
and evaluation of goodness of fit of models.

Identification of a model implies that it is possible to determine a unique set of values for the unrestricted parameters. For a
model to be identified, there have to be restrictions on B, $\Gamma$ and
$\Psi$, for instance that some parameters have an equal value or are
constraint to be one or zero. It is often difficult to prove identification. Identification should be proven algebraically by solving
the parameters of the model in terms of the variances and covariances of the model. As a shortcut the matrix of second partial deri-
vatives is often checked for being positive definite, which is a
necessary and sufficient condition for identification.

A model being identified, its parameters can be estimated. Most often, <u>estimation</u> is done using the maximum likelihood method. When $\xi$ and $\eta$ are multivariate normally distributed, the ML-estimators are unbiased and are asymptotically normal distributed. This last property makes it possible to test individual parameters to be different from zero. Very little is known about the robustness of these estimators against violations of the normality assumption in LISREL (see Boomsma, 1983). Using ML-estimation, a <u>chi-square test</u> can be performed to measure the difference between the hypothesized model (yielding $\hat{\Sigma}$) and the unrestricted model (yielding S). The number of degrees of freedom for this test is equal to the difference between the number of independent parameters in the restricted and the unrestricted model. When the test value is significant, the hypothesized restricted model should be rejected. However, when the assumption of normality is not met, the test statistic should be used with caution. It is possible to test the difference between nested restricted models, using the difference in test values and the difference in degrees of freedom. This difference in chi-square is also chi-squared distributed.

Only when a model fits the data, it makes sense to start <u>interpretation</u> of the unrestricted parameter values. When variables are standardized, and thus the covariance matrix is equal to the correlation matrix, interpretation is straightforward : when we find for $\beta_{21}$ a value of e.g. .30, this indicates that a change in value of 1 standard deviation in variable 1 results in a change of .30 in variable 2. Indirect effects can be evaluated by the product of direct effects : the indirect effect of exogenous variable 1 on endogenous variable 2, via endogenous variable 1, is equal to $\gamma_{11}\beta_{21}$. Summarizing, LISREL seems to be a useful tool to test theories that can be expressed in arrow-diagrams. This property is even more pleasant in our case, since we have two theories trying to explain the same set of data in different ways. Drawbacks of the current approach are the assumptions of multivariate normality and linearity

of relations, but work seems to be done on relaxation of these assumptions (see Bentler, 1980, 1983; Brown, 1982, 1984; De Leeuw,
1983; Mooijaart, 1985; Muthen, 1983).  In the sequel we will first
describe the variables and the theories to be tested, secondly two
problems we met using LISREL, and their solutions, and finally we
discuss some models we fitted with LISREL.

## 3. Data and theories

### 3.0. Introduction

    In Junger-Tas and Junger (1984) a description can be found of
the theories to be tested, and the variables used in testing these
theories.  In this section we will rely heavily upon their paper.
First we discuss the variables, secondly the social control theory,
thirdly the labelling theory, and finally we mention some other
possible relations, and relations that are impossible from a theoretical point of view.  We follow Saris and Stronkhorst (1984), who
make a point of it that arrows should be theoretically justified
before drawing them in a diagram.

### 3.1. The variables

    The R.D.C. (Research and Documentation Centre, Ministry of
Justive, The Hague) conducted a study on juvenile delinquency, from
which three variables were issued :
    1. Social integration.  Questions were asked on social inte-
gration in family life, school, leisure time activities and relationships with friends.  From these questions a 10-point scale index
is constructed, called social integration (SI).
    2. Self-reported delinquency.  Youngsters were asked whether,
and how often they performed each of seven prespecified offenses
last year.  The total number of offenses was computed for each
subject, taking frequency into account (1 = once, 2 = 2 or 3 times,
3 = more).  This variable will be called 'reported delinquency' (DEL).

3. Judicial contacts.  Information was gathered concerning the number of recorded contacts the youngsters had last year with the police and the prosecutor.  This variable will be called 'recorded contacts' (RC).  In total 332 youngsters were interviewed in 1981 as well as in 1983.  For the computation of the correlation matrix we only used those youngsters who did not have any missing values (n = 325).

## 3.2. Social control theory

The social control theory stems from Hirschi (1968).  The theory states that if a youngster is well integrated in society he will not commit offenses, and therefore will not have contacts with the judicial system.  When the youngster is integrated badly, the opposite will be the case.  Graphically this can be represented as in Figure 1. Furthermore, it seems logical that there is a strong relationship between social integration in 1981 and 1983.  So far for the social control theory.  Using LISREL it is possible to allow errors of different structural equations to be correlated.  Correlation of errors makes sense for the variables DEL81 and DEL83, and RC81 and RC83, since it is possible that there are other (unknown) variables, not included in the equations, which influence these variables at both time points in the same way.  It is not possible to correlate errors of SI, because SI81 is not to be predicted from a structural equation.

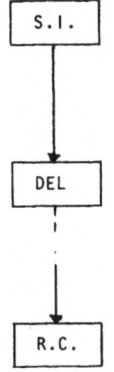

Fig.1. Social integration theory

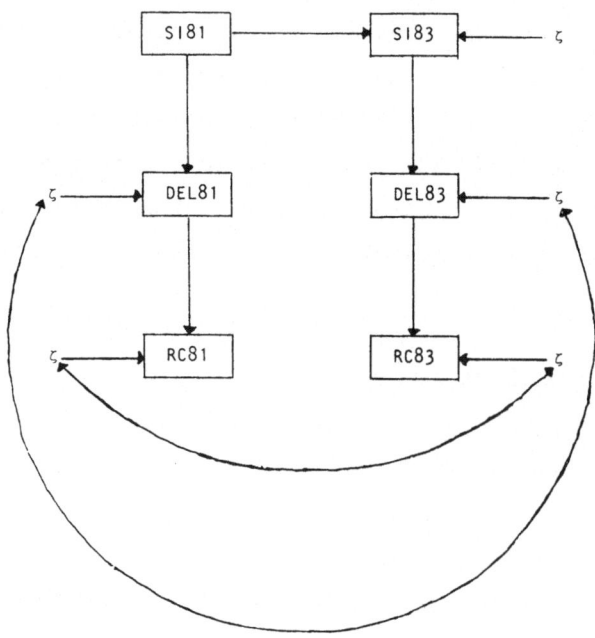

Fig.2. Scheme of relationships according to Social Control Theory

Summarizing the social control theory, one can make a scheme of relationships as is shown in figure 2.

## 3.3. Labelling theory

According to Junger-Tas & Junger, the labelling theory tries to explain delinquent behavior stressing other relationships. If a person has had judicicial contacts in the past, the theory predicts that this person will show more delinquent behavior and subsequently will have additional judicial contacts in the future. The relationships are represented in figure 3 (In Junger-Tas and Junger (1984) this figure is different : there is no arrow from DEL83 to RC83, but an arrow from RC81 to RC83. This seems logically impossible, since a person can only have contacts because he is delinquent. See also section 3.4 and section 4.2.), where the influence of social integration has disappeared. It can be argued that it

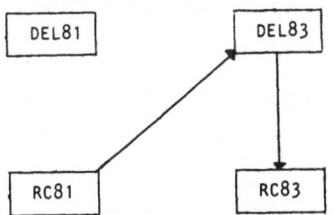

Fig.3. Labelling Theory

is in the spirit of the labelling theory to say that the relation
from RC81 to DEL83 can go over SI83.  We shall make use of this in
section 5.4.

## 3.4. Other possible relationships

So far relationships have been mentioned which fit rather
straightforward in the social integration theory or the labelling
theory.  But looking at the two schemes, other relations catch the
eye which seem appealing.  In this section these relations are discussed.

### 3.4.1. Influence of delinquency over time

It seems possible that there is a relationship between delinquency in 1981 and social integration in 1983.  An explanation for
this could be that a person who has a lot of delinquent behavior,
probably also chooses suitable friends, thereby changing his own
social integration in a certain direction.

Besides this relation it seems also possible that there is a
direct relation between delinquency in 1981 and 1983.  A lot of psychological research seems to indicate that the best predictor of
future behavior is past behavior.

### 3.4.2. Influence of recorded contacts over time

It is possible that the number of recorded contacts in 1981 has an influence on the social integration in 1983. For instance, a recorded police contact is often reported to the parents, which could cause the parents to change the social integration of the youngster.

Furthermore, a note should be made on the relation between the number of recorded contacts in 1981 and the delinquent behavior in 1983. Labelling theory postulates a positive relation. However, it is also possible that a negative relation is found. This could be interpreted as a result of deterrence.

## 3.5. Relations which are left out

So far relations between concepts have been discussed which fit either in the social integration theory, the labelling theory, or the ideas of the researchers. But besides these relations there are a number of relations which are excluded by us. These relations are discussed in this section.

1. It is impossible that arrows go from 1983 to 1981.

2. We assume that SI81 cannot influence DEL83 directly, but that this effect is mediated by SI83.

3. SI81 cannot influence RC83, because this effect is mediated by SI83 and DEL83.

4. DEL81 cannot cause RC83, because a person cannot have contacts with the police in 1983 for delinquent behavior in 1981, except in very serious cases. It is more reasonable to assume that this effect goes via SI83 and DEL83, or only via DEL83.

5. RC81 cannot cause directly RC83, because it is only possible to get into contact with the judicial system by behaving delinquently.

6. Both SI81 and SI83 cannot have a direct influence on RC81 and RC83. The relation between social integration and recorded contacts should go via delinquency.

## 3.6. Conclusion

In this section we have discussed which variables are thought to be related, following both theories and our own ideas. Before we can start with LISREL analyses, two more problems have to be solved, one concerning linearity of relations between the variables, and one concerning conditional causal relationships. This will be the subject of section 4.

## 4. Two problems to be solved before using LISREL

### 4.1. Problem 1 : non-linearity of relations

In section 2 it has been discussed that one of the assumptions of the LISREL-approach is linearity of relations among variables. Both in 1981 and 1983 Junger-Tas & Junger have measured three concepts : social integration (SI), delinquency(DEL), and recorded contacts (RC). The matrix of correlations between these six variables (see table 1) contains two relations which can not be linear from a theoretical point of view. The two relations we have in mind are the relation between DEL en RC both in 1981 and 1983 : a low score on DEL can only lead to a low score on RC, whereas a high score on DEL can result in a low and a high score on RC. Constructing two cross-tables, it can be seen that only in a triangle of these tables the frequencies are higher than zero. In case of a

Table 1. Original correlation matrix

```
SI    1.00
DEL   -.49 1.00
RC    -.36  .41 1.00
SI     .57 -.36 -.30 1.00
DEL   -.23  .36  .17 -.45 1.00
RC    -.18  .27  .35 -.26  .20 1.00
       SI  DEL   RC   SI  DEL   RC
```

Table 2. Correlation matrix based on PRINCALS-solution

| | | | | | |
|---|---|---|---|---|---|
| SI | 1.00 | | | | |
| DEL | -.37 | 1.00 | | | |
| RC | -.31 | .51 | 1.00 | | |
| SI | .57 | -.29 | -.25 | 1.00 | |
| DEL | -.28 | .30 | .20 | -.46 | 1.00 |
| RC | -.21 | .27 | .32 | -.29 | .31 | 1.00 |
| | SI | DEL | RC | SI | DEL | SI |

linear relationship large frequencies are to be found around the diagonal of the matrix (i.e. only when the correlation differs considerably from zero). We have two solutions for this non-linear relationship.

The first solution is to use a method that diagonalizes the two cross-tables as much as possible by rescaling the categories, under the restriction that the original order of the categories should remain intact. This can be accomplished with the program PRINCALS (De Leeuw & Van Rijckevorsel, 1981). Two PRINCALS-solutions were obtained, one for each cross-table. From these solutions the rescalings of the categories were used to recode the original categories of the variables, and compute a new correlation matrix (table 2). The difference with table 1 is large. The correlations between DEL and RC is considerably higher for both time-points. In fact, they are maximal under all possible monotone requantifications of the variables. The requantifications of the four variables are shown in figure 4.

The second alternative we used in order to cope with the non-linear relation, consists of treating all variables as if they are discreticized measurements of latent continuous normal variables. This results in a matrix of so-called polychoric correlation coefficients, which can be obtained using the LISREL-V program (see Jöreskog & Sörbom, 1984). Table 3 shows the matrix with polychoric

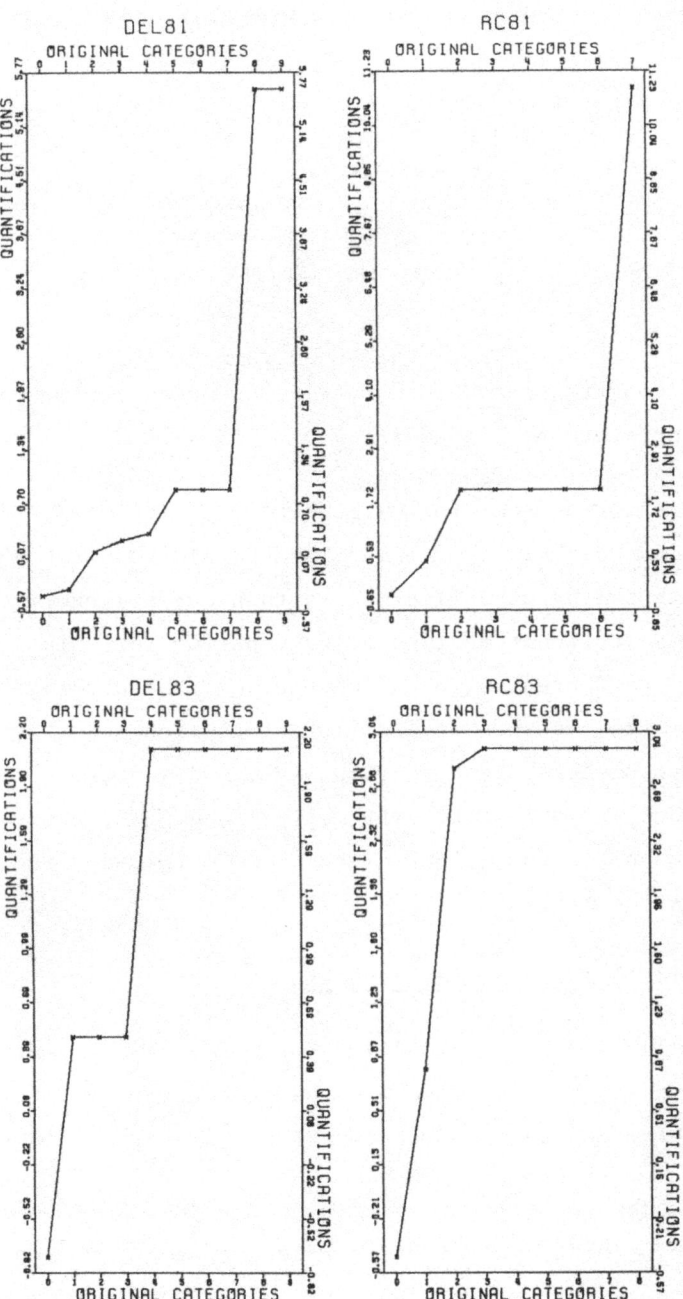

Fig.4. Requantifications of original category scores for four
       variables

Table 3. Polychoric correlations

```
SI    1.00
DEL   -.50 1.00
RC    -.36  .42 1.00
SI     .57 -.37 -.30 1.00
DEL   -.21  .33  .13 -.43 1.00
RC    -.20  .30  .40 -.27  .19 1.00
       SI  DEL   RC   SI  DEL   RC
```

correlation coefficients.  This matrix resembles the matrix of numerical correlations (table 1).

In all analyses that follow, the correlations computed with the quantifications derived from the PRINCALS solution were more in accordance with the hypothesized models than the polychoric correlations.  It is difficult to find an explanation for this.  Maybe it is caused by the unrealistic assumption that the observed variables are discretisized measurements of latent continuous variables : since most youngsters are not delinquent, and thus have no recorded contacts, four variables are extremely skewed.  Therefore, in the following we only report LISREL-results we obtained using the PRINCALS correlation matrix.

## 4.2. Conditional causal relationships

Junger-Tas and Junger (1984) pay considerable attention to considerations that the police and the prosecutor have for recording a contact, provided that the contact takes place.  We quote :

"Given schoollevel, social economic status (SES), age, sex and father's employement, police officers and prosecutors react differently to youngsters.  This means that boys whose father is unemployed or has a low SES will have a greater chance to be prosecuted. Furthermore, boys with a lower education level will also be prosecuted sooner".

Some of these variables are contained in the social integration index, or influence this index. Furthermore, when a contact takes place, the probability for this contact to be recorded is much larger when there are preceding recorded contacts (M.Junger, pers.comm.). How should we deal with these relations in a model ?

It seems that we have to do with conditional causal relationships (see Saris and Stronkhorst, 1984), i.e. <u>only when</u> a contact takes place, there is a direct relation between SI and RC, and RC and former RC. With the variables in this research, this can be graphically represented as in figure 5.

In figure 5 the circles denote the not-measured, or latent variable 'contacts' (CON). From CON81 an arrow points to the arrow from SI81 to RC81, indicating a conditional causal relationship. From CON83 two arrows point to other arrows. Conditional causal relations cannot be tested in a standard way. One way to get rid of conditional relations is by selecting the subgroup for which the conditional relations holds. However, here the subgroups should be

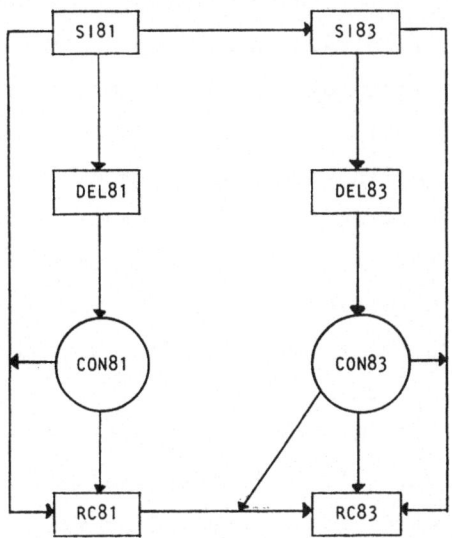

Fig.5. Conditional causal relationships

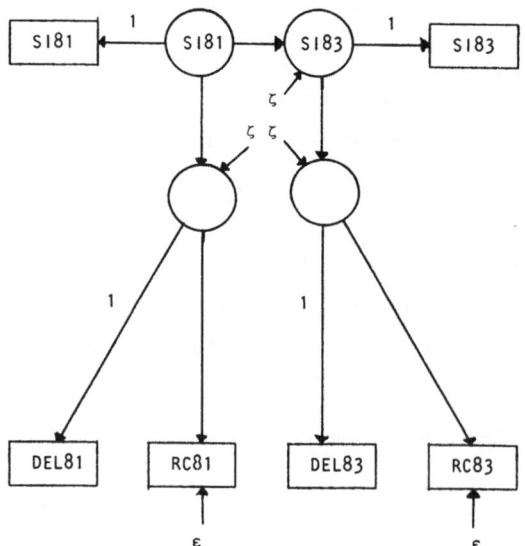

Fig.6. Theoretical solution conditional relationships

selected on the basis of their score on the variable CON, which is
not measured.  Therefore, we have to solve this problem in a different way.

   A first solution is to redefine the problem substantially, and
to work with latent variables.  One way to do this is with a model
as is represented in figure 6.  In this model the latent SI (circle),
is measured by one variable (square), which is confounded with error
(circle).  The latent delinquency is measured by two variables, i.e.
self-reported delinquency and recorded contacts.  A rationale for
this redefinition is that it is recommended in criminological research to use these two measurements for delinquency, because using
only recorded contacts one is measuring in particular the more
'serious' cases (Steinmetz, R.D.C., pers.comm.).  Redefining our
problem, we have omitted the step from DEL to RC, and therefore the
latent variable 'contact' also.  Furthermore, this solutions makes
it also possible to let SI to be measured by more than one variable.
This seems a very elegant way to solve the problem.  However, for

the labelling theory there is a crucial difference between self-
reported delinquency and recorded contacts (see section 3.3).  Therefore we had to find a solution keeping the order of DEL-RC intact.
Such a solution is to work with models containing only manifest variables.  More specifically, we let it be an empirical question
whether we have to draw arrows from SI to RC, and from RC81 to
RC83.  For this reason the restriction 5 and 6 of section 3.5 are
overruled.  When these arrows have to be drawn, we interpret this
as indicating that the influence of the variables SI and RC on the
decision of the police officer or the prosecutor cannot be neglected.

## 5. The LISREL solutions

### 5.1. Introduction

In this section we are testing the social integration theory
described in section 3.2, and the labelling theory discussed in
section 3.3.  We will see that, when testing the social integration
model, some extra relations will have to be postulated in order to
fit the data.  The rationale for these relations is described in
section 4.2.  We start with the discussion of results concerning the
social control theory.  The labelling theory is discussed in the
following section in terms of its further contribution to the goodness of fit already obtained for the social control theory.

### 5.2. Social integration theory

The social integration model represented in figure 2 is the
first model we are going to test using the LISREL V program.  Figure
7 once again shows the model, but now supplemented with the LISREL
ML-estimates of the paramters placed next to the arrows.  To repeat
shortly, the uni-directional arrows refer to the effect of an exogenous variable upon an endogenous variable ($\gamma$-parameter) or the
effect of an endogenous variable upon another endogenous variable
($\beta$-parameter).  In our data the only exogenous variable is SI81.

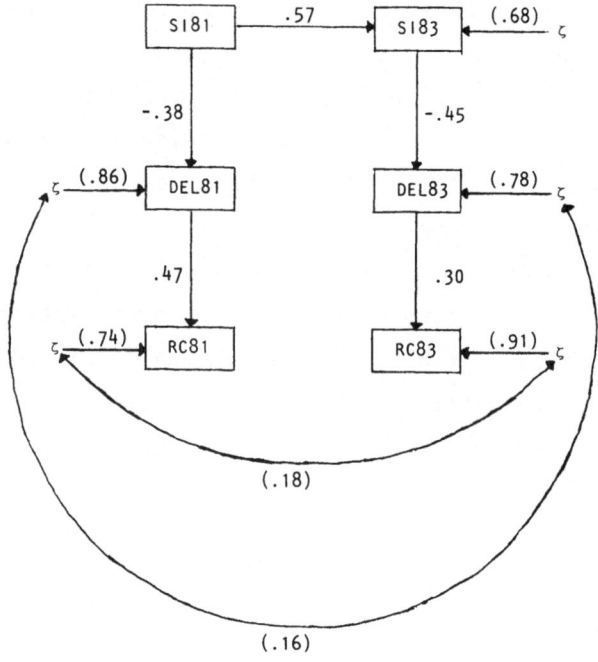

F: Fig.7. Model 1; df is 8, $\chi^2$ = 30.6, p = .000

The bidirectional arrows indicate the correlation of the errors of two endogenous variables.

The fit of model 1 is indicated by the chi-square value, which is 30.62, df is 8, p = .000. Obviously in this case the model does not fit the data well. Therefore we do not have to interpret the estimates of the parameters of the model. (Interpreting the chi-square values we should keep in mind that our data do not meet the LISREL assumptions of multi-variate normality and linearity of relations. Therefore we only use the test values as an indication of fit, not as a measure which can be used to accept or reject models).

Summarizing, social control theory alone does not provide an adequate fit to the data.

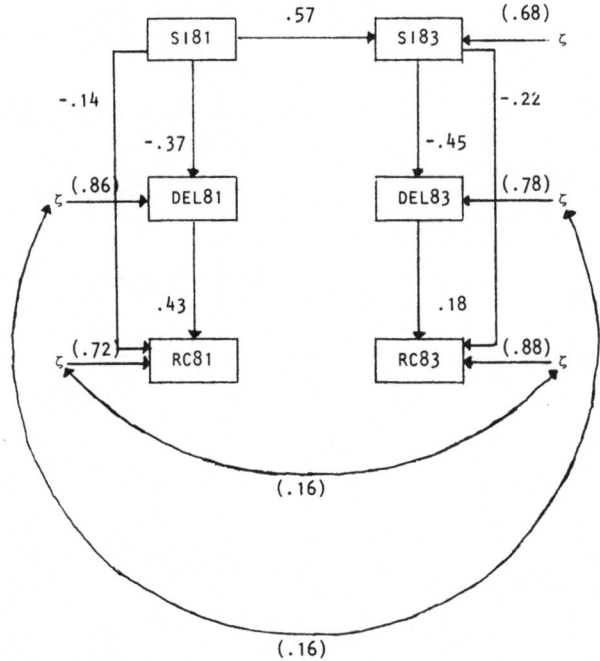

Fig.8. Model 2,   df  is 6, $\chi^2$ = 15.5, p = .017

## 5.3. The influence of SI and RC during contacts

By relaxing some of the constraints in the model, for instance by relaxing a parameter which was constrained to be zero (no effect), the fit of the model usually improves.  A measure for the improvement in fit is given by the so-called modification index, provided by the LISREL V program.  The modification index for a specific parameter indicates the minimal reduction in chi-square obtained by relaxing that specific parameter (all other constraints remaining the same).  Relaxing a parameter results in a loss of one degree of freedom.  The modification indices for model 1 (figure 7) indicate that a substantial reduction in chi-square is obtained by relaxing the two parameters between SI and RC (5.9 for 1981, and 5.2 for 1983).  Model 2 (figure 8) shows the result of the analysis with these two parameters relaxed.  Model 2 gives part of the solution to the conditional causal relationships using manifest variables

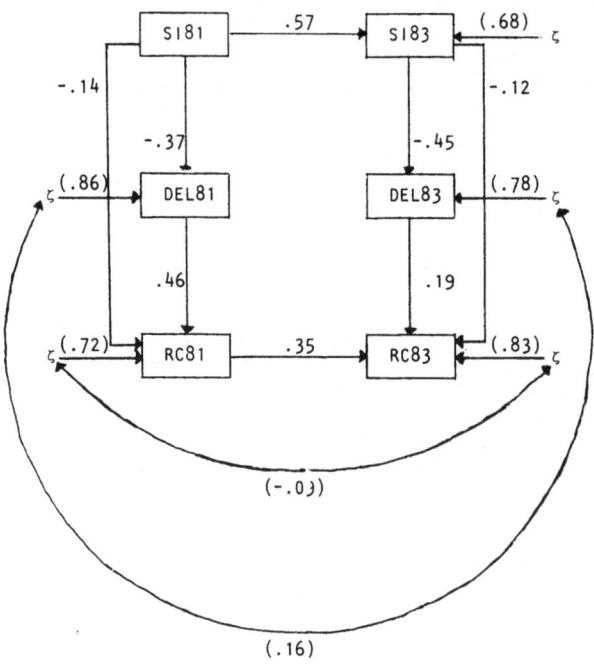

Fig.9. Model 3;  df  is 5, $\chi^2$ = 5.4, p = .37

(see section 4.2).  The chi-square value has dropped considerably :
this value is 15.50, df  is 6 and  p = .017.  Obviously this model
fits the data better.  Figure 9 shows model 3 : the results of the
analysis where we relaxed the parameter for the effect from RC81
to RC83.  First it should be noted that model 3 fits the data
rather good (chi-square is 5.38, df  is 5, p = .37), which means
that this model is a good explanation for the observed correlation
matrix.  Secondly, there is a strong effect from RC81 to RC83  (.35).
Thirdly, by postulating an effect from RC81 to RC83, the correlation
between the errors on these variables changes from positive to negative.  We come back to this in section 5.4.  A last point to be
noticed is that the influence of SI83 to RC83 has decreased somewhat.

We can conclude from model 3 that the data seem to be explained by the social control theory (cf. section 3.2), together with consi- derations police officers and the prosecutor have to record a contact, once the contact is taking place (cf. section 4.2).

## 5.4. The labelling theory

Although the fit of model 3 is good, it is possible that the labelling theory still contributes significantly. To investigate this, we relaxed the restriction on the effect from RC81 to SI83. The result is shown in model 4 (figure 10). As is to be expected this model also fits the data well (chi-square is 2.02, df is 4, p = .733). Comparing model 4 with model 3 one finds that relaxing the parameter for the effect from RC81 to SI83 gives a significant contribution to the model (chi-square is 3.36, df is 1, .10 > p > .05). According to the data, a part of the effect of judicial con-

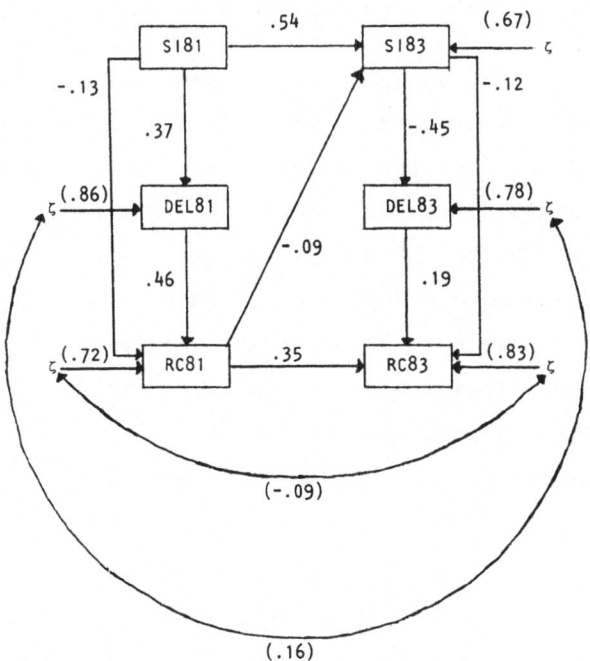

Fig.10. Model 4; df is 4, $\chi^2$ = 2.0, p = .733

tacts on delinquency goes via social integration. However, this
effect is very small, namely - .09 × - .45 = .04. This effect is
not large enough to be of theoretical significance. (The direct
relation from RC81 with DEL83 was found to be only .02. For reasons
of space we do not show the corresponding model). Therefore we
conclude that, given social control theory and the effect of SI and
RC on the decision of the prosecutor, the influence of the labelling
theory can be neglected. Model 3 is our final model.

## 5.5. The variable age

It is possible that the variable age plays a special rôle in
our models. Because we did not take this variable into account in
the analyses, the influence of age (if there is any) has to be found
in the error of the structural equations. This could possibly form
an explanation for the fact that the correlation between the errors
for RC81 and RC83 is negative : perhaps the influence of age has
reversed from 1981 to 1983. However, we have not checked this.

The influence of the variable age is also interesting in another way. If there would be no systematic effect for age, we could
assume that the effects for the different time-points would be equal,
e.g. the effect from SI81 to DEL81 could be hypothesized to be equal
to the effect from SI83 to DEL83. By supposing such an effect to
be equal, one would gain one degree of freedom. However, as a difference of two years can have an important influence for youngsters,
we have not done so. This makes it possible to study the difference
between corresponding parameters for the two time-points to age.
Perhaps these differences are caused by the age of the youngsters.

## 6. Discussion and conclusions

The problem to be solved was which theory, social control theory
of the labelling theory, provides the best explanation for the relationships in the data. Using LISREL, and more specifically structu-

ral equation models, this question could be answered : social control
theory together with the influence of SI and RC on the decision of
the prosecutor whether to record a contact or not, seem to provide
a reasonable explanation of the data. The influence of labelling
theory can be neglected.

Something has to be said about the reliability of our results,
since we have modified our model 1 on empirical rather than theoretical grounds. In general the best way to assess the reliability
is by using new data; in other words, the best way is by replica-
ting the study. Another way is by means of cross-validation : in
cross-validation the data are splitted randomly in two independent
samples. One sample is used to find which parameters are to be
included of excluded, the other sample is used to evaluate the ap-
propriateness of the inclusion and exclusion of these parameters.

Our solution is imperfect for the following reasons. First,
the relation between some of the variables is not linear : therefore a LISREL assumption is violated. We tried to solve this as
good as possible by rescaling the categories of the concerning variables. However, it is possible that this linearization is still
not sufficient. Secondly, LISREL presupposes that our variables
follow a multi-variate normal distribution. However, four of our
variables are extremely skewed. Not much is known about the effects
of violation of this assumption. However, generalizations of the
LISREL approach on aspects such as linear relationships and the
presupposed multi-variate normal distribution is currently being
worked on. Although our solution of the problem can be criticised
for these and perhaps also other reasons, we think it is one of the
best solutions that could be achieved with the current state of
affairs, i.e. knowledge, viable computer programs, etcetera.

The concept of causality has led to great confusion in the
context of covariance structure modelling (Cliff, 1983). These

models are often used in cases that researchers want to make statements that go further than that some variables are merely related.
The confusion still has to be resolved.  We take the stand that it
is not possible to prove causality of relations using these models,
but on the other hand, that is possible to prove that hypothesized
causal relations are non-existent.  So the fact that a hypothesized
causal model cannot be rejected merely indicates that there is some
further evidence this model might be true.  However, there are probably also other causal models that cannot be rejected, given the
same dataset.  It will be clear that in this context the rôle of
substantive theory cannot be underestimated.

As an example, as an alternative for model 3 (figure 9) also
the model holds in which the arrow for correlated errors between
DEL81 and DEL83 is replaced by a direct arrow from DEL81 to DEL83
(coefficient is .18; $\chi^2$ = 5.21, df  is 5, p = .391).  It will be
clear that the substantial·difference of model 3 and this model can
have important repercussions for governmental policy.  In this
respect it is perhaps better, for the time being, by not assuming
a causal relation between DEL81 and DEL83, but by adopting our
model 3.  It seems to us that a proper choice between these models
can only be made after one has tried to get rid of the correlation
between the errors by tracing the variables that cause this correlation.  When this is possible, model 3 should be preferred above
the model with direct relation between DEL81 and DEL83;  when it
is not, model 3 should be rejected.  However, this work is beyond
the scope of this paper.

## References

P.M.Bentler (1980). Multivariate analysis with latent variables :
    causal modelling. Annual Review of Psychology, 31, 419-56.
P.M.Bentler (1983). Simultaneous equations systems as moment struc-
    ture models : with an introduction to latent variable models.
    Journal of Econometrics, 22, 13-42.

A.Boomsma (1983). On the robustness of LISREL (maximum likelihood
     estimation) against small sample size and non-normality.
     Amsterdam : Sociometric Research Foundation.

M.W.Browne (1982). Covariance structures. In D.M.Hawkins (ed.),
     Topics in applied multivariate analysis (72-141). London :
     Cambridge University Press.

M.W.Browne (1984). Asymptotically distribution-free methods for the
     analysis of covariance structures. British Journal of Mathema-
     tical and Statistical Psychology, 37, 62-83.

N.Cliff (1983). Some cautions concerning the application of causal
     modelling methods. Multivariate Behavioral Research, 18,
     115-128.

J.Junger-Tas & M.Junger (1984). Juvenile delinquency, backgrounds
     of delinquent behaviour. The Hague : R.D.C., Ministry of
     Justice.

J.de Leeuw (1983). Models and methods for multivariate analysis.
     Journal of Econometrics, 22, 113-38.

J.de Leeuw & J.van Rijckevorsel (1980). HOMALS & PRINCALS, some
     generalizations of principal components analysis. In : E.Diday
     et al. (eds), Data analysis and informatics. Amsterdam :
     North Holland.

J.S.Long (1983a). Confirmatory factor analysis. Beverly Hills, C.A. :
     Sage.

J.S.Long (1983b). Covariance structure models : an introduction to
     the LISREL approach. Beverly Hills, C.A. : Sage.

A.Mooijaart (1985). Factor analysis for non-normal variables.
     Psychometrika, 50, 323-342.

K.G.Jöreskog & D.Sörbom (1981). LISREL V user's guide. Chicago :
     National Educational Resources.

B.Muthén (1983). Latent variable structural equation modelling with
     categorical data. Journal of Econometrics, 22, 43-66.

W.Saris and H.Stronkhorst (1984). Causal modelling in non-experimen-
     tal research : an introduction to the LISREL approach.
     Amsterdam : Sociometric Research Foundation.

SECTION 6.5


SOME CONCLUSIONS


M. Junger, P.G.M. van der Heijden and J.A. Hagenaars



The research question posed in Junger-Tas and Junger (this book)
was : "Which of both theories - social control theory of labelling
theory - is right, or are they both predicting existing relationships equally well ?" Both theories describe relations between
concepts in causal terms, while they have some concepts in common.

## 1. Causality

The three solutions that are proposed by Junger, van der Heijden
et al., and Hagenaars (all in this book) have some aspects in common,
which can be found in the literature under problems concerning causality (see for example Blalock, 1972). All solutions agree that
causality cannot be proved. What one can do is setting up a model
of the supposed causal relations between the variables and investigate whether the empirical predictions are valid or not. If these
predictions are invalid, the model is (at least partially) rejected.
If all the empirical outcomes are as predicted, the model is confirmed, not definitely proved. Other models may be found that explain
the data equally well. A crucial part of such a causal model is the
causal order of the variables : does A causes B or vice versa.
In this respect, substantial theory should help in determining the

direction of the arrows between variables.  Secondly, a time lag can
be helpful : in the research under study all variable are measured
twice, and arrows can only go from concepts in 1981 to 1983, and not
the other way round.  And thirdly, the researcher can have extra
information which can be helpful in determining the direction of
arrows (and which is not always available to the methodologist).

An example of an important problem in the determination of the
direction of arrows - which is a point in which the solution proposed from Hagenaars differs from those of Junger and van der Heijden
et al. - is the causal order between Social Integration 1981 and
Delinquency 1981.  From other parts of the study, published in
Junger-Tas and Junger (1984) Junger has the extra information that
teachers and parents almost never have knowledge concerning the
existence of delinquent behavior in their youngsters.  Considering
that social integration is to a large extent determined by relations
in school and in the family, it seems unlikely that delinquent
behavior does influence social integration - which is for the largest
part determined by school and family.

This example illustrates the problem with the concept of causality : causality is not visible, neither can it be proved, it can
only be made plausible.  The direction of arrows can be determined
using not only statistical information, but also theoretical considerations, the presence of a time-lag, and eventually some extra
information concerning the data.  Without the latter elements, a
researcher cannot give guidance to the analyses.  Furthermore, given
one correlation matrix a lot of models are possible, a lot of which
are difficult to interpret.  Such empiricism seems to be unfruitful.

## 2. The methods used

In the solution of Junger, path analysis is used, while in the
solution of Hagenaars and van der Heijden et al. the LISREL-approach
is used.  The latter approach should be evaluated as the better,

because it provides the researcher with an overall test for the model,
instead of one test for each arrow.  The resemblance between the
solutions, however, is large.  A large number of relations is found
in all three solutions (SI81 to SI83; SI81 to DEL81; SI83 to DEL83;
SI81 to JC81; SI83 to JC83; DEL81 to JC81 and JC81 to JC83), while
some relations are absent in all solutions (for instance the important one JC81 to SI83).

An important difference between the solutions of Junger and
Hagenaars versus van der Heijden et al. is the arrow from DEL81 to
DEL83.  van der Heijden et al. have used the possibility of LISREL to
hypothesize correlated errors on the equations, and they provide the
reader with an acceptable justification.  Indeed, in econometrics
the postulation of correlated errors is something that is often met:
in this tradition researchers should prove that errors are not correlated.

A problem in the field of path analysis and LISREL is the assumption of multivariate normality (see van der Heijden et al.).  van der
Heijden et al. try to solve this problem by using a program for monotone regression.  This results in higher coefficients for the arrows
from DEL to JC (but lower coefficients for some other relations).
However, the stability of their solution is unknown, and it would
have been better to investigate that stability.

## 3. Conclusion

Junger-Tas and Junger are very satisfied about the extra information that is provided by the solutions of Hagenaars and van der
Heijden et al. Indeed, the LISREL approach seems much better than the
path analysis approach.  The absence of the arrow from DEL81 to
DEL83, proposed in the solution of van der Heijden et al., may be a
better solution from a theoretical point of view, but, as they note
in their contribution, further research seems to be necessary to be
more sure of the correctness of this absence.  The solution of

Hagenaars covers a wider range because he also relates some extra
variables (socio-demographic variables) to the six variables used
in both Junger and van der Heijden et al. Furthermore, he stresses
the difficulties that can be met in the interpretation of relations
as being causal relations. To us this last aspect seems one of the
most difficult, but also one of the most important aspects of research
in the social sciences.

## References

H.M.Blalock Jr (1972). Causal inferences in non-experimental research.
    New York, Nortan and Company.
J.Junger-Tas and M.Junger (1984). Juvenile delinquency. Backgrounds
    of delinquent behaviour. The Hague, R.D.C., Ministry of Justice.

KEYWORDS

INDEX

Non-linearity, 345

Operator, 13
Orthogonal
  basis, 13
  polynomial of Legendre, 123
  projection, 31
  supplement, 49
Overdiagonal minimization, 283

Paired comparison rôle, 213
Panel study, 313, 331, 345
Parabolic curve, 123
Parameters
  length, 13, 31, 49, 67, 89
  skeletal, 13, 31, 49, 67, 89
  stoutness, 13, 31, 49, 67, 89
  weight, 13, 31, 49, 67, 89
Part families, 273
Partial order, 255
Partition, 213, 273
Percentage of variance, 177
Period
  epipaleolithic, 115
  mesolithic, 115
  of growing, 3
Pharmacology, 31
Physical anthropology, 7
Political
  families, 177
  trends, 173
Polynomial, 67, 123
Posac diagrams, 255
Prediction, 255, 313, 331, 345
Prehistoric
  assemblages, 115
  settlements, 115, 123, 153
Princal solution, 345
Problem of coding, 233
Production
  management, 273
  subsystem, 273

Quadratic form, 283

Radiocarbon dating, 115, 123
Regression, 49, 123, 313, 331, 345
Representation
  barycentric, 233
  simultaneous, 13, 123, 177

Selfreported delinquency, 305
Sequential information, 10
Seriation
  approximation, 153
  chronological, 123
Similarity index, 209
Singular value description, 89
Social
  control theory, 313
  integration, 305
Somatic characters, 49
Space diagram, 255
Spatial organization, 273
Spectramap, 31, 153, 247
Standardization of data, 7, 13,
    49, 89
Statis, 153
Stratigraphy, 123
Structuples, 255
Sum of squares, 89
Supplementary
  individuals, 7, 49, 177, 233
  variables, 7, 49, 177, 233

Table
  Burt, 233
  contingency, 233
  disjonctive, 233
  juxtaposition of, 177
  mean, 177
  presence-absence, 123
Taylor series, 67
Test
  Beales f, 67
  chi-square, 345
Three
  dimensional portrayal, 193
  way structure of data, 7, 31
Time trends, 89
Trajectories, 13, 49, 89, 177
Transnational study, 177
Triangular
  chart, 31
  map, 31, 153, 247
Triordonnance
  aggregation, 213
  coding, 213
Triple comparison, 213
Tuckals, 7

Typology, 115, 283
Two-way data, 7

Variable
  dependent, 313
  endogeneous, 313, 331, 345
  errors in, 345
  exogeneous, 313, 331, 345
  latent, 345
  qualitative, 13, 213, 233
  quantitative, 13, 213, 233
  sociodemographic, 305
  space, 13

Workshop organization, 283